SUSTAINABILITY SCIENCE AND TECHNOLOGY

AN INTRODUCTION

SUSTAINABILITY SCIENCE AND TECHNOLOGY

AN INTRODUCTION

Edited by
Alejandro De Las Heras

CRC Press
Taylor & Francis Group
Boca Raton London New York

CRC Press is an imprint of the
Taylor & Francis Group, an **Informa** business

CRC Press
Taylor & Francis Group
6000 Broken Sound Parkway NW, Suite 300
Boca Raton, FL 33487-2742

First issued in paperback 2017

© 2014 by Taylor & Francis Group, LLC
CRC Press is an imprint of Taylor & Francis Group, an Informa business

No claim to original U.S. Government works

Version Date: 20140207

ISBN 13: 978-1-138-07569-6 (pbk)
ISBN 13: 978-1-4665-1808-7 (hbk)

Library of Congress Cataloging-in-Publication Data

Sustainability science and technology : an introduction / edited by Alejandro De Las Heras.
    pages cm
    Summary: "This book builds a bridge between sustainability science and engineering. It provides the latest information on sustainability issues and solutions by discussing them from basic concepts to scientific consensus to technical solutions. It identifies and addresses the major issues related to sustainability with an environmental science perspective, and presents the connections and interactions between science and technology through a very practical approach and easy understanding. "-- Provided by publisher.
    Includes bibliographical references and index.
    ISBN 978-1-4665-1808-7 (hardback)
    1. Sustainability. 2. Sustainable development--Technological innovations. 3. Environmental responsibility. I. De Las Heras, Alejandro, editor of compilation.

HC79.E5S86164 2014
338.9'27--dc23   2013049500

Visit the Taylor & Francis Web site at
http://www.taylorandfrancis.com

and the CRC Press Web site at
http://www.crcpress.com

Contents

Preface

The intention of this book is to promote the widest possible dialog around sustainability issues and their solutions. This called for a book written without assumptions of a disciplinary field but rather with cultivated and intelligent people in mind, who are mostly busy professionals who need to engage in multidisciplinary work, decision makers who need to understand the possible unintended consequences of their resolutions, and undergraduates who have a heavy study load.

The perspective of the book is that natural and human subsystems interact in the planetary system. The global view is rendered based on examples from around the world.

The story is told one bite-sized chapter at a time, about the size of a scientific journal article. The chapters are self-contained, each grappling with a large topic. This provides more in-depth coverage of a topic than a standard encyclopedia article. The assertions made are backed by references in the text. Similar to a textbook, the 17 nonoverlapping but linked chapters can help higher education teaching staff and independent students cover the main sustainability issues and currently proposed solutions.

No exercises are provided in this edition to avoid the impression that sustainability issues are well-formulated problems in real life. Instead, the reader is given the option to disagree and build upon the solutions proposed in each chapter. In fact, this critical-but-constructive approach is essential where work on sustainability is concerned.

The mathematical and statistical knowledge to understand this book probably does not go beyond the fundamentals. However, the definition of sustainability that is needed to focus the discussion has to be operative (measurable) and actionable (it has to lend itself to decision making and enforcement) from the individual to the global level. No background is provided in this Preface on the emergence of the concern for sustainability, other than mentioning that the global debate had started in earnest by 1987, and that by 1992, dozens of definitions were available. To date, however, scarcely any definition is operative and actionable; most definitions fairly adequately deal with the relationship between the environment and human consumption, but they fail to encompass social sustainability

Defining Sustainability

An operative and actionable definition of sustainability can be expressed as follows. Conceptually, C is the amount of human consumption of natural matter and energy (the extraction of natural resources). R is the amount of natural regeneration, or biomass production. S is the amount of restoration that humans have to invest in nature to return to a more predictable state of the planetary system. Under natural conditions, $C \leq R$ (a species cannot consume more than the planet can produce). Also, the species on the planet may want to use as much of the planet resources as possible and save for crises, hence they have to care for the full replenishment of planet resources; this justifies S in terms of allowing for maximum natural production.

Operatively,

$$C < R + S$$

which means that at any time in the future, consumption must not exceed natural regeneration and human restoration of nature. Whenever this expression is not verified, sustainability is not happening. Equivalently, human populations and wildlife cannot last.

Graphically, the most salient features are two periods of sustainability. First, a period of necessary decrease in consumption and second, after the point where regeneration newly exceeds consumption, the costs incurred by humankind for the restoration of their impacts become negative and translate into savings. Fluctuations in this second period may loosely respond to cyclical and random variations in natural output and future human population. Scarcity risks are limited by the fact that consumption is well below the natural output. Dematerialization may still allow for (potentially unlimited) wealth generation since the latter is decoupled from natural resource extraction.

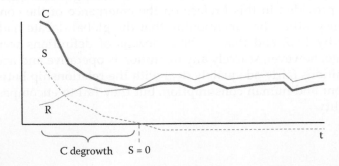

Turning to social sustainability, it amounts at the very least to the initial natural condition, $C = N_b$ (consumption is limited to biological needs N_b) for everyone, which should be recognized as a human right. But stone technology that began 2.5 million years ago, and art more than 100,000 years ago, indicates that human endeavors go beyond N_b. The possibility to exploit one's individual potential N_c (cultural needs) means that $C = N_b + N_c$ under sustainability.

N_b are the necessary conditions (food, health services, housing, clothing, in this order) for maximum healthy survival; they are produced with ever-more natural resource efficiency as science and technology evolve. N_c are nonmaterial or nonconsumptive satisfactors, which include enjoying nature and pursuing art, science, sports, and technology endeavors. Equal access to N_c satisfactors should be considered a right, while the utmost exertion of capabilities should be considered a duty. The distribution of N_c is as unequal as individual capabilities are, after equalizing the opportunities to contribute to human endeavors. Operationally,

$$N_b + N_c < R$$

and graphically

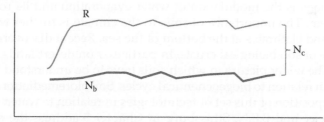

Under environmental and social sustainability, R is maximal and C is minimal. The reason for C's material degrowth and a global postmaterial economy as soon as possible is that the cost of S has to be as small as possible.

Outline of the Book

Chapter 1 by A. de las Heras and T. Macagno underlines the uneven distribution of freshwater in land environments, coupled with population growth and urbanization in arid areas, mostly in the United States, China, and India, which are enduring climate change in the form of prolonged droughts. By 2030, it is expected that global demand will exceed supply. In the arid portion of the United States, the limits of water extraction seem to already have been reached. Food production is still the leading reason for water exploitation, but industrial pollution probably explains most of the decrease in water quality, a form of progressive loss of water availability. Whereas the more developed countries are facing industrial pollution only, middle-income countries face a double sanitary threat of infectious and industrial freshwater pollution.

At the large scale of human societies, Chapter 2 by M. Wexler and J. Burgess highlights the need for humans to help nature itself clean up both freshwater and marine waters. Water bioremediation uses widespread microorganisms that have evolved detoxification capabilities and directs their activity to anthropogenic pollution. A wastewater perspective is developed in detail. Oil spill disasters are compared. And the more extreme radioactive and heavy metal pollution cases show the extent of bioremediation capabilities.

Far easier to solve but somehow resisting human ingenuity or global political will are clean water and sanitation issues claiming millions of children's lives every year. The workaround, called water adequate technologies, has been known since the 1970s but has been suppressed by the privatization trend that relies on large urban facilities and networks. A catalog of the recent water adequate technologies is presented in Chapter 3 by M. Islas-Espinoza and A. de las Heras.

References to water are also made in Chapter 4 by M. Islas-Espinoza, J. Burgess, and M. Wexler, which is dedicated to soils. Among the vast array of soil functions is the modulation of water evaporation and its role in the climate system. The importance of soils in the climate is further tracked to permafrost and clathrates at the bottom of the sea. Recent discoveries of the essential role of soil biological crusts, in particular on desert lands, are also pointed to. The wider picture in which soils have to be understood is finally expounded in relation to biogeochemical cycles. Soil bioremediation complements the exposition of this set of technologies in relation to water.

A.E. Latawiec and B.B.N. Strassburg in Chapter 5 address the emergent land issues relevant to both soil and vegetation. Land grab and biofuel crops are perhaps the most critical instances of competition for land. The former is most acute in Africa and represents an attempt to cultivate solely for exports to affluent countries, leaving the nutrition issues affecting local people unsolved. As for biofuels, not only do they compete for space with staple crops and wildlife, they also resort to Green Revolution technologies, such as pesticides, herbicides, fertilizers, and monocultures, which have brought about pollination and seed dispersal crises.

The solutions proposed by A.E. Latawiec and B.B.N. Strassburg in Chapter 6 can be summarized under three headings: sustainable intensification, carbon capture markets, and organic agriculture. The precise mechanisms to avoid pitfalls in implementing these solutions are discussed in detail.

Following up on soil and climate, the use of biochar by L. Peake, A. Freddo, and B.J. Reid in Chapter 7 illustrates possibly the sole geoengineering technology that has garnered substantial support. Biochar is a product of biomass pyrolysis or low-oxygen thermal decomposition. This is a means for durable sequestration of carbon, which might add substantial soil organic matter that is favorable to plant life. Biochar production is also part of a process that generates bioenergy.

Delving deeper into energy, in Chapter 8, D. Durán and E.A. Rincón-Mejía introduce the key notions of entropy, exergy, and emergy to manifest the methods whereby energy sources and setups are compared from an efficiency and sustainability viewpoint. They also show that zero-carbon solutions and carbon-negative solutions cannot be achieved independently from carbon sequestration, and are only possible using restoration of natural carbon sequestration and renewable energy sources, which the authors list as close-at-hand solutions in the impending transition from fossil fuels to wind energy first, and other clean energy sources thereafter.

In Chapter 9, I. Martínez-Cienfuegos and E.A. Rincón-Mejía show that solar energy is similar to wind energy, a soaring market. Furthermore, solar concentrated energy systems have very high exergy and power, which makes them suitable for industrial applications. New materials, geometries, and the hybridization of solar concentration and geothermal energy highlight the vast array of combinations that will support solar energy growth.

M. Islas-Espinoza and B. Weber pinpoint the bioenergetic principles shared by all living beings in Chapter 10. They then explore advances in bioenergy solutions, identifying sustainable energy sources. Foremost among these is biomethane, which is generated from microbial anaerobic decomposition of waste biomass and whose combustion is the cleanest of all the hydrocarbons in terms of carbon release and absence of sulfur and nitrogen compounds. Recycling of waste and multifarious applications of biomethane make it highly exergetic and emergetic, respectively.

Renewable energy sources, as well as reductions in energy consumption, constitute the chief solutions to climate change and air pollution whose intertwinement is presented in Chapter 11 by A. de las Heras. This is because fossil fuels egress greenhouse gas and sulfur and nitrogen compounds to the atmosphere, among other pollutants. Chlorine compounds are also anthropogenic pollutants. Together they are the main causes of climate change, the depletion of the ozone layer, the Asian brown clouds, acid rain, and smog (the most relevant global and regional atmospheric issues to date). Ancillary solutions are monitoring, modeling of possible effects of new manmade compounds, and mitigating (with a strong emphasis on co-benefits of measures, taking advantage of the fact that the same pollutants are attributable to the same sources and participate in different pollution phenomena).

Added to their global and regional roles, air pollutants have toxic effects on the human organism. M. Islas-Espinoza in Chapter 12 reviews indoor and outdoor exposures and effects of the global and regional pollutants, and stresses the effects of organometal(loid)s, radiation, and endocrine disrupting chemicals. The conventional recourse to filters is a sink of pollutants needing disposal and energy to be operated; this is why they are very seldom used in the Third World. Appropriate solutions have to do with the transition to renewable energy sources and green roofs to reduce the generation of photochemical smog.

Chapters 13 through 17 explore the context in which sustainable science and technology can develop. Chapter 13 is concerned with the evolutionary origin of human features linked to current unsustainability. Neither our survival nor our biological and technological inheritance seemed to depend on or be directed to large-scale human encroachment. However, dominance present in all primates transformed into gender and generational inequalities. Power leverage, owing to language and objects symbolizing prestige, led to the emergence of elites. Technological advances in domestication led to surplus food production, which was controlled by elites.

Contemporary unsustainability has been explained by a combination of factors. Among these, consumption per capita (rather than simply affluence) and technology dominate. Chapter 14 points to evidence that the human population affects the environment through the food system, which is then analyzed. This system is at the core of global social unsustainability: excess food production and hunger coexist. A model of population stabilization is grounded on demographic theory but excess food production has to be curbed and redistributed.

Consumption and property, pivotal in current economic systems, make unsustainability deeper: consumer goods tamper with some human biological functions and seldom contribute to happiness (Chapter 15). Strong economic sustainability and social sustainability based on the development of individual human capacities are conditions of socioeconomic sustainability, without which market mechanisms (eco-innovation, ecodesign, and corporate valuation of ecosystem services) result pointless. Dematerialization of the economy is a favorable trend that can be harnessed to create value without exhausting nature. Another favorable trend is the evolution of macroeconomic indicators that may guide changes in taxes and subsidies toward sustainable technological change and social equity.

Chapter 16 examines material waste and transport dysfunctions in the built environment. Solutions have to account for regional and local environmental conditions (biourbanism) and create urban–rural flows of biodegradable waste recyclable into soil amenders and biomethane. These will gradually replace costly construction, operation, and maintenance of large waste facilities. Solutions to transport issues should aim for very low energy consumption in ecocities, reduction of miles traveled via mixed uses, and more compact built environments. Renewable materials, long-lived buildings, and adaptively reusable buildings are solutions to material waste.

Chapter 17 takes a step back and examines the sustainability of science and technology themselves. The gap between knowledge generation and usable knowledge affects science and has been dealt with in networks of actors linking knowledge and know-how, called knowledge systems. Sustainable technology (re)training is the main issue for engineering associations, which has called for the introduction of sustainability skill sets in undergraduate and postgraduate training. An urgent matter is harnessing ongoing scientific and technological revolutions. Essential progress has been made by green chemistry and sustainable nanochemistry. Computing, however, warrants considerable efforts emulating pioneering companies. The latter have powered data centers using renewable energy sources or provided end-user products that offset carbon dioxide emissions via tree plantation or wetland conservation. Green and sustainable computing seem to hold the key to the participation of citizens in sustainability decision making, emulating existing online social networks.

The scope of this book leaves out biological systems per se, although the technological applications of microbiology are mentioned, as are the underpinnings of human evolution and biology that directly impinge on sustainability. More potential applications and discussions on conservation and marine ecosystems warrant special treatment in a forthcoming volume. In the perspective of this book, however, sustainability can do without the plundering of marine fisheries, the equivalent of hunted meat on land.

This sets the stage for discussion and sustainability (re)training across professional divides. We wish to acknowledge the enthusiasm of Irma Britton at CRC Press/Taylor & Francis, thanks to whom this book came to life.

Editor

Alejandro de las Heras (DEA in demography at Paris Sorbonne, Ph.D. in environmental sciences at the University of East Anglia) has worked with grassroots movements and nongovernmental organizations in Mali, France, and Mexico. He has also taught in higher education in England and Mexico. He is currently involved in oasis restoration in the Baja Peninsula desert. E-mail: aheras38@hotmail.com

Contributors

Jeremy Burgess
Retired Scientific Researcher and
    Photographer
Norfolk, United Kingdom

Dolores Durán
College of Engineering
University of the State of Mexico
State of Mexico, Mexico
mddg_2210@hotmail.com

Alessia Freddo
University of East Anglia
Norwich, United Kingdom
a.freddo@uea.ac.uk

Marina Islas-Espinoza
Inter-American Center of Water
    Resources
University of the State of Mexico
State of Mexico, Mexico
marinaislas@ymail.com

Agnieszka Ewa Latawiec
International Institute for
    Sustainability
Rio de Janeiro, Brazil
and
Department of Production
    Engineering and Logistics
Opole University of Technology
Opole, Poland
and
School of Environmental Sciences
University of East Anglia
Norwich, United Kingdom
a.latawiec@iis-rio.org

Thomas Macagno
University of East Anglia
Norwich, United Kingdom
tmacagno@gmail.com

Iván Martínez-Cienfuegos
College of Engineering
University of the State of Mexico
State of Mexico, Mexico
igmartinezc@uaemex.mx

Lewis Peake
University of East Anglia
Norwich, United Kingdom
l.peake@uea.ac.uk

Brian J. Reid
University of East Anglia
Norwich, United Kingdom
b.reid@uea.ac.uk

Eduardo A. Rincón-Mejía
College of Engineering
University of the State of Mexico
State of Mexico, Mexico
rinconsolar@hotmail.com

Bernardo B.N. Strassburg
International Institute for
    Sustainability
Rio de Janeiro, Brazil
and
Department of Geography and the
    Environment
Pontificia Universidade Catolica,
Rio de Janeiro, Brazil
b.strassburg@iis-rio.org

Bernd Weber
College of Engineering
University of the State of Mexico
State of Mexico, Mexico
bweber@uaemex.mx
and
Technische Hochschule Mittelhessen
Hessen, Germany

Margaret Wexler
University of East Anglia
Norwich, United Kingdom
margaretwexler@ymail.com

1

Freshwater Today: Uses and Misuses

Alejandro de las Heras and Thomas Macagno

CONTENTS

Provision of Water by Ecosystems

Freshwater is scarce on Earth, unevenly distributed, and sometimes too heavily loaded with salts and minerals to sustain land life (Table 1.1). Water is mostly present in the oceans where salt involves a large osmotic pressure (high salt contents inside living cells), an additional pressure of one atmosphere for every 10 meters of additional depth, and a rapid dimming of solar energy. Deserts have vast water reserves accumulated over millennia, which are seldom accessible to life. Freshwater is found beyond the polar circles where cold imposes adaptation constraints on living organisms (such as elevated metabolisms to avoid water from freezing inside tissues). Finally, water vapor in the atmosphere is scarce and only available to life in land ecosystems (Figure 1.1) as a result of precipitation.

TABLE 1.1

Terrestrial Water Stocks/10^3 km^3

Total water	1386 million
Oceans	1340 million
Total freshwater	35,000
Saline groundwater and lakes	13,085
Total glaciers, arctic islands, permafrost	24,364
Freshwater lakes	91
Wetlands	11
Rivers	2
Biological matter	1
Atmosphere	13

Source: Gleick PH, Palaniappan M, 2010, *PNAS* 107:11155–11162.

Total precipitation (mm)
High: 2353
Low: 0

FIGURE 1.1 (See color insert.)
The main climatic limits for life on Earth's continental masses in the early Anthropocene. Water is delivered through precipitation (rainfall, snow, hail, fog, and dew) with very large differences: dry and cold deserts are defined by unpredictable rainfall (rain can go unrecorded there for years). Polar areas and continental areas (far from the sea) receive 80 to 150 mm per year. Meanwhile, temperate coastal areas receive 300 to 1500 mm and tropical forests might receive 3000 mm. Not surprisingly, in the early Anthropocene (the era dominated by mankind), human life is mostly concentrated in areas well endowed with solar energy and water (the Tropics) where crops can be harvested at low costs. (Data from the Intergovernmental Panel on Climate Change [IPCC], 2005, Climate Research Unit [CRU] high resolution climate data, version 2.1, http://www.ipcc-data.org/observ/clim/cru_ts2_1.html.)

Water Scarcity: Human Appropriation and Competing Uses

The arid and semiarid regions are experiencing long rows of seasonal droughts, such as the Sahel in Africa (Chapter 14). Australia has recently faced a decade with three of the harshest droughts on record. Southwestern United States, Northern China, and India are poised to go through climate-change-driven extreme droughts, whereas other regions of said countries have to brace themselves against floods. Floods and storm runoff transport pollutants and add to already prevalent river contamination in cities like Sydney (Norris and Burgin 2009).

Water pumping, especially in desert urban settlements, has soared. Water extraction from surface water bodies has doubled since 1960, with long-term consequences on regional water cycles and freshwater (Millennium Ecosystem Assessment [MEA] 2005). The United Nations nevertheless estimate 1 billion people lack access to sufficient water supply, defined as a source likely to provide 20 liters per person per day at a distance no greater than 1000 meters. This water stress is correlated to poverty, the most severe consequence of which is an inability to bridge intraseasonal dry spells and droughts. It could be avoided with better water management rather than additional extraction (Rockström and Karlberg 2010).

Human water appropriation is most notable in surface water bodies (lakes and rivers) and groundwater, leading to blue scarcity (Figure 1.2). Blue water scarcity hotspots are densely populated basins that straddle the tropics or are heavily industrialized. Human appropriation diverts water away from wildlife habitats. Green water scarcity derives from rainwater appropriation and mostly occurs in rain-fed, subsistence agriculture.

The aggregate level (total water demand) is a translation of both economic production in and exports from a region. On a per capita basis, water demand translates the efficiency of economic and population sectors. Developing countries rely on subsistence and exports agriculture. Industrialized countries predominantly use water in the industrial sector (Figure 1.3). Agricultural and industrial uses compete with the urban lifestyle for scarce water resources, which often undergo deep seasonal changes as a result of human pumping. This competition makes water more expensive, which has made it an important business for large corporations that rely on clean water scarcity for profits.

Months in a year with scarcity > 100%

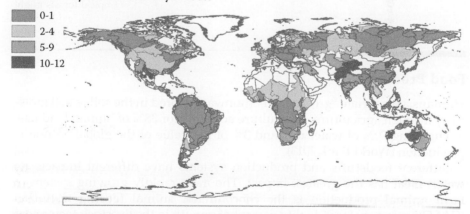

0-1
2-4
5-9
10-12

FIGURE 1.2 (See color insert.)
Blue water scarcity in the most important drainage systems of the world. Blue water is surface water and groundwater. Scarcity in the majority of the basins responds to human appropriation. Biological and chemical pollution come from domestic, agricultural, and industrial uses. Lack of wastewater treatment makes water a single-use commodity driving extraction and blue water scarcity upward. (Data from Hoekstra AY et al., 2012, *PLoS ONE* 7: e32688.)

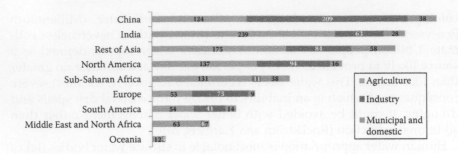

FIGURE 1.3
Water demand per water use in 2010 in the regions of the world (billion m³). (Data from the 2030 Water Resources Group, 2009, Charting our water future, http://www.2030waterresourcesgroup. com/water_full/Charting_Our_Water_Future_Final.pdf.)

As a result of increasing demand, the global freshwater supply in 2030 is expected to be 40% smaller (4200 km³/year) than demand (6900 km³/year) (The 2030 Water Resources Group 2009). Agriculture in all regions will account for a majority of future water use growth, except in China and Europe, where industrial uses are forecasted to continue predominating. Conversely, in Africa and the Middle East industrial uses are expected to play a minor role. The probability of water scarcity is more than 25% in all settled areas or agricultural breadbaskets (Global Water Intelligence 2011). Scarcity will befall the most densely populated areas and their agricultural hinterlands with a probability higher than 75%, as urban lifestyle and agriculture—both traditional and agroindustrial—make poor use of water. In many countries, water subsidies encourage squandering.

Food Production

Humans appropriate "green water" (rainwater stored in the soil as soil moisture) through agriculture. Agriculture accounts for 38% of human land use, a whopping 70% of water use, and 3% of the value of the global economic production (World Bank 2013).

Different foodstuffs and production systems have different impacts on water; meat has the greatest impact. The most rapidly growing system of farm animal production is the concentrated animal feeding operation (CAFO), or factory farm. Eighty percent of growth in the livestock sector now comes from CAFOs. Globally, they already account for the production of 72% of poultry, 55% of pork, and 43% of eggs. CAFOs have massive impacts on ecosystems: 4 million acres are producing vegetables for human consumption in the United States, while 56 million acres are producing hay for animals only. Waste from livestock pollutes more than 27,000 miles of rivers and

TABLE 1.2

Water Footprint of the Five Most Important Crops and Meats

Crop	Water (L per kg)	Meat	Water (L per kg)
Rice	1900–5000	Beef	15,000–70,000
Soybeans	1100–2000	Sheep	6100
Corn/maize	1000–1800	Chicken	3500–5700
Alfalfa	900–2000	Goat	4000
Wheat	900–2000	Eggs	3300

Sources: Mekonnen MM, Hoekstra AY, 2010, The green, blue and grey water footprint of crops and derived crop products, Value of Water Research Report Series No. 47, Delft: UNESCO-IHE; Worldwater.org, 2011, Water content of things: The world's water 2008–2009, http://www.worldwater.org/data20082009/Table19.pdf.

groundwater in many American states and is responsible for algal blooms contributing to a "dead zone" in the Gulf of Mexico reaching 7700 square miles during the summer of 1999 (WorldWatch Institute 2003).

At the global scale, cereal products are the largest contribution to the water footprint of the average consumer (27%), followed by meat (22%) and dairy (7%) (Hoekstra and Mekonnen 2012, Table 1.2). The water footprint of meat and dairy also has to include the use of water in feed crops (9% of the total crop's footprint). Increasing water scarcity raises the issue of future food security and hunger eradication.

Industries

Physicochemical properties have made water desirable in every industrial application, used as solvent (to separate different compounds), container sanitization, coolant, lubricant for boring tools, or simply to transport material from one place to another. Ensuing water pollution is varied and widespread.

Heavyweight Champion of Manmade Industrial Pollutants: Dioxins

Dioxins, a family of 210 which are among the most toxic chemicals, are not produced intentionally but are byproducts of manufacture and combustion. Such industrial processes have led to major well-known pollution accidents since the late 1950s; to name only a few: the Chicken Edema, Agent Orange, and Seveso incidents in the United States, Vietnam, and Italy, respectively. As to combustion as a source of dioxins, it has been traced to around 1935 when the chemical industry introduced massive consumption products made of plastic, some of which were chlorinated phenols. These, upon disposal and combustion, produced dioxins and their furan relatives. They were released into the atmosphere, deposited in water, and accumulated in lake sediments.

Hormone Disruptors

Many compounds found in water are chemically similar to natural hormones and so can be confused by an organism's hormone receptors and assimilated. The effects of these pollutants, called hormone or endocrine disruptors, are multifarious, as recently expounded by an official Endocrine Society Statement (Diamanti-Kandarakis et al. 2009). Among a number of conclusions therein are, first, that low level but chronic exposure leads to health effects in humans, and second, that "evidence for adverse reproductive outcomes (infertility, cancers, malformations) from exposure to endocrine disrupting chemicals is strong, and there is mounting evidence for effects on [other human biological systems]." These compounds have been monitored across many industrial countries since at least 1999 and come from human urine, household products, agricultural waste, the industry, and are also found in the vicinity of landfills and wastewater treatment plants (Wise et al. 2011). The effects have also been known to be underestimated in laboratory assays as compared to effects occurring in the more complex environmental wastewaters.

A Loophole in Pollution Control: Toxicity of Mixtures

Environmental regulations rely on scientific information. But the latter narrowly focuses on one pollutant at a time in 95% of the studies. It is only in the last decade that concern has emerged (Monosson and Lincoln 2006) and that scientific societies have formulated research programs focusing on lower doses than for a single pollutant (even below observable damage), complex histories of mixtures for different individuals including to short-lived pollutants (very different from lab settings), and synergies (more potent effects or difficult to predict from just two pollutants). One feature of the toxicology of mixtures is that environments are multimedia, that is, comprised of air, soil, and water, and so the entry points of pollutants into organisms include all systems in direct contact with the environment (skin, respiratory, digestive, and mucose systems). The importance of mixtures can be illustrated by the exposure to hormones such as mentioned earlier and fertilizers that are known as hormone agonists (they enhance the effects of hormone exposure).

Other pharmaceutical products, such as antibiotics and polluting byproducts of the industry and agroindustry, also circulate in the environment, are often degraded, and their degradation products can be more toxic than the original pollutant. All these products can interact among themselves and with living organisms. With regard to water and pharmaceutical products alone, a series of compounds including hormones, antibiotics, blood lipid regulators, nonsteroid analgesics and anti-inflammatory agents, beta-blockers, antiepileptics, antineoplastics, tranquilizers, and diagnostic contrast media are more likely to interact in mixtures because they are more soluble and not very volatile (Buzby n.d.), but besides forerunning studies (such as Brain et al. 2004, who used eight pharmaceuticals in mixtures) to date, few mixture studies in water are available.

Water Trade

Trade, the information and communication technologies, along with financial services are rising in economic importance. This means that water use in wealthy countries is no longer a main factor of economic growth. However, it has been suggested that water extraction capacity has reached its limits, explaining the plateau in U.S. water withdrawals (Figure 1.4). Most U.S.

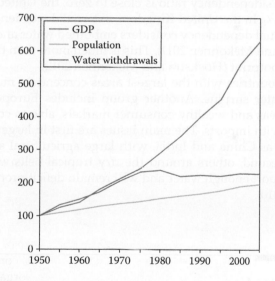

FIGURE 1.4 (See color insert.)
Peak water and decoupling of economic output, population, and water withdrawals after 1975 in the United States (1950 = 100). Water withdrawals stabilized their increase thereafter independently from socioeconomic factors. The explanation may be that maximum water extraction capacity has been reached either in technical or ecological terms, as suggested by Gleick and Palaniappan (2010). This runs counter to the more optimistic decoupling interpretation whereby GDP could grow without exhausting water resources. This decoupling also means that reducing water extraction may not jeopardize the evolution of well-being. GDP in U.S. dollars, billions; water withdrawals in billion gallons per day. (Data from Kenny JF et al., 2009, Estimated use of water in the United States in 2005, U.S. Geological Survey Circular 1344; U.S. Census Bureau, Population Division, 2011, Intercensal estimates of the resident population by sex and age for the United States: April 1, 2000 to July 1, 2010; U.S. Bureau of Economic Analysis, 2013, U.S. economic accounts, http://www.bea.gov/.)

watersheds experience water scarcity at least 4 months a year, especially in the dry southwest, which is growing demographically. A recent survey found that 40% to 50% of the water bodies in the United States could have impaired or threatened water quality (Environmental Protection Agency [EPA] 2005).

Virtual Water

Trade contributes to water shipping across borders as water-rich produce and embodied water in manufactured products (see Figure 1.3). Virtual water looks at the water footprint of a country generated by the goods it imports or exports. A country can have a large external footprint, meaning it is dependent upon the freshwater resources of other countries (Hoekstra and Mekonnen 2012).

The water dependency of a country is the ratio of external water availability to overall uses. It is an indicator of future political risks and an impulse to innovate. China's dependency ratio is close to zero, the United States is 8%, and India's is 31%. In the United Kingdom, external dependence on water is 1.4%, but its virtual dependency considers embodied water and is as high as 75% (Hoekstra and Mekonnen 2011). This is much more than the global 20% average water footprint (Hoekstra and Mekonnen 2012).

Overall, the countries with the largest areas concentrate more water and export their water surplus. Another group includes European countries with smaller areas and wealthy consumer markets, able to compensate for water scarcity with imports. The main issues are first in large but populous countries, such as China and India, with large agricultural and industrial exports. And second, others around the dry tropical belts will likely lack resources to import enough water and will remain deficit prone (Rockström and Karlberg 2010).

Domestic Water Use

Household water consumption is a modest part of water consumption (0.9%) compared to thermoelectric power (47.9%), irrigation (33.6%), or industries (4.8%) (for the United States in 2000) (Mines and Lackey 2009). However, households egress a wide array of pollutants via wastewater. An example is sucralose contained in diet beverages, which is 800 times sweeter than sugar and is excreted in human urine. Wastewater systems cannot filter it out. Many other drugs are found in watersheds. Even the chemical that has for the most part stopped waterborne diseases in urban areas, namely chlorine, has byproducts such as trihalomethane and haloacetic acids, which are suspected carcinogenic agents (Mines and Lackey 2009).

Water Quality Transition

Global water scarcity is around the corner but in developing countries it already combines with the single most important water issue: unsafe drinking water with waterborne pathogens. In addition, agricultural, industrial, and urban pollution sources are increasingly affecting richer areas. Thus, a water quality transition from biological to chemical pollution is underway whereby socioeconomic development under the current models leads from one malady to the other. This is an instance of the already known epidemiological transition (Omran 1974).

Middle-income countries are affected by an overlapping period during which waterborne biological issues linger in the poorest areas, while industrial pollution augments. The more developed countries have reduced industrial water pollution through "delocalization" (migration) of heavy and polluting industries to countries with lower wages. Affluent countries have as yet failed to reign in acid rain and the concomitant use of water and pollutants in the manufacture of consumer goods. Also, the chemical industry supplies agroindustrial inputs leading to eutrophication risks (excess of nutrients such as nitrogen in water) and cancer hazards in humans via nitrite pollution.

One feature of the transition is the unpredictability of interactions. Pollution from industrial estates often percolates (infiltrates) the soil and permeates into groundwater, which nowadays tends to be overexploited; and so, the concentration of pollutants increases in groundwater for human consumption. The pervasive use of chemicals augments the probability of interactions. For example, acid rain seems to increase the toxicity of chromium VI to organisms in ponds and channels in the tropics (Abbasi et al. 2009). Human uses competing for water frequently lead to irrigation of food crops with industrially contaminated water (with, for instance, heavy metals) in densely populated areas; wheat grain thus irrigated presents a health hazard to human populations (Si et al. 2011). Interactions with airborne and foodborne pollution occur within organisms. A second feature is a feedback loop: as population grows, safe water becomes scarce and water of poorer quality must be used.

References

Abbasi T, Kannadasan T, Abbasi SA (2009) A study of the impact of acid rain on chromium toxicity. *International Journal of Environmental Studies* 66:765–771.

Brain RA, Johnson DJ, Richards SM, Hanson ML, Sanderson H, Lam MW, et al. (2004) Microcosm evaluation of the effects of an eight pharmaceutical mixture to the aquatic macrophytes *Lemna gibba* and *Myriophyllum sibiricum*. *Aquatic Toxicology* 70:23–40.

Buzby ME (n.d.) Pharmaceuticals in the environment: A review of PhRMA initiatives. Pharmaceutical Research and Manufacturers of America. Merck & Co. Inc.

Diamanti-Kandarakis E, Bourguignon J-P, Giudice LC, Hauser R, Prins GS, Soto AM, Zoeller RT, Gore AC (2009) Endocrine-disrupting chemicals: An Endocrine Society scientific statement. *Endocrine Reviews* 30:293–342.

Environmental Protection Agency (EPA) (2005) Handbook for developing watershed plans to restore and protect our waters. Vol. EPA 841-B-05-005-2005. http://www.epa.gov/owow/nps/pubs.html (accessed April 14, 2012).

Gleick PH, Palaniappan M (2010) Peak water limits to freshwater withdrawal and use. *PNAS* 107:11155–11162.

Global Water Intelligence (2011) Global water risk index. http://www.water-rick-index.com/index.html (accessed March 17, 2012).

Hoekstra AY, Mekonnen MM (2011) National water footprint accounts: The green, blue and grey water footprint of production and consumption. Value of Water Research Report Series No. 50. Delft: UNESCO-IHE.

Hoekstra AY, Mekonnen MM (2012) The water footprint of humanity. *PNAS* 109:3232–3237.

Hoekstra AY, Mekonnen MM, Chapagain AK et al. (2012) Global monthly water scarcity: Blue water footprints versus blue water availability. *PLoS ONE* 7: e32688.

Intergovernmental Panel on Climate Change (IPCC) (2005) Climate Research Unit (CRU) high resolution climate data, version 2.1. http://www.ipcc-data.org/observ/clim/cru_ts2_1.html (accessed December 14, 2012).

Kenny JF, Barber NL, Hutson SS, Linsey KS, Lovelace JK, Maupin MA (2009) Estimated use of water in the United States in 2005. U.S. Geological Survey Circular 1344.

Mekonnen MM, Hoekstra AY (2010) The green, blue and grey water footprint of crops and derived crop products. Value of Water Research Report Series No. 47. Delft: UNESCO-IHE.

Millennium Ecosystem Assessment (MEA) (2005) Ecosystems and human well-being: Synthesis. Washington DC: Island Press. http://www.millenniumassessment.org/documents/document.356.aspx.pdf (accessed October 12, 2012).

Mines R, Lackey L (2009) *Introduction to environmental engineering*. New York: Prentice Hall.

Monosson E, Lincoln D (2006) Comparison of PCBs, organochlorine pesticides, and trace metals in cold liver from Georges Bank and Stellwagen Bank, USA and Canada. *Marine Pollution Bulletin* 52:572–597.

Norris A, Burgin S (2009) Rhetoric and reality surrounding water quality issues in a peri-urban western Sydney community. *International Journal of Environmental Studies* 66:773–783.

Omran AR (1974) Changing patterns of health and disease during the process of national development. In *Community medicine in developing countries*. New York: Springer.

Rockström J, Karlberg L (2010) The quadruple squeeze: Defining the safe operating space for freshwater use to achieve a triply Green Revolution in the Anthropocene. *AMBIO: A Journal of the Human Environment* 39:257–265.

Si W, Ji W, Yang F, Lv Y, Wang Y, Zhang Y (2011) The function of constructed wetland in reducing the risk of heavy metals on human health. *Environmental Monitoring and Assessment* 181:531–537.

U.S. Bureau of Economic Analysis (2013) U.S. economic accounts. http://www.bea.gov/ (accessed June 13, 2012).

U.S. Census Bureau, Population Division (2011) Intercensal estimates of the resident population by sex and age for the United States: April 1, 2000 to July 1, 2010 (US-EST00INT-01).

The 2030 Water Resources Group (2009) Charting our water future. http://www.2030waterresourcesgroup.com/water_full/Charting_Our_Water_Future_Final.pdf (accessed May 14, 2012).

Wise A, O'Brien K, Woodruff T (2011) Are oral contraceptives a significant contributor to the estrogenicity of drinking water? *Environmental Science and Technology* 45:51–60.

World Bank (2013) World development indicators: Agriculture and rural development. http://databank.worldbank.org/data/views/reports/tableview.aspx/ (accessed June 30, 2013).

WorldWatch Institute (2003) *Vital signals: The trends that are shaping our future 2003–2004*. London: Earthscan Publications.

Worldwater.org (2011) Water content of things: The world's water 2008–2009. http://www.worldwater.org/data20082009/Table19.pdf (accessed July 3, 2013).

2

Water Bioremediation

Margaret Wexler and Jeremy Burgess

CONTENTS

The Context

Bioremediation is the use of microorganisms and plants to clean up pollutants in the environment. All human life causes pollution to some degree. Only in very low-density agrarian societies, or among nomadic peoples, can the natural pace of environmental processes overcome the polluting effects of human societies. For such societies, pollution may eventually "go away" on its own. But today humans live increasingly in large cities; Tokyo has 30 million citizens; Seoul, Mexico City, and Mumbai each hold 20 million or so. City dwellers are not farmers or nomads; they are consumers and workers, often in industrial plants. The variety and amount of pollution we all produce and the rate at which we produce it mean that it is no longer a problem that merely goes away on its own; it must be tackled.

One of the foundations of sustainability is the harnessing of ecological processes to achieve desirable ends. Bioremediation is precisely such a process. In this chapter we shall see how both plants and microorganisms can be used to ameliorate the effects of pollution in a wide variety of environments. The pollutants include natural sewage waste from the mere process of being alive, to waste from factories and mines, together with the results of accidental events such as oil spills. For all these problems, bioremediation techniques are already well on the way to providing a sustainable solution.

Bioremediation: An Overview

Contamination of soil, groundwater, and sediments is a major problem facing the industrialized world. Frequently encountered contaminants include chlorinated solvents, aliphatic and aromatic hydrocarbons, heavy metals, polychlorinated biphenyls, pesticides and radioactive compounds. Bioremediation can play a role in removing or detoxifying all these, whether by treatment *in situ* (at the site) or *ex situ* (elsewhere). Bioremediation may involve the use of plants, a process known as phytoremediation, although most methods employ microorganisms able to convert pollutants into harmless compounds through oxidation–reduction (redox) reactions. In most cases, energy in the form of adenosine triphosphate (ATP) is generated from these reactions. In some instances pollutants themselves may be utilized as carbon substrates.

Biodegradation can take place in the presence or absence of free or bound oxygen. It is important to grasp the terminology of these different processes.

During *aerobic* biodegradation oxygen is the final electron acceptor or oxidizing agent, and is reduced to water. In most cases, the pollutant is mineralized—converted to carbon dioxide—and microorganisms increase in number.

In the absence of oxygen redox reactions are anaerobic or anoxic. In such cases the terminal electron acceptor is a species other than molecular oxygen. In *anoxic* conditions this is commonly nitrate or sulfate. Ferric ions or methane are widely used under conditions of *anaerobic* degradation. In practice, both aerobic and anaerobic reactions may be required to detoxify a pollutant.

In rare cases, microorganisms can degrade pollutants without obtaining carbon or energy. This arises when an enzyme whose usual biological function is to degrade a different compound, is also able to break down a pollutant, a process known as cometabolism. An example of this is the degradation of trichloroethylene (TCE) by methane monooxygenase, an enzyme made by methane-utilizing methanotrophic bacteria (Little et al. 1988). This enzyme normally converts methane to methanol. TCE is taken up by methanogens and methane monooxygenase converts it to TCE epoxide, which is extruded from the cell and readily degraded by a range of bacteria (Alvarez-Cohen and McCarty 1991).

Intrinsic bioremediation occurs naturally in most polluted environments, because microorganisms able to degrade pollutants are generally present. Microorganisms may be said to have been dealing sustainably with natural levels of pollution for millions of years. Where appropriate microorganisms are not present, or when the scale of the polluting event means that degradation would occur too slowly, bioaugmentation (the addition of specific microorganisms) or biostimulation (the stimulation of microorganisms already present) may be used (see Table 2.1 for examples).

Water Treatment

Wastewater Perspective

Wastewater is a general word that broadly means water that is—maybe temporarily—of no use or even hazardous. In physical terms, it is a combination of sewage, water from agricultural and industrial processes, and stormwater (the runoff from buildings and streets). Sewage mainly contains human excrement; wastewater may also contain chemical pollutants, rubbish, stones, and soil. In former times sewage was redirected to cesspools, natural water bodies, or onto fields. With the growth of large cities, these methods have become unsustainable, due both to the risk of widespread pollution and the fact that they are themselves wasteful of a precious resource. The wide-scale treatment of domestic sewage in specially designed treatment plants began in the late 19th century. It probably represents the greatest advance in human health of the last 200 years.

Biochemical Oxygen Demand

The objectives of wastewater treatment are to reduce organic and inorganic materials within it to a level that no longer supports microbial growth, to remove toxic materials, and to reduce levels of pathogens. If these objectives are attained, the water becomes usable once more.

TABLE 2.1

Microbial Bioremediation Strategies

Type of Bioremediation	Mechanism of Bioremediation	Example	Reference
Bioattenuation (intrinsic bioremediation)	Natural population of microorganisms degrades pollutant	Degradation of benzene, toluene, ethylbenzene-xylenes (BTEX) in aquifers as a result of accidental petrol leakage	Salanitro 1993
Biostimulation	Addition of nutrients provides the optimum nutrient ratio for growth of degrading microbial population	Addition of Inipol EAP22 to shoreline following *Exxon Valdez* oil spill in Prince William Sound, Alaska	Atlas and Hazen 2011
	Addition of cosubstrate enables degradation of pollutant through the action of enzyme whose target is the cosubstrate	Addition of methane to trichloroethylene (TCE)-contaminated aquifer to stimulate methane monooxygenase activity of methanotrophs	Semprini et al. 1990
	Addition of electron acceptor enables growth of degrading microbial population	Addition of oxygen to TCE-contaminated aquifer to stimulate aerobic growth of methanotrophs	Semprini et al. 1990
	Addition of surfactant improves bioavailability and dispersal of pollutant	Injection of surfactant COREXIT 9500 into *Deepwater Horizon* wellhead to reduce formation of surface oil slicks	Atlas and Hazen 2011
Bioaugmentation	Addition of pure culture(s) of microbial degrading species to a site not previously exposed to pollutant or to a site that requires enhanced biodegradation	Addition of commercially available strains of the bacterium *Dehalococcoides ethenogenes* for degradation of chlorinated solvents in soil and groundwater	Hazen 2010
	Addition of preadapted microbial community to a site not previously exposed to pollutant or to a site that requires enhanced biodegradation	Addition of mixed culture to wastewater to enhance nitrification	Leu and Stenstrom 2010
	Addition of genetically engineered microorganisms	Addition of *Pseudomonas fluorescens* HK44 to naphthalene-contaminated soils; HK44 carries a *lux* reporter gene fused to polyaromatic hydrocarbon degradation genes, allowing this bacterium to bioluminesce as it degrades specific polyaromatic hydrocarbons such as naphthalene	Ripp et al. 2000

One of the most important tests used to assess the efficiency of treatment is biochemical oxygen demand (BOD). This is a measure of the amount of readily oxidizable organic matter present in a water sample. To determine BOD, diluted samples of wastewater are seeded with a mixed microbial culture, saturated with oxygen, and incubated in the dark at 20°C. Dissolved oxygen is measured at the start and end of the incubation period, normally, after 5 days (referred to as BOD_5). Wastewater treatment of domestic sewage typically reduces BOD_5 from around 250mg/l to 25mg/l (European Commission 1991).

Stages of Wastewater Treatment

The treatment of wastewater involves a series of stages designed to produce an end product—the effluent—that can safely be discharged to surface water bodies or used as the input for drinking water purification facilities. It is an example of bioremediation, since it involves industrial-scale use of microorganisms.

The preliminary stage of the process is the physical removal of coarse materials (e.g., stones and rags) that would otherwise interfere with subsequent processes. This material is normally disposed to a landfill. Primary treatment employs screening and grit removal to further clarify the wastewater, followed by a period in a primary sedimentation tank. During this phase, sludge—the biotic and abiotic solids that passed through the screening process—settles to the bottom of the tank. This sludge can be removed and further treated.

Primary treatment is mainly a physical process, although some fermentation reactions occur in the primary sedimentation tank, and BOD_5 is typically reduced by 30% to 40%. Early wastewater treatment plants used primary treatment only.

Secondary treatment exploits microorganisms to accelerate the degradation of organic and inorganic matter present in the wastewater. BOD_5 is typically reduced by 80% to 90%. A variety of secondary treatments is available, with the method of choice depending on such factors as population size and density, environmental constraints, and the availability of finance. The most widely used treatment method throughout the world is the activated sludge process, developed by Arden and Lockett in Manchester, United Kingdom, in 1914. This is discussed in more detail later. Other commonly used processes include trickling filters and rotating biological contactors. In the former, wastewater is trickled through a permeable medium (for example, a filter bed of graded stones). Rotating biological contactors comprise circular discs, often of PVC, which are rotated slowly as wastewater flows over them. Both of these processes aerate the wastewater. This encourages the degradation of organic matter as the wastewater passes mixed populations of microorganisms growing as biofilms on a solid support, whether stones or PVC discs.

Nonindustrial Methods of Treatment

The most ancient method of treating wastewater involves the use of stabilization ponds or lagoons. This relies on natural biological activity and is generally a low-cost method with low energy inputs. It corresponds to a secondary treatment process.

The most common type of stabilization pond is the facultative pond, which relies on aerobic and anaerobic processes (Bitton 2010). Photosynthetic algae take up nutrients including phosphate and nitrogen. The oxygen from photosynthesis enables aerobic bacteria to grow and degrade organic matter. Solid wastes, including dead bacteria and algae, settle to the bottom of the pond and are broken down by anaerobic bacteria. Zooplankton within the pond feed on living bacteria and algae.

Facultative ponds are particularly suitable for use in warm climates with intense sunlight. Their disadvantages include the need for large areas of land, with attendant odor problems, particularly from hydrogen sulfide, and the possibility of mosquito infestation.

Other types of seminatural secondary treatment include artificial wetlands and soil aquifer treatment. Maturation ponds can be used in addition to stabilization ponds as a form of tertiary treatment. The maturation process occurs in shallow aerobic water; it helps to remove excess nutrients and to further reduce BOD_5. Important pathogens such as nematodes, helminthic eggs, and protozoan cysts may also be inactivated.

Individual on-site sewage treatment systems often use septic tanks in which microorganisms anaerobically degrade waste. Effluent flows through perforated piping into a drainage field and is further degraded by soil microorganisms. Increasingly, dry toilet systems are being used, especially at roadside or remote locations. These function with very little water and rely on aerobic decomposition, so tend to work faster than septic systems.

Activated Sludge Process

Activated sludge is the most effective secondary wastewater treatment process available (Figure 2.1). It is versatile and may be reconfigured to enable biological nutrient removal.

As we have seen, primary treatment is largely a physical process involving some anaerobic fermentation. Following primary treatment, the wastewater is agitated and aerated in large tanks. This causes the growth of microorganisms as a particular form of mixed biomass known as a biological floc. Flocs are comprised predominantly of bacteria and contain a wide range of species, including slime-forming *Zoogloea* and filamentous bacteria, protozoa, viruses, and some fungi, embedded in an extracellular polymeric matrix. Important, flocs are large aggregations of microorganisms within a gelatinous mass that is bulky enough to settle under gravity. This is essential for the eventual clarification of the processed wastewater.

FIGURE 2.1 (See color insert.)
The aerobic stage of an activated sludge wastewater treatment process. (© Photo by Margaret Wexler.)

The microbiology of activated sludge is complex. Within the flocs, the microorganisms break down carbonaceous matter (see Equation 2.1). In addition to flocs, other microorganisms are important in the ecology of activated sludge. These include free-living protozoa and ciliates, feeding on free-living bacteria; flagellates, which feed on organic particles: and rotifers, which ingest pathogenic protozoan oocytes (ova) and nematodes.

The wastewater passes from the aeration tank into a settlement tank. At this stage—clarification—the flocs sink to the bottom to produce a layer of activated sludge.

Following clarification, a proportion of the activated sludge (typically 20% to 40%) is recycled back into the aeration tank. A well-performing activated sludge plant will generate good effluent quality and remove at least 90% of the dissolved organic matter. As biomass is continually being produced, some of the sludge must be removed as waste, otherwise it would eventually choke the aeration tank. An activated sludge plant inevitably generates a large amount of sludge. This is a disadvantage of the process, but as explained later, this excess sludge can be put to good use.

$$\text{Organic matter} + O_2 \rightarrow CO_2 + NO_3 + SO_4 + PO_4 + H_2O \tag{2.1}$$

Sludge Stabilization

Wasted-activated sludge is disposed of by incineration or landfilling; it may also be recycled to agricultural use. Prior to landfill disposal, it is dewatered and then treated using heat or chemicals. Sludge destined for agricultural use as a fertilizer is first dewatered and then chemically treated, usually with lime in order to raise its pH and destroy pathogens. Alternatively, it may be composted or digested in an aerobic or anaerobic digester. The residue from digestion is further dewatered before being used as fertilizer. Wasted-activated sludge can also be stabilized in lagoons.

Anaerobic Digestion

Anaerobic digestion has many advantages as a method for stabilizing activated sludge. It generates the biogas methane, which can be used to power the wastewater treatment plant or for other energy requirements. It uses less energy than aerobic processes and it results in the biodegradation of many recalcitrant compounds. It can also be used as a way of treating sludge from the primary sedimentation tank and in the processing of agricultural slurries. A disadvantage of anaerobic digesters is they generate effluent with high N and P, which may require further treatment.

Anaerobic digestion is carried out in large enclosed tanks, and involves a series of digestive and fermentative reactions carried out by different bacteria and archaea under anoxic conditions. Anoxic conditions occur when molecular oxygen is absent and other oxygen species such as carbon dioxide are present. Complex polymers including polysaccharides, lipids, and proteins are hydrolyzed by extracellular microbial enzymes to monosaccharides, fatty acids, and amino acids. These are fermented by acidogenic bacteria to organic acids, alcohols, and acetate. Acetogenic bacteria convert organic acids and alcohols to acetate and methanogenic archaea convert acetate to methane and carbon dioxide, or carbon dioxide and hydrogen to methane and water. The final gaseous product comprises approximately 50% to 75% methane, 25% to 35% carbon dioxide, together with other gases (Equation 2.2).

$$\text{Organic matter} \rightarrow CH_4 + CO_2 + H_2 + NH_3 + H_2S + H_2O \qquad (2.2)$$

Nutrient Removal from Effluent: The Context

Eutrophication of receiving waters with phosphorus and nitrogen represents a potentially serious problem that may occur following wastewater effluent discharge (Figure 2.2). Eutrophic waters can produce overgrowth of toxin-producing Cyanobacteria and phytoplankton. This overgrowth causes oxygen depletion, to the detriment of aquatic life. Water containing ammonia and nitrites is especially toxic to fish. Freshwater environments are highly impacted by phosphorus, whereas marine environments are more sensitive to high levels of nitrogen.

Wastewater treatment plants are legally obliged to remove these micronutrients prior to effluent discharge. Nitrogen and phosphorus can be

FIGURE 2.2 (See color insert.)
Eutrophication of a lake as a result of sewage contamination. (© Photo by Jeremy Burgess.)

removed from wastewater by chemical or physical means. These techniques damage the environment and are costly, since they may involve the use of ferric salts, chlorine, or the venting of ammonia into the atmosphere following treatment with lime. Biological nutrient removal relies on the action of microorganisms, achieved by manipulating environmental conditions during wastewater treatment. Most biological removal processes involve modifications of activated sludge treatment systems, although reduction may be achieved using rotating biological contactors and trickling filters. Lagooning and reed beds can also be effective. Well-managed biological nutrient removal processes do not produce environmental damage, and despite relatively high setup costs, are cheaper to run than chemical and physical processes (Wexler 2011).

Biological Nitrogen Removal

Nitrogen in sewage occurs principally as urea from urine and organic nitrogen from feces. Both are readily converted to ammonium salts.

The nitrogen is removed in two stages, involving nitrification and denitrification. The simplest configuration uses an aeration stage and an anoxic stage, after which the wastewater is recycled for a further passage through the aeration stage, resulting in carbonaceous oxidation. During the initial aerobic stage, ammonia-oxidizing bacteria such as *Nitrosomonas* species convert ammonium to nitrite, and nitrite-oxidizing bacteria (often *Nitrobacter* species) oxidize the nitrite to nitrate. Nitrifying bacteria are autotrophs, using carbon dioxide as their carbon source. They grow slowly, which means that to be effective passage through the aeration stage must be slow enough to produce a long sludge age and adequate hydraulic retention time. Bioaugmentation—the addition of bacteria adapted to specific conditions— may be used to enhance nitrification.

The denitrification stage takes place in an anoxic tank. Denitrification is catalyzed by a wide range of bacteria using an organic carbon substrate such as methanol as an electron donor. Suitable carbon substrates must be added to the wastewater if they are not already present. Treatment plants serving industrial premises will often receive suitable substrates as factory waste, via the sewage network. The rate of denitrification depends on the concentration of organic substrate and of biomass. Under anoxic conditions, different bacteria catalyze a series of reductive reactions; nitrate to nitrite, nitric oxide, nitrous oxide, and finally nitrogen gas. Nitrous oxide, a potent greenhouse gas, is inevitably released during this process. Nitrification/denitrification requires considerable energy for aeration and generates large sludge volumes. The need to add an external carbon substrate may incur further cost.

A more recently established biological nitrogen removal method is the SHARON/Anammox process, which is used to treat effluent from sludge dewatering facilities and other ammonia-rich wastewater with low organic carbon content. During the SHARON (*s*ingle reactor system for *h*igh activity *a*mmonium *r*emoval *o*ver *n*itrite) process, around half the ammonium present is oxidized to nitrite, aerobically at 30°C to 40°C. Anammox (*an*aerobic *ammo*nium *ox*idation) combines nitrite with ammonium forming nitrogen gas and water. This stage selects for autotrophic, anaerobic ammonia-oxidizing bacteria, including *Candidatus* "Brocadia anammoxidans," an unculturable species belonging to the Planctomycetales.

SHARON/Anammox operates with a shorter sludge retention time, approximately 60% less O_2, requires no organic carbon, and converts an estimated 90% of the incoming nitrogen into nitrogen gas, leaving approximately 10% as nitrate (Kumar and Lin 2010).

Biological Phosphorus Removal

Phosphorus removal can be achieved when activated sludge wastewater treatment plants are operated with alternating anaerobic and aerobic reactor zones, a process known as enhanced biological phosphorus removal (EBPR), or *luxury uptake* of phosphorus. This process transfers phosphate from wastewater to sludge by incorporating it into bacterial cells as polyphosphate.

The initial stage is anaerobic and occurs as the wastewater moves slowly in a channel from the primary settlement tank toward the aeration tank. Bacteria known as polyphosphate accumulating organisms (PAOs) synthesize carbon storage molecules called polyhydroxyalkanoates (PHAs) and glycogen from short-chain volatile fatty acids present in the influent. The energy source for this is stored polyphosphate within the bacteria. These anaerobic reactions result in the release of phosphate into the wastewater. When the wastewater reaches the aeration tank the transformations are reversed. The PAOs accumulate polyphosphate and glycogen, and they degrade PHAs. This results in a transfer of phosphate from the wastewater to the sludge. The wastewater then moves to a settlement tank and is clarified as the sludge falls to the bottom. The clarified effluent contains low levels of P and can be safely discharged. The P-rich sludge is either returned to the anaerobic phase, or collected and further treated for eventual use as fertilizer.

Combined biological phosphorus and nitrogen removal can be achieved using various configurations including the University of Cape Town (UCT) and modified Bardenpho processes, which involve an initial anaerobic stage followed by various anoxic and aerobic stages. The anaerobic zone is maintained free of oxygen by, in this case, recycling the sludge to the anoxic zone.

Treatment of Other Types of Polluted Water

Bioremediation of Oil Spills: The Context

Every year millions of tons of oil are released into the environment. Natural seepage occurs from underground reservoirs both on land overlying large oil deposits and under the oceans. Over long periods of time this process has given rise to microorganisms that are adapted to use oil products as an energy or carbon source. Crude oil is today perhaps the most reviled of pollutants; but as we have suggested before, at natural rates of production, the ecosphere can, and did, manage the problem "sustainably."

In the last 150 years, the amount of oil released to the environment has increased vastly due to human activity. Accidents involving oil tankers and oil rigs are among the most high-profile, damaging, and expensive pollution events in today's world. Bioremediation is increasingly used to deal with the consequences of these accidents, exploiting the microorganisms that have gained the ability to metabolize oil over evolutionary time.

Nature of Crude Oil

Crude oil is a complex mixture of more than 17,000 compounds. Most components are biodegradable hydrocarbons including alkanes, cycloalkanes, and volatile aromatics. Heavy metals, such as Ni and V, heterocyclic compounds, asphaltenes, and resins make up only a small proportion of crude oil. These have very slow degradation rates and are generally very toxic (Head et al. 2006).

Oil deposits originated from algae and other marine plankton laid down under the seafloor millions of years ago. Today, some marine microorganisms synthesize long-chain hydrocarbons similar to those found in crude oil, for example, the green algae *Botrycoccus braunii* can synthesize C_{30}-C_{36} chain length hydrocarbons (Madigan et al. 2011).

Oil and Microorganisms

Crude oil degradation by microorganisms is ubiquitous. Species of bacteria, fungi, and some green algae can degrade hydrocarbons found in oil. The most significant are the hydrocarbonoclastic or oil-degrading bacteria, many of which are members of the gamma proteobacteria, to which *E. coli* is also affiliated. Probably the most important oil degrader in the marine environment is *Alcanivorax borkumensis*, which degrades alkanes. This species is an early colonizer following marine oil spills and has a highly restricted carbon substrate spectrum that includes hydrocarbons, fatty acids, and pyruvate. Later colonizers include *Cycloclasticus* species, which degrade polycyclic aromatics, and catechol-degrading *Pseudomonas* species. Many of these synthesize surfactants which help break up the oil slick and make it easier to biodegrade. Normally, the bacterial population of marine environments contains less than 1% of oil degrading species, but following an oil spill this proportion may increase to over 10% (Head et al. 2006).

Oil-degrading bacteria are found worldwide, in all waters, sludge, soil, and sand. Many factors are important in oil degradation including the availability of inorganic nutrients, notably nitrogen and phosphorus. Physical parameters, microbial ecology, and bioavailability of hydrocarbons are also significant.

Degradation of oil begins with the evaporation of volatile hydrocarbons followed by the degradation of medium and longer chain aliphatic compounds. Branched chain and polycyclic compounds are slower and more difficult to degrade.

A diverse range of enzymes is involved in oil degradation although most degradation uses monooxygenases or dioxygenases. Hydrocarbons are utilized as electron donors and as carbon sources. Most degradation is aerobic; however, anaerobic degradation is also important, and in this case sulfate and nitrate are commonly used as electron acceptors.

Case Studies of Two Oil Spills: Exxon Valdez *and* Deepwater Horizon

Exxon Valdez *Oil Spill*

The *Exxon Valdez* disaster has two dubious claims to fame in the field of sustainability studies. It was the first time contingent valuation was tried as a way of assessing the monetary level of damage compensation from an oil spill. It was also the first time bioremediation was used to clean up an oil spill on such a large scale.

On March 24, 1989, the oil tanker *Exxon Valdez* was en route from Valdez, Alaska, to Los Angeles. It ran aground on Bligh Reef, releasing 42 million liters of crude oil into Prince William Sound, causing 2000 km of shoreline to be contaminated.

The initial response involved the use of booms and skimmers, physical methods designed to limit the spread of the oil and to allow some of it (a valuable commodity) to be recovered, or if recovery was not economically worthwhile, to be burnt at sea. Unfortunately, in the days after the spill, wind-driven mixing of oil and seawater resulted in an emulsion, which was no longer conducive for burning, and difficult to remove from the surface or shore. Another physical part of the early response at sea was to apply dispersants, the effects of which were reduced due to lack of wave action. Prevailing currents meant oil was able to reach the shoreline. The use of dispersants was discontinued within days.

On the shoreline, hoses spraying seawater were used to flush oil away from the shore. Manual cleaning and mechanical means were also employed to clean the oil on the shoreline. Such physical, even manual, methods are slow, inefficient, and involve unpleasant work that may cause additional mechanical damage to delicate shore habitats.

Bioremediation efforts began with the application of dispersants and bio-augmentation, using *Pseudomonas,* but treatments were ineffective. However, biostimulation, using Inipol, containing nitrogen, phosphorus, surfactant, and Customblen, a granular slow release fertilizer, was successful in treating 120 km of shoreline over a period of 2 months. Within 2 to 3 weeks, oil on the shore was visibly degraded. Concerns that these fertilizers might cause eutrophication or toxicity to marine species later transpired to be unfounded. Further studies demonstrated nitrogen was the main component enabling enhanced degradation.

Prince William Sound was a pristine environment and the environmental consequences were devastating. Thousands of sea otters, a quarter of a million sea birds, and hundreds of harbor seals died immediately after the spill. The fishing and tourism industries were seriously impacted. Most of the oil was broken down within a few years after the spill. The slowness of this process reflects the low temperatures at this site. Toxic subsurface oil remains present, and the chronic exposure to its components, particularly polycyclic aromatics, continues to affect wildlife (Peterson et al. 2003).

Macondo (Deepwater Horizon) *Blowout: Overview*

In popular memory, the *Exxon Valdez* oil spill remained the iconic oil accident for two decades. However, on April 20, 2010, it was dwarfed when the Macondo exploration well connected to the *Deepwater Horizon* drilling rig suffered a blowout following a catastrophic explosion and fire on the rig. Consequently, 770 million liters of crude oil were released into the Gulf of Mexico. This was the largest marine oil spill in history. Eleven workers died and the flow of hydrocarbons continued for 87 days.

The release of oil occurred 1500 m below the ocean surface, 50 km from the shore. Six thousand sea turtles, 400 dolphins, and 82,000 birds died as a result. The Gulf of Mexico is a warm water environment with important fishing and tourist interests, and represents a valuable economic resource to the coastal populations of five U.S. states. At the peak of the disaster, 230,000 sq km of federal fisheries were closed. In this highly political environment, the response to the disaster had to be immediate and on a very large scale. The site of the leakage was at an unprecedented depth below the ocean surface. This led to the adoption of some untested techniques, both in engineering and bioremediation (Atlas and Hazen 2011).

Macondo (Deepwater Horizon) *Blowout: The Response*

The initial response was of a physical nature, deploying booms and skimmers, and controlled burning. To try to prevent shoreline contamination, 4000 km of booms were eventually used, and 130 million liters of oil were surface burned. The physical cleanup involved over 460,000 people.

For the first time dispersants (COREXIT) were used underwater, in an attempt to prevent flammable oil from reaching the surface. This resulted in a deep sea cloud of oil, found to contain high concentrations of psychrophilic bacteria, initially dominated by members of the Oceanospirillales, then *Cycloclasticus* and *Collwellia*. Alkanes and aromatics were rapidly degraded by these consortia. The well leaked methane in addition to oil. Thousands of tons of methane were shown to have been degraded by numerous methane-utilizing methanotrophs. Approximately 40% of the oil reached the sea surface. Dispersion, along with evaporation and photooxidation, helped to break down this surface oil. Three weeks after the leak was capped oil slicks were no longer visible on the sea surface and comparatively little oil had reached the shore. Underwater plumes continue to persist and the most serious damage is likely to have occurred in the deep ocean (Atlas and Hazen 2011).

Disaster Comparison

Compared to the *Exxon Valdez* spill, the response to the *Deepwater Horizon* spill was faster, as it had to be. A tanker disaster can only spill the entire contents of the vessel once; an oil well will continue to gush until capped. Unlike Prince William Sound, the environment of the Gulf of Mexico was not pristine. A history of oil seepages and spills meant that a microbial population well adapted to degrading oil was already present. The weather conditions and currents were more favorable in the Gulf than in Prince William Sound, facilitating oil removal by skimming, although storms did eventually impede the engineering work at the wellhead. The Gulf has warmer waters, increasing evaporation and enabling more rapid degradation. A far higher proportion of oil from the Gulf accident did not reach the shore and was widely dispersed. By February 2011, 99.6% of federal fisheries were open to fishing again. The *Exxon Valdez* was carrying heavy crude oil, which is more difficult to degrade than the light crude oil and methane that leaked from the seafloor following the *Deepwater Horizon* event. The long-term effects of the use of underwater dispersants in the Gulf are as yet unknowable.

Bioremediation of Heavy Metals and Radioactive Compounds

The contamination of soil, sediments, and water by heavy metals is a world-wide problem. Its impact is most significant in the case of drinking water due to the acute toxicity of heavy metals. Ubiquitous elements such as aluminum can be mobilized by leaching ("acid rain"), causing damage to trees and polluting waterways. Mining and other industries contribute to the load of contaminants in soils and sediments, which may also then be leached into water supplies. Heavy metals with the highest toxicity have stable oxidation states, for example, Cd^{2+}, Pb^{2+}, Hg^{2+}, Ag^+, Zn^{2+}, and As^{3+}. These react with biomolecules to form stable biotoxic compounds. If heavy metals are not removed or neutralized from the environment, they will bioaccumulate in lower-form plants and animals, and may eventually be ingested by mammals and humans.

Bioremediation strategies are available to deal with metal contamination based on the ability of both bacteria and plants to concentrate metals within their cells or to adsorb metals on their surfaces. Such methods are in increasingly common use, since the physicochemical alternatives are expensive and less suited to large volume remediation.

One strategy, known as biotransformation, exploits the ability of certain microorganisms to change the oxidation state and solubility of the metal. Often bacteria do this when they use heavy metals as terminal electron acceptors. In other instances, metal reduction occurs as a means of detoxification using enzymes that may be excreted. Through these activities, many different heavy metals present at toxic concentrations in the environment can be reduced from a soluble to an insoluble form (Gadd 2010). Reduced metals all precipitate as oxides, hydroxides, or coprecipitate with ferric minerals that form during ferrous reoxidation. Metal-contaminated environments may also contain high concentrations of suspended solids; oxidation of organic compounds is often coupled to metal reduction. One example is remediation of chromium-contaminated waters. Chromium is found in the effluent of cooling water from power stations and as a result of industrial processes, often resulting in groundwater contamination. The reduction of chromate (CrO_4) in contaminated waters to chromium hydroxide [$Cr(OH)_3$], which precipitates in the sediment, is carried out by various microorganisms, including the soil bacterium *Pseudomonas putida* (Cervantes et al. 2001). This results in the conversion of Cr^{6+}, which is highly toxic and carcinogenic, to nontoxic Cr^{3+}.

Other types of microbial biotransformation solubilize heavy metals, increasing their mobility and therefore facilitating their removal from the environment. An example of this is the solubilization of copper in copper ores by *Acidithiobacillus thiooxidans*, discussed later in the section titled "Bioleaching."

Heavy metals may be converted to a less toxic form by volatilization. Several heavy metals, for example mercury, are converted from an ionic (Hg^{2+}) to a nontoxic gaseous elemental form (Hg^0) through biological action. However, it should be noted that elemental mercury may also be reoxidized to mercuric ion by photooxidation. Mercury in its ionic form is very toxic even at low concentrations and due to its widespread industrial use (for example, in batteries, pesticides, and electrical equipment) environmental contamination is common. It may originate from the mining of mercury, its industrial use, or photooxidation of elemental mercury from burning of fossil fuels and wastes. Microorganisms readily methylate mercuric ion to methyl or dimethyl mercury, which has even higher toxicity. Bacterial reduction of the mercuric ion to elemental mercury, through the action of mercuric reductase, is catalyzed by some bacterial species, often pseudomonads (Gadd 2010). Certain plants, including tobacco, also catalyze this transformation, enabling volatilization of elemental mercury (Roane et al. 1995).

Another type of bioremediation used for metal-contaminated soil and water is biosorption, and bioaccumulation by intracellular or extracellular sequestration of metal. Intracellular sequestration, by microorganisms or plants, uses metallothioneins or phytocheletin, proteins that bind to heavy metals such as mercury and cadmium (Roane et al. 1995; Cobbett and Goldsbrough 2002; Gadd 2010). Metals remain immobilized inside the cell or within vacuoles. Passive binding of metals to biological material, such as polysaccharides on bacterial cell surfaces or to algae cell walls, is another means of metal remediation. The biological material need not be living and may be recycled following metal recovery. An example is the use of a commercially available immobilized nonliving microalgae to remove cadmium (Monteiro et al. 2009).

Bioremediation of Radioactive Compounds

Radioactive waste originates from uranium mines, nuclear power stations, or weapons production facilities. Uranium, plutonium, cesium, and strontium are the most common radioactive contaminants. Phytoremediation by bioaccumulation of radionuclides can help restore contaminated sites, although the plants remain toxic and biological material may require further treatment. A number of bacterial species including *Geobacter sulfurreducens* are able to reduce soluble uranium U^{6+} to insoluble U^{4+}. Reduction requires an electron donor (for example, organic acids) and recovery of insoluble uranium may be problematic.

Recently, research efforts have centered on the development of bioelectrical systems to treat radioactive and heavy metal contamination of subsurface environments. These systems are based on the ability of some microorganisms including *Geobacter* to accept electrons from electrodes; such bacteria are known as electrode-oxidizing bacteria. Immobilized uranium can be removed from the subsurface by pulling up the electrode following a period of treatment and stripping off the U^{4+} with bicarbonate (Gregory and Lovley 2005).

Acid Mine Drainage: The Context

Acid mine drainage is a metal-rich sulfuric acid solution released from mine tunnels, open pits, and waste rock piles. It is generated by the oxidative dissolution of pyrite (FeS_2) and other sulfide minerals present in coal and metal ores. It occurs following exposure of underground ores to water and oxygen. When a mine is abandoned, pumping of excess water ceases, causing the mine to be flooded by groundwater, resulting in contamination of the local watercourse. Paradoxically, the closure of numerous coal mines has exacerbated the problem of acid mine drainage.

Oxidative dissolution of sulfide minerals proceeds spontaneously, and is accelerated by acidophilic bacteria and archaea, including the iron and sulfur-oxidizing bacterium *Acidothiobacillus ferrooxidans* and iron-oxidizing archaean *Ferroplasma acidarmanus*. The overall reaction may be summarized as

$$FeS_2 + 14Fe^{3+} + 8H_2O \rightarrow 15Fe^{2+} + 2SO_4^{2-} + 16H^+ \qquad (2.3)$$

Heavy metals leach into solution and precipitate into sediments of rivers and lakes. Jarosite ($HFe_3(SO_4)_2(OH)_6$), which is poisonous to aquatic life, is frequently generated and appears as an orange-yellow precipitate. In addition, the low pH and high concentration of metals are in themselves highly toxic to aquatic species.

Treatment Methods for Acid Mine Drainage

Treatment of acid mine drainage uses active or passive techniques. Active treatment involves the use of highly engineered water treatment systems, which require constant maintenance, whereas passive methods require little maintenance and are self-contained in regard to treatment and waste.

The most common abiotic remediation is an active treatment that uses lime to neutralize acid, and aeration and oxidizing agents to accelerate the rate of ferrous iron oxidation and precipitation of iron and other metals as hydroxides. A metal-rich toxic sludge is generated, which must be treated further prior to disposal (Johnson and Halberg 2005). This treatment is neither economically nor environmentally sustainable.

Bioremediation of acid mine drainage is based on the abilities of microorganisms to generate alkalinity and immobilize metals, and of plants to bioaccumulate metals.

Passive methods use natural or constructed wetland ecosystems. One treatment system that has been used since the 1980s and has shown good long-term performance and high metal removal rates is that of combined anoxic limestone drains and aerobic wetlands. Initially, water flows through an anoxic limestone drain that is impervious to air and water, and which contains gravel. The minewater is acidic. As it flows through, limestone dissolves, which raises the pH while maintaining iron in the ferrous form. The minewater is then directed to shallow, aerobic wetlands where abiotic and microbiologically catalyzed oxidation reactions precipitate the metals as oxides and hydroxides. Plants stabilize the precipitates, which are retained within sediments. One disadvantage to this system is the eventual coating of the limestone with aluminum and iron hydroxides, necessitating its replacement (Johnson and Halberg 2005).

Active bioremediation involves continuous application of alkaline materials to neutralize acidic minewaters and precipitate metals. Sulfidogenic bioreactors allow performance monitoring and efficient recovery of metals for economic gain. However, they are costly to construct and operate. Acid mine drainage enters the bioreactor, in which an organic carbon and energy source has been added, and sulfate-reducing heterotrophic bacteria such as *Desulfovibrio* catalyze the reduction of sulfate to sulfide. This reduces acidity by transforming a strong acid (sulfuric) into a weaker one (hydrogen sulfide) (Johnson and Halberg 2005). The reduction of sulfate to sulfide removes toxic metals from the minewater, since many metals form highly insoluble sulfide precipitates, which may be recovered and reused.

Effluent originating from Iron Mountain, a former metal and pyrite mine in California, was previously responsible for significant pollution of the Sacramento River. Drainage from the mine is now treated at the source, using a lime neutralization/high-density sludge water treatment plant (U.S. Environmental Protection Agency 2006). Metals are precipitated from solution and recovered using bioreactors. This highly polluted natural environment was the largest source of surface water pollution in U.S. history and a financial settlement worth $950 million has been committed to cleaning up this site. It is estimated that Iron Mountain will continue to produce acid mine drainage for 2500 to 3000 years (U.S. Environmental Protection Agency 2006).

Bioleaching

Microorganisms, responsible for acid mine drainage, such as *Acidothiobacillus ferrooxidans*, may also be used to recover metals such as copper from mines, in a process known as bioleaching (Gadd 2010). Metal-containing ore is dumped in a leach dump and sulfuric acid is added to maintain a low pH. Covellite (copper sulfide) is oxidized by oxygen and ferric iron, which generates solubilized copper (Cu^{2+}). Aerobic oxidation of covellite and pyrite, also present in metal-rich ores, may be spontaneous or mediated by *A. ferrooxidans*. The liquid from the pile containing soluble copper is transported to a precipitation plant containing scrap steel (Fe^0). The steel reduces the copper ions in solution to metallic copper (Cu^0), which can be recovered. Instead of copper ions, the solution now contains ferrous iron (Fe^{2+}). It is sent to an oxidation pond where ferrous iron is oxidized by *A. ferrooxidans* to ferric iron, which is recirculated and used once again to oxidize more covellite. Biological oxidation is essential for regeneration of ferric iron, as under acidic conditions, it is not spontaneously oxidized. Bioleaching is also used to recover iron, zinc, gold, and uranium from low-grade ores and mine tailings.

Concluding Remarks

Bioremediation is a very good illustration of a sustainable solution to current environmental problems. It makes use of natural processes and for the most part increases their efficacy without causing additional long-term problems.

It is not above criticism. For example, the treatment of human waste is regarded by critics as involving an unnecessarily large use of water, particularly in the context of a future in which the supply of potable water is likely to become increasingly problematic. It is certainly rational to point out the folly of using potable water for, for example, garden watering and toilet flushing. But this particular criticism can be countered by pointing to the enormous amount of engineering work that would be required to separate gray and potable water supply to domestic premises in the world's megacities. And it is not mere complacency to restate that the benefits of the present system have been and continue to be immeasurably great. However, where new urban developments are built, this differentiation of potable and greywater is feasible, as is illustrated by the adoption of an integrated urban water management system at a number of developments in Australia, the United States, and elsewhere. A common solution is to arrange for reuse of gray water by means of dual reticulation; where two separate pipes deliver water of different qualities into homes, commercial premises, or industries. One pipe carries potable water and the other nondrinking water. This may be recycled water or obtained from alternative sources such as groundwater. Nondrinking water is used for toilet flushing, watering gardens, clothes washing or car washing, and for industrial processes.

In general, then, bioremediation is a very good example of a sustainable solution to current environmental problems. It makes use of natural processes and for the most part increases their efficacy without causing additional long-term problems.

A more general weakness of bioremediation has been hinted at: the unavailability in nature of organisms able to degrade novel chemical compounds. In such instances *in situ* bioaugmentation might appear to be the most likely solution. However, numerous trials have demonstrated better success is usually achieved using biostimulation. This may be partly due to the inability of introduced microorganisms to survive in their new environment caused by, for example, competition from indigenous species and lack of required nutrients. It is also probable that in many cases only complex microbial consortia can carry out the multistep degradation processes necessary to mineralize or detoxify the xenobiotic.

One notable success of bioaugmentation is the use of commercially available strains of the bacterium *Dehalococcoides ethenogenes* for degradation of chlorinated solvents in soil and groundwater (Hazen 2010). This bacterium uses chlorinated compounds as terminal electron acceptors in a process known as reductive dechlorination. It is one of only a small number of species from the phylum Chloroflexi, shown to fully degrade chlorinated ethenes such as trichloroethylene to ethene (a nontoxic gas). In this particular case, biostimulation alone is often unsuccessful as *D. ethenogenes* is not ubiquitous or present in low numbers (Bedard 2008).

An alternative approach is product substitution, the use of biodegradable plastics in food packaging being a clear example. However, biodegradability is not an unalloyed good. It is clear that no one would wish to have the plastic materials in their various "durable" consumer items to be liable to degradation by microorganisms or sunlight. A possible solution to this might be the development of organisms genetically modified to specific tasks of chemical degradation, but this would have to be undertaken with the greatest care to ensure that such organisms were safely confined. Some introductions have taken place, for example, *Pseudomonas fluorescens* HK44, the first genetically engineered microorganism approved for field release in the United States. This bacterium carries a *lux* reporter gene fused to polyaromatic hydrocarbon degradation genes, resulting in bioluminescence during degradation of specific hydrocarbon compounds such as naphthalene (Ripp et al. 2000).

Although bioremediation can be a cost-effective solution, it is often expensive and slow. One pragmatic solution is intrinsic bioremediation—letting the intrinsic microorganisms deal with the contamination problem undisturbed. In such cases the degradation of the pollutant needs to be monitored. Bacterial biosensors, often genetically engineered to respond to the presence of a specific contaminant, are being increasingly used to monitor the disappearance of toxins.

A final thought on this topic: notice the word *remediation*. Bioremediation does offer a solution to a problem. But it does not remove the source of the problem.

References

Alvarez-Cohen L, McCarty PL (1991) Effect of toxicity, aeration and reductant supply on trichloroethylene transformation by a mixed methanotrophic culture. *Applied and Environmental Microbiology* 57:228–235.

Atlas RM, Hazen TC (2011) Oil biodegradation and bioremediation: A tale of the two worst spills in U.S. history. *Environmental Science and Technology* 45:6709–6715.

Bedard DL (2008) A case study for microbial biodegradation: Anaerobic bacterial reductive dechlorination of polychlorinated biphenyls—From sediment to defined medium. *Annual Review of Microbiology* 62:253–270.

Bitton G (2010) *Wastewater microbiology*. Hoboken, NJ: John Wiley & Sons.

Cervantes C, Garcia JC, Devars S, Corona FG, Tavera HL, Torres-Guzman J Carlos, et al. (2001) Interactions of chromium with micro-organisms and plants. *FEMS Microbiology Review* 25:335–47.

Cobbett C, Goldsbrough P (2002) Phytochelatins and metallothioneins: Roles in heavy metal detoxification and homeostasis. *Annual Review of Plant Biology* 53:159–182.

European Commission (1991) European Commission Urban Wastewater Directive 91/271/EEC.

Gadd GM (2010) Metals, minerals and microbes: Geomicrobiology and bioremediation. *Microbiology* 156:609–643.

Gregory KB, Lovley DR (2005) Remediation and recovery of uranium from contaminated subsurface environments with electrodes. *Environmental Science and Technology* 39:8943–8947.

Hazen TC (2010) *In situ*: Groundwater bioremediation. In *Handbook of hydrocarbon and lipid microbiology*, Timmis KN (ed), 2584–2596. Berlin: Springer-Verlag.

Head IM, Jones DM, Roling WFM (2006) Marine microorganisms make a meal of oil. *Nature Reviews, Microbiology* 4:173–182.

Johnson DB, Halberg KB (2005) Acid mine drainage remediation options: A review. *Science of the Total Environment* 328:3–14.

Kumar M, Lin J-G (2010) Co-existence of anammox and denitrification for simultaneous nitrogen and carbon removal: Strategies and issues. *Journal of Hazardous Materials* 178:1–9.

Leu SY, Stenstrom MK (2010) Bioaugmentation to improve nitrification in activated sludge treatment. *Water Environment Research* 82:524–535.

Little CD, Palumbo AV, Herbes SE, Lidstrom ME, Tyndall RL, Gilmer PJ (1988) Trichloroethylene biodegradation by a methane-oxidizing bacterium. *Applied and Environmental Microbiology* 54:951–956.

Madigan MT, Martinko JM, Stahl DA, Clark DP (2011) *Brock biology of microorganisms*. San Francisco: Pearson Benjamin-Cummings.

Monteiro CM, Castro PML, Malcata FX (2009) Use of the microalga *Scenedesmus obliquus* to remove cadmium cations from aqueous solutions. *World Journal of Microbiology and Biotechnology* 25:1573–1578.

Peterson CH, Rice SD, Short JW, Esler D, Bodkin JL, Ballachey BE, Irons DB (2003) Long-term ecosystem response to the *Exxon Valdez* oil spill. *Science* 19:2082–2086.

Ripp S, Nivens DE, Ahn Y, Werner C, Jarrell J, Easter JP, et al. (2000) Controlled field release of a bioluminescent genetically engineered microorganism for bioremediation process monitoring and control. *Environmental Science and Technology* 34:846–853.

Roane TM, Pepper IL, Miller RM (1995) Microbial remediation of metals. In *Bioremediation: Principles and applications*, Crawford RL, Crawford DL (eds), 312–340. Cambridge: Cambridge University Press.

Salanitro JP (1993) The role of bioattenuation in the management of aromatic hydrocarbon plumes in aquifers. *Ground Water Monitoring and Remediation* 13:150–161.

Semprini L, Roberts PV, Hopkins GD, McCarty PL (1990) A field evaluation of in-situ biodegradation of chlorinated ethenes: Part 2, Results of biostimulation and biotransformation experiments. *Ground Water* 28:715–727.

U.S. Environmental Protection Agency (2006) Iron Mountain mine case study. www.epa.gov/aml/tech/imm.pdf (accessed June 14, 2013).

Wexler M (2011) Microbiology and biochemistry of biological nutrient removal. *Water and Sewerage* 2:39–41.

3

Water Appropriate Technologies

Marina Islas-Espinoza and Alejandro de las Heras

CONTENTS

In response to the overwhelming challenges facing conventional water and sanitation technologies (Chapter 1), the appropriate technologies movement has looked for a way to do things differently and rethink the technology in order to improve water and sanitation; reduce the impact on the environment; and cut equipment, operation, and maintenance costs. Appropriate technologies improve environmental and economic sustainability in the water sector in developed and developing countries, but in the latter they also contribute to social empowerment and set the scene for autonomous technological development.

Principles of Appropriate Technologies

Technically, two principles underpin these technologies. The first one is that externalities, that is, importing water, impeding its natural course, or transporting pollutants beyond the point where water is used, have to be avoided. This can be expressed as saying that a watershed or a household is as close to a closed system as possible. Water is recycled and recirculated within these places, thereby reducing consumption.

The second principle is to emulate nature and means that humans have to use the same physicochemical and biological tools as nature to achieve sustainability. It also means that humans have to act within the biophysical limits of nature and cater to the needs of other organisms and natural abiotic mechanisms using water. In particular, solar energy in different forms, natural attenuation, and bioremediation have to be resorted to (Chapters 2 and 9).

In addition, communities and their culture have to be factored in. This is best done through participatory planning and design involving the end users as well as DIY (do-it-yourself) construction, operation, and maintenance of appropriate technology devices by community and household members. DIY is the best way to learn and uses local human, material, and technical resources. On the contrary, devices obtained for free may be subject to inadequate use and early failure. Participatory planning and design contest the separation between scientific and traditional knowledge, as well as conventional boundaries of scientific disciplines, that fail to link people and technology in situations of technology adoption and lifestyle change (Sas 2004). These gaps help explain why several United Nations Decades devoted to water and sanitation have not yielded adequate results in the Third World. Specific socioeconomic, energy, and matter components are needed to alter this historical trend (Table 3.1).

The systematic application of water and sanitation appropriate technologies helps diminish the impacts of irrigation, the single most water-consuming human use, while contributing to food safety (see Box 3.1).

The efficiency of water appropriate technologies is clearly expressed in an example of greywater reuse in a constructed wetland. The end result is that organic matter (including pathogens) and inorganic constituents of wastewater (which include nitrogen and phosphorus) decrease to below levels required by Health Canada (Table 3.2).

Rainwater Harvesting

Rainwater often has a very good biological quality and requires very little work to be collected. This distinguishes rainwater from well water (often saline), spring water (which needs to be transported), and river

TABLE 3.1

Appropriate or Intermediate Solutions Differ from Conventional and High-Tech Solutions

System Components	Appropriate Solutions	Conventional and High-Tech Solutions
Logic	Technology fitted to local human basic needs, minimal cost and throughput, health goals (human, environment) approved by local population	Maximum profit, linked to increase in supply
Social	Household or community based	Government or corporation led
Capital	Social, not for profit	Economic, for profit
Investments	Disperse, autonomous, self-funded	Heavy, complex, centralized
Energy	Human/animal power, gravity, biogas, wind, nonphotovoltaic solar	Fossil fuels, electric grid or high-tech renewable
Materials	Locally available, nonpolluting disposal, salvage value and reuse	Imported technology, complex breakdowns
Production, distribution	Point of use, off the grid	Centralized production, network distribution

Sources: Diwan I, Rodrik D, 1989, *Patents, appropriate technology, and North-South trade*, Washington DC: World Bank; Montgomery MA, Bartram J, Elimelech M, 2009, *Environmental Engineering Science* 26:1017–1023; Hashmi F, Pearce JM, 2011, *Sustainable Development* 19:223–234; De Luca MJ (2012) Appropriate technology and adoption of water conservation practices: Case study of greywater reuse in Guelph, Master's dissertation, University of Guelph, Ontario, Canada.

BOX 3.1 WATER APPROPRIATE TECHNOLOGY FLOW CHART

A systematic use of water-appropriate technologies (Figure 3.1) should minimize extraction, transport, storage, desalination (carried out only for drinking and cooking), and disinfection. And conversely, the reuse of plant nutrients (e.g., nitrogen and phosphorus) should be maximized. The World Health Organization and the United Nations Environment Programme recommend avoiding sanitation systems where excreta and wastewater are mixed; the reuse of both and greywater is encouraged in agriculture and aquaculture, as exemplified in the United States, China, and other countries (WHO 2006). One important windfall is that less water is needed when enough nutrients are available to plants. Disgust is a natural reflex that can be overcome with education on the role of beneficial microorganisms and how to avoid contact with pathogens.

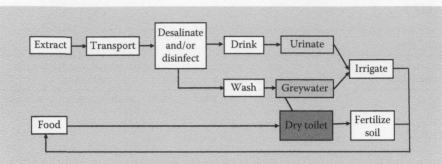

FIGURE 3.1
Water appropriate technology flow chart. The chief features of an integrated system are separation of solid and liquid waste, and water recycling. Protection of water from contamination starts at the source (spring, dug well, or borehole). Potable water has to be dedicated to washing vegetables, kitchenware, and body and hands only. To minimize pollution, water must not be used as a flushing and transport fluid. Dry-toilet composting systems are essential and double-vault or Muldrum setups are acceptable in most cultural settings.

TABLE 3.2

Efficiency of One Appropriate Technology in Greywater Pollutant Removal

Water Parameters	Influent (High-Strength Greywater)	Effluent (Vertical Flow Constructed Wetland)	Health Canada Guidelines for Reclaimed Water Used in Toilet and Urinal Flushing
DBO$_5$ (mg/L)	466	0.7	≤20
Fecal coliforms (cfu/100 mL)	5×10^7	2×10^5	≤200
TSS (mg/L)	158	51	≤20
N (total, mg/L)	34.3	10.8	NA
P (total, mg/L)	22.8	6.6	NA

Source: De Luca MJ (2012) Appropriate technology and adoption of water conservation practices: Case study of greywater reuse in Guelph, Masters dissertation, University of Guelph, Ontario, Canada.

Notes: A colony-forming unit (cfu) is an estimate of viable microbial cell numbers. Since a colony may comprise 1 cell or 1000 cells, the results are given per milliliter or per gram. Total suspended solids (TSS) is the weight of solid particulate matter in a water sample. N is total nitrogen, P is total phosphorus. NA: not applicable. The rationale is that these nutrients must be fully recycled in gardening or agriculture.

water (often contaminated). Apart from rainwater collection, fog and dew collection (Figure 3.2), air wells promoting condensation, condensation bags, and pits have been successfully used in communities with specific needs and supply from nature. The world's largest fog collection project provides 130 persons with 6000 L per day thanks to 30 large fog collectors (Rosato et al. 2010). Therefore, atmospheric water should be considered a prime source of potable water. The technical issues to be solved are, however, collection and storage in good conditions over dry spells or seasons. A fail-safe approach

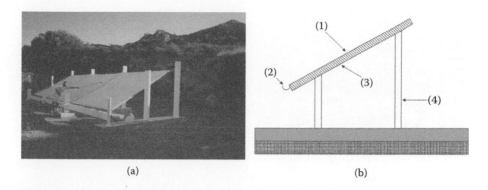

(a)　　　　　　　　　　　　　　　　　(b)

FIGURE 3.2
Dew condensers. (a) Dew condenser in Corsica, France. (b) Diagram of a radiative condenser designed to collect dew (commons.wikimedia.org 2009). (1) Radiating/condensing surface (the radiating surface allows for cooling it faster than surrounding air making it a condensing surface for water vapor), (2) collecting gutter, (3) backing insulation, (4) stand (en.wikipedia.org 2009).

is to disinfect water stored for long periods using SODIS (solar disinfection) containers (see the section, "Solar Ultraviolet Radiation").

Drilling, Pumping, Qanats, and Wadis

Because soil is a thorough natural filter, wells provide very good biological water quality. Drilling and pumping (Figure 3.3) are the main operations involved and disinfection is only necessary when contamination of the well has occurred. Manual drilling can use a Rota-Sludge system (for hard strata, with grinding teeth and vertical and rotation movements). The Stonehammer system, also for hard strata, uses a heavy element hitting upon the drill, and is adequate for perforations of 5 to 12.5 cm in diameter and up to 40 m deep.

Pumping may use the basic Baptista method, U.S. Patent 6283736B1, which requires a very limited force, or the Emas system, a public domain alternative to the Baptista system. Other pumping systems include piston and "rope" pumps. South African roundabout playpumps activated by children have hitherto been too expensive for most places. Mostly, hand pumps and treadle pumps are destined to mantle depths less than 10 m deep, except for the Flexi-Pipe Pump, which can reach 25 m. Deeper water tables can be reached using underwater pumps. Hippo water roller and Q-drum mechanisms are also well adapted to bringing in water from remote locations in somewhat flat terrains.

In the past, frequent failures have troubled hand pumps, especially in Sub-Saharan Africa, as shown in relevant literature (Sanders and Fitts 2011). This makes it important to involve traditional knowledge about alternative

(a)   (b)

FIGURE 3.3
Wind-driven and manual pumps. (a) A windmill-activated pump in San Juan, Utah, in the 1970s. (From Usna P, 1972, Power without pollution, Wind driven water pump 05/1972, http://www.dijitalimaj.com/alamyDetail.aspx?img=%7B2CDCF7B7-EEBA-45EB-BE25-DF651D09F350%7D.) (b) A hand-operated, reciprocating, positive displacement pump in Pakistan. (From commons.wikimedia.org 2007a.)

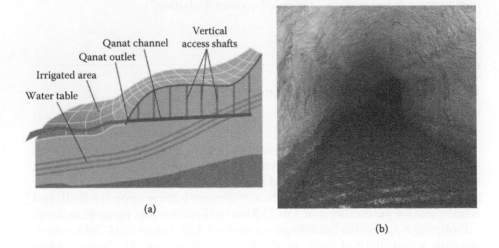

(a)   (b)

FIGURE 3.4
Qanats. (a) The qanat channel intercepts the water table but instead of spending energy pumping water, gravity pushes water down the slope and out to the surface. (From es.wikipedia.org 2009.) (b) Flooded tunnel section of the Firaun Qanat in Jordan. (From commons.wikimedia.org 2007b.)

water supply. Wadis are used to date in Sub-Saharan Africa: they are seasonally waterlogged areas or land flooded by water intrusion. Wastewater reuse there is crucial for fertilization and to avoid the contamination of wadis. Both measures tend to preserve shallow wells (Mohamed 2011) for the most critical needs.

Qanats are another system that accesses groundwater for a year-round supply (Figure 3.4). Developed in Persia more than 2600 years ago, it spread as far as China and the Roman Empire in Europe, and even to former Spanish colonies in America. Although an important social capital is required (i.e., organization of collective work) and seasonal water supply may vary, the advantages of this system are that after the initial investment is performed, the system works as a horizontal well for domestic and irrigation uses.

Disinfection

Disinfection can only be performed when water has low turbidity. Otherwise, microorganisms and vira may shelter in large suspended solids. Suspended matter has to be removed from water prior to disinfection (Adams et al. 2009) by sedimentation, or more commonly filtration.

Boiling

Bringing water to a rolling boil has long been recognized as the safest way to kill pathogens in water. But this implies energy as well as exposing people to combustion smoke. So synergies are being sought between cooking stoves and water disinfection to inactivate thermotolerant coliform bacteria. One such tested device is a stove that allows for pasteurization of water using the stove's chimney (Figure 3.5). This also drastically improves indoor air quality in kitchens otherwise burdened with noxious smoke. This has resulted in

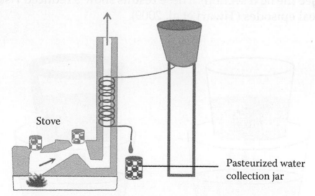

FIGURE 3.5
Integrated stove-water pasteurizing appliance. This allows for improved efficiency of combustibles compared to boiling water (although it may not replace it altogether), as well as improved indoor air quality. (Modified from Singh A, et al., 2010, An investigational research on the synergy of water pasteurization and improved cook stove for providing safe drinking water and improved indoor air, in *Regional Conference on Appropriate Water Supply, Sanitation, and Hygiene (WASH) Solutions for Informal Settlements and Marginalized Communities*, Sabitri T, Shrestha HK, Rai YM, Rajiv C (eds), 33–43, Kathmandu: Nepal Engineering College, Imperial College London and Preston University, Pakistan.)

the absence of *Escherichia coli* bacteria in the treated water, and a 3.5 to 15.0 L daily water output, depending on the model. Users also welcomed the integrated technology (Singh et al. 2010).

Filtration

Filtration in appropriate technologies currently uses porous ceramics (clay), often treated with silver to add antimicrobial properties. Such a setup has successfully eliminated all bacteria from water, while using locally available materials in South Africa, and is easy to operate and maintain. The tested device, called a silver-impregnated porous pot (SIPP, Figure 3.6), seemed to perform better than other well-known devices (a biosand filter, a zeolite-enhanced biosand filter, a bucket filter, and a ceramic candle filter) (Mwabi et al. 2012).

Another type of device is the intermittently operated slow-sand filter. In Kenya, replicating a study in the Dominican Republic, slow sand filtration (also known as a Biosand or Manz filter) reduced the concentrations of fecal coliforms and so contributed to reduced probability and duration of diarrheal episodes in children up to 15 years old. This can be achieved at a low cost (US$15–25/unit), with a durable design (Figure 3.7), allowing for a high flow rate (3–60 L/h), and an ability to process very turbid waters. Other studies have shown that biosand filtration eliminated all *Giardia lamblia* cysts, 99.98% of *Cryptosporidium* oocysts, 95% to 99% of bacteria, variable amounts of viruses, and reduced turbidity (enough to use SODIS treatment afterward) (see the next section). These results show a reduced risk and gravity of diarrheal episodes (Tiwari et al. 2009).

(a)   (b)

FIGURE 3.6
Ceramic filter options for water disinfection. (a) Ceramic candle filter. (b) Silver-impregnated porous pot. (Modified from Mwabi JK, Mamba BB, Momba MN, 2012, *International Journal of Environmental Research and Public Health* 9:139–170.)

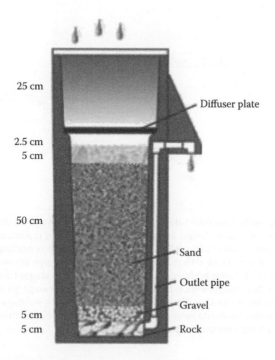

FIGURE 3.7
Biosand filter. (Modified from Tiwari et al., 2009, *Tropical Medicine and International Health* 14:1374–1382.)

The aforementioned filters can best be used at the household point-of-use, although biosand filters are often found to be shared by several households. In addition, community-scale technologies are increasingly found, such as "multistage filtration" (Figure 3.8), which implies important investments. These filtration utilities are made of a sequence of increasingly fine filters prior to disinfection. This setup is adequate for removing high turbidity, while remaining relatively simple to maintain and operate.

Solar Ultraviolet Radiation

One of the simplest and cheapest disinfection solutions according to the World Health Organization was discovered by women in India who had stored transparent water bottles in the sun and realized that the quality of the water had improved. Nowadays, this solution is called a Solar Disinfection System (SODIS) (Figure 3.9) and only requires transparent bottles and solar light. A series of studies in laboratories and in operating conditions have shown the synergy between solar ultraviolet A (UV-A) radiation and heat (>45°C). Boiling water is not necessary: bringing it to 70°C for 6 minutes is sufficient but the World Health Organization suggests a vigorous boil to be on the safe side (Akinjiola

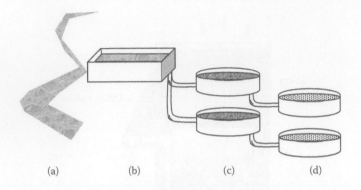

(a)   (b)   (c)   (d)

FIGURE 3.8
Similar to biosands, multistage filters develop a biologically active layer on top of the filtration substrates. This particular design is from a utility built by a community in Honduras. (a) Influent. (b) Dynamic coarse filter. It is called dynamic since it is gradually obstructed by solids unless a sudden flow peak arrives, in which case a return pipe allows the excess water and solids to return to the water body (a) to prevent disruption of stages (c) and (d). (c) Coarse filter. (d) Slow filter where the biological filter is most active. After stage (d), disinfection takes place. (Modified from Martínez-Castillo A, 2009, Filtración en múltiples etapas como tecnología apropiada para la potabilización de agua en una comunidad rural de Honduras. Análisis FIME El Naranjal. Master en Ingeniería del agua.)

FIGURE 3.9
Solar energy disinfection solutions (SODIS). SODIS in PET bottles filled with water to treat. Solar radiation and heat are needed for at least 6 hours. A solar concentration device, such as a solar cooker, can be used to bring water to 70°C in about 20 minutes at altitudes of 2700 m above sea level.

and Balachandran 2012). Research is also currently underway to remove suspended matter prior to SODIS disinfection using salt, with promising results obtained on removing bentonite clay (Dawney and Pearce 2012).

As stated earlier with respect to boiling water, fuelwood has detrimental environmental effects (deforestation, CO_2 emissions, and indoor air pollution).

Solar energy is an adequate alternative if high-tech solutions are sidelined in favor of cheaper investments that are also easier to operate and maintain, such as concentrated solar thermal systems based on mirrors that focus on solar radiation (Chapter 9). These systems can replace fuelwood, fossil fuels, and electricity as well as provide potable water. Designs have greatly evolved but the general principles were discovered more than 2000 years ago. Local manufacture is also quite possible (Akinjiola and Balachandran 2012).

Chlorination

The World Health Organization considers chlorination the most cost-effective disinfection technique, despite increasing evidence of its unsafe side effects. As a nonpermanent or emergency measure, however, chlorination is worth considering along with similar chemical techniques such as colloidal silver. The many commercial presentations of chlorine include bleaching powder, liquid bleach, calcium hypochlorite powder, sodium hypochlorite solution, and sodium dichloroisocyanurate (NaDCC) tablets. Perhaps more interesting from an appropriate technology standpoint, sodium hypochlorite can be obtained by electrolysis in communities with a 12V, 50Ah, car battery; a carbon electrode such as a pencil lead; and a copper cable. Water and sodium chloride are dissolved in a water bottle. The lid of the bottle is punctured so the lead is partially in the bottle. The cable is fixed to the lead outside the bottle and connected to the battery. After 2 or 3 h the solution becomes sodium hypochlorite, which can be filtered to eliminate carbon particles (Garfi 2007). Via electrolysis, 30 g/L of salt produce 0.5% sodium hypochlorite. Electricity is needed and also an electrolysis cell, amounting overall to a US$28 investment per family, and a $1 expense every 4 months (i.e., 1 L disinfectant).

Effective chlorination takes more than 30 minutes between chlorine addition and the moment water is drunk. Safe levels of free chlorine residual (the chlorine free form in the water after 30 mn) should be 0.5 to 1.0 mg/L, which can be measured with simple methods such as a color comparator and diethyl-p-phenylenediamine (DPD) tablets (Adams et al. 2009).

Vegetal Treatments

The biochemical properties of some plants make them useful, specific, water disinfectants. This appears to be the case with *Alternanthera sesselis*, an African plant whose extracts are toxic to the mollusk *Bulinus* (phy) *globosus*, the intermediate host of schistosomiasis parasites (Azare et al. 2007). Similarly, antibacterial and coagulant (bulk-making) properties of *Moringa oleifera* are useful in both turbidity remotion and disinfection. Cacti and potatoes have similar properties.

Desalination

With 96.5% of Earth's water located in oceans, tapping this source would be a major breakthrough in terms of human access to safe water while allowing for wildlife access to freshwater sources hitherto appropriated by humans. This is especially important as two-thirds of the human global population live near the seashores. Pumping groundwater near the seashores is increasing intrusions of salty seawater into fresh groundwater mantles. Excessive irrigation has also augmented evaporation and capillarity, creating an upward movement of water, which in turn results in soil surface increases in salinity, eventually hampering vegetation growth. And finally, groundwater in several regions is loaded with metal or metalloid salts, such as arsenic.

With regard to seawater, an increasingly favored desalination method is reverse osmosis, which is cheaper than thermal (evaporative) systems, especially since crude oil prices have gone up from US$20 to US$70 per barrel from 1997 to 2007. With increasingly large reverse osmosis desalination plants from 1989 onward, the price of desalinated seawater has gone down from nearly US$2 to US$0.50 today. Of special interest are brackish waters since they have lower salt concentrations. Analyses seem to show that virtually all costs are lower for brackish water than for seawater; the total cost per cubic meter of desalinated brackish water would be US$0.13 as compared to US$0.53 for seawater (Greenlee et al. 2009).

Solar stills (Figure 3.10) are even more energy-efficient than reverse osmosis systems. They have had good acceptance among the local population in the state of Rajasthan in India to evaporate brackish water to obtain distilled

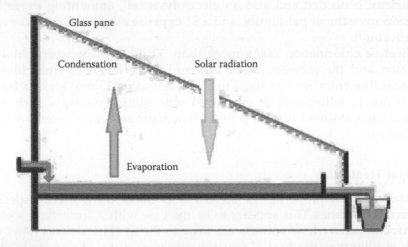

FIGURE 3.10
Solar stills. The brine is optimally 1.5 cm deep. Improved systems feed water onto the glass plate so as to preheat the water and also cool the glass to enhance condensation. Some models have siphons that allow for a continuous flow.

water (Khanna 2008). Distilled water, however, is arguably not apt for long-term human consumption as it is devoid of necessary minerals. Another challenge for the technique is also to obtain a continuous feed mechanism to avoid burdensome work.

With regard to arsenic, the exposed population is 30 million in Bangladesh (with arsenic levels of 2500 parts per billion, i.e., 2500 micrograms per liter); 6 million in India and West Bengal (3200 ppb); 5.6 million in China; 2 million in Argentina (5300–7800 ppb); and several hundreds of thousands in Nepal, Chile, Mexico, and southwest United States (with concentrations in the range of 620 in Mexico and 2600 ppb in the United States) (Espinoza and O'Donnell 2011).

The Kanchan Arsenic Filter (KAF) was developed in 2002 to comply with Nepalese official guidelines of 50 micrograms per liter. The system uses a Biosand setup but adds iron nails so that arsenic binds to iron rust and rust binds to sand. To date, this system has been deployed in around 25,000 households in Nepal and a sample indicated that poor performance came from special chemical conditions in some localities. Removal efficiency was 99% for arsenic groundwater below 100 micrograms per liter but this decreased to above the official guideline with higher arsenic concentrations and after 3 years of use decreased to 75% efficiency, calling for maintenance. Also, some people have removed the nails and sand to speed up the process. Lack of maintenance and erroneous modifications were more frequent when people obtained a KAF for free or through subsidies. People with arsenicosis or people who bought the filter took better care of the KAF. Since groundwater was a constant 20°C throughout the year, filtering it cooled it off and so people were less inclined to use the KAF in winter (Espinoza and O'Donnell 2011).

The Sono filter is not only an alternative to the KAF (Figure 3.11). These two filters can work in sequence in places where high arsenic concentrations are found. The Sono filter consists of a removal material (composite iron matrix, CIM), sand, and charcoal layers in three vertical buckets. Charcoal removes bad tastes. Maintenance implies replacing the upper sand layers as flow rate decreases; the sand can be washed and used again. The CIM removes arsenic at very high concentrations (6000 ppb). Its longevity is 11 years or more. In addition to arsenic, 23 metals, bacteria, and viruses are removed (arsenic and bacteria achieve levels below WHO guidelines). The Sono filter was developed in Bangladesh and was awarded the Golden Prize in the Grainger Challenge by the U.S. National Academy of Engineers for the World's Best Household Arsenic Removal System. In Nepal, total coliform reduction was 92.4%. However, out of 15 filters, bacteria were present in 6 filters and high bacteria counts in 4 filters. Lack of filter maintenance increased metal concentrations (Tuladhar and Shakya 2010).

Cost analyses show that in Pakistan, the best economic solutions for arsenic removal are the Stevens Institute Method and the KAF, ranging from

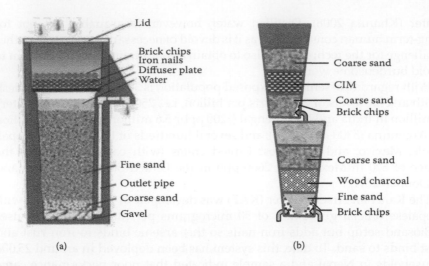

FIGURE 3.11
Arsenic filters. (a) Kanchan Arsenic Filter. Water with arsenic goes through the rusting iron nails layer. Iron is oxidated from Iron(II) to Iron(III), the latter binds to arsenic(V) and iron binds to sand. Water free from arsenic and iron is obtained. (Modified from Espinoza CM, O'Donnell MK, 2011, Evaluation of the Kanchan™ Arsenic Filter under various water quality conditions of the Nawalparasi District, Nepal, Masters dissertation, Massachusetts Institute of Technology, Cambridge.) (b) Sono filter. CIM is made from iron turnings from local foundries or workshops, which are then prepared with food-grade acids to enhance chemical bindings; 93% iron, 4.5% carbon, 1.5% silicium oxide, 1.5% magnesium, and 1.5% sulfur and phosphorus are processed into CIM. (Modified from Hussam A, Munir AKM, 2007, *Journal of Environmental Science and Health Part A* 42:1869–1878.)

US$0.03 to US$0.05 per liter, which also are readily applicable (Hashmi and Pearce 2011). However, as seen earlier, arsenic removal efficiency and maintenance must also be considered. The Stevens Institute Technology mixes iron sulfate and calcium hypochloride into one bucket and in another bucket, flocs (lumps) produced by the chemicals settle down while agitated with a stick, and are filtered in a sand layer. However, flocs bung the sand filter, which requires washing several times a week.

Overall, reverse osmosis seems to remove all trace of arsenic but this method requires electricity, equipment, and training, and so may not operate properly in every setting. As an alternative, CIM is said to be excellent in removing arsenic below 10 ppb as per WHO guidelines, especially at slow flows (Ongley et al. 2012). Although the Sono filter is very effective, some drawbacks are the proprietary CIM, which makes it necessary to buy replacements, a cost that may bring about lack of maintenance. Also, the media may host microorganisms (WS Atkins International Ltd 2001). Finally, an inconvenience inherent to all filters is that they accumulate increasing arsenic concentrations over time; and most often will affect the environment upon disposal, as arsenic is released from the filter. This calls for bioremediation, or microbial change of one arsenic form to another, less toxic one. Promising research is underway.

Greywater

Greywater comes from kitchens, washing basins, showers and bathtubs, washing machines, and dishwashers. Greywater issues could have more to do with boron contained in current detergents, or food grinding and flushing. But even the latter concern could be overestimated.

A decade of experience in greywater treatment systems in Palestine has shown that greywater can substantially improve irrigation and reduce wastewater disposal costs, thereby improving food security and household economy. Reused greywater is more culturally acceptable than other wastewaters. Acceptance is related to knowledge of these benefits. In Lebanon and Jordan, greywater reuse has highlighted the role of women in the implementation of a household reuse project, compared to male members of the communities. In turn the project led to large benefits for women in terms of reducing the daily workload and the associated back pains. Moreover, women were empowered and became more involved in local decision making. Public perception is an important barrier to widespread greywater reuse. In surveys conducted in Oman, Palestine, Jordan, and Ethiopia, most people (76%–89%) either agreed with or used untreated greywater irrigation in their gardens but they were unjustifiably concerned with safety, environmental, or cultural issues. Participatory design and implementation with the community, government and nongovernment participants, and researchers in some settings or media participation in other places have helped dispel these concerns. Literature on illnesses caused by greywater irrigation is scant. There could be more concern on the effects of increase in salts (in particular boron), grease, surfactants and bacteria in some greywater, soil, or vegetable samples in Jordan or Israel. Some soil types showed less permeability. But chemically the crops were unchanged, or were highly tolerant (olive trees), salts were not above levels attained by conventional fertilized-water irrigation and did not affect plants. Boron, which can be toxic to plants, was due to detergents in greywater. Overall, there is a debate where some recommend greywater treatment prior to irrigation while others find more impacts in treated greywater.

In the United States, 13% of the people in Tucson, Arizona, used greywater without a permit. But studies in preparation for the new law revealed that permitting was bureaucratic and expensive. Kitchen-sink water was possibly the greatest issue and was recommended to be removed from greywater definition even though the systems already in use were deemed devoid of health risks (Allen et al. 2010).

Dry Toilets

Right after potable water, sanitation is healthwise the most important water necessity. Interestingly, all the solutions were already known 30 years ago, and classified by Kalbermatten et al. (1980). Foremost among those solutions are dry toilets because they avoid soiling potable water for such nonessential functions as flushing, odor seals, and transport to centralized fecal matter treatment plants. Dry toilets are also important sources of fertilizers, which can be microbiologically disinfected at high temperatures via aerobic composting.

They are not only rural toilets, as exemplified by the CK Choi Building at the Institute of Asian Research of the University of British Columbia in Vancouver, which is devoid of sewer connection since composting toilets and urinals are waterless and greywater is recycled and used for irrigation. This won the building recognition as a 2000 Top Ten Green Project from the American Institute of Architects. But as the designers themselves acknowledge, the toilets were the most surprising feature of the building in the eyes of users (Williams 2007).

As part of ecological sanitation, water must be limited to hand washing and to anal cleansing, wherever this practice replaces toilet paper. Urine diversion is also essential as this is a readily usable fertilizer (direct use, however, is not recommended in areas with high schistosomiasis prevalence). Multrum models (Figure 3.12a) and double vault composting toilets are the most frequent models. The latter originated in Vietnam and is a very good option in urban housing: one container is used at a time; when full, a slab (which includes a urine separation pipe) is moved to the other container while the first one is sealed for 1 year for facultative (aerobic and anaerobic) composting. A variation of the double vault is the Arborloo, a lightweight

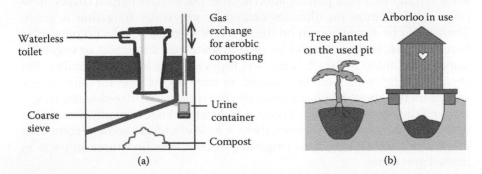

(a)   (b)

FIGURE 3.12
Urban and rural dry toilets. (a) Toilet in the CK Choi Building in Vancouver. The vent fan, liquid removal, and moisturizing systems ensure optimal aerobic composting. (Drawing modified from Clivus Multrum, 2008, Composting dry toilet. www.clivusmultrum.com.au.) (b) Arborloo system design that moves a lightweight shed around when the pit is full. (Modified from Morgan P, 2008, The Arborloo, http://en.wikipedia.org/wiki/File:Arborloo-en.svg, 2012.)

shed that moves around when the pit is nearly full and is then planted with a tree that will have sufficient fertilizer for years (Figure 3.12b).

Septic Tanks

Because they run counter to ingrained hygiene principles as well as microbiological misconceptions, dry toilets may not be easily accepted and alternatives must be deployed. Among these are septic tanks (Figure 3.13), which are only adequate when water is not a scarce resource, for instance, in Canada. This also rules out many locations in the United States, China, and India, to name a few.

An innovative septic tank design uses more vertical baffles than the usual design in Figure 3.13. By doing so, the wastewater goes up and down, and mixes with the resident microbial biomass, without needing for stirring devices. Residence time for the biomass also increases. Improved and longer contact means enhanced biodegradation of wastewater organic matter and nutrients. Aerobic pathogens also die in this anaerobic system. Sludge is also supposed to decrease compared with traditional settings. The system may also be resilient to load changes as a result of a diversity of adapted microbial communities separated by baffles (Foxon et al. 2004).

Soakaway pits or infiltration trenches should only be used if they irrigate plants. These pits, trenches, and drainfields should be more than 1.5 m above the groundwater table, more than 30 m from any well, pond, or spring

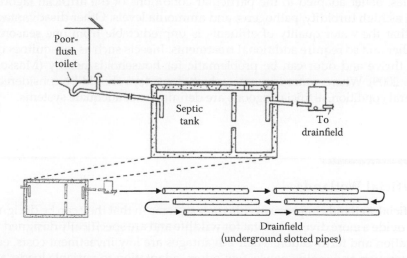

FIGURE 3.13
Septic tank and drainfield. Septic tanks are designed to anaerobically degrade wastewater organic matter, use up excess nutrients, and diminish the presence of aerobic pathogens. The effluent of the septic tanks irrigates drainfields planted with useful, native, noninvasive species. Drainfields can feed artificial wetlands (see Figure 3.14).

(Adams et al. 2009). These distances depend on the permeability of the soil and should be considered as minima.

Lagoon Systems

Artificial lagoons have been around for some four centuries. They have been used to degrade the organic matter coming from domestic sewage. They rely on a symbiosis between photosynthetic algae providing oxygen to aerobic bacteria who metabolize organic matter (they degrade it to a mineral, i.e., bioavailable compounds, which can be taken up by the algae). A byproduct of aerobic metabolism is carbon dioxide, which is used by algae in photosynthesis. Most artificial lagoons are facultative: the upper layer stirred by the wind is aerobic and harbors algae. The sediment bottom layer is anaerobic and bacteria there metabolize organic matter from wastewater at lower rates, releasing methane, a greenhouse gas. One advantage of artificial lagoons is that they can adapt to alternating heavy and light load periods if their effluents are used for irrigation because of high nutrient content. However, nutrients do not remain suspended and tend to sediment, thus requiring dredging to be used. The accumulation of nutrients requires maintenance lest the lagoon eutrophicates (becomes even shallower until it dries out as a result of land plants colonizing it and boosting evapotranspiration). They may provide habitat for some wild species, better adapted to the particular conditions of the artificial lagoons, such as high turbidity, pathogens, and ammonia levels. Other disadvantages are that the water quality of effluents is unpredictable from one season to another and so require additional treatments. Insects such as mosquitoes can also thrive and odor can be problematic for households nearby (Massoud et al. 2009). Where water is scarce, which nowadays should be considered a general condition, artificial lagoons are definitely not adequate systems.

Artificial Wetlands

Artificial wetlands improve over lagoon systems in that they can be designed to provide a more diverse habitat for wildlife and are specifically designed for irrigation and fertilization. Their advantages are low investment costs, ease of operation and maintenance, few odors, adaptation to variable loads, and smaller land requirements than lagoons (Massoud et al. 2009). Esthetically, artificial wetlands are among the best available water treatment solutions.

A key variable to enhance their usefulness is the selection of plants fertilized by an artificial wetland. These should be native, possibly staple crops, or provide prime material for handicrafts, have commercial value, be included in the local pharmacopeia, or a combination of these. Land vegetation can be combined with macrophyte aquatic plants. If the wetland is located in a gully, erosion can be controlled by a judicious terrace design and plant selection.

Constraints for sustainable artificial wetlands are as follows. Maintenance is needed in designs that incorporate subsurface piping. Plants sensitive to water shortages should be avoided to allow for maintenance downtimes and droughts. Excess ammonia may become toxic if uncontrolled. Finally, the systems seldom supply solid fertilizers, which are easier to transport: fertilization and irrigation are mostly *in situ*. As for all water technologies, fly tipping and solid waste may alter the system. To avoid this, vegetation fences and hedges can be required (using cacti in semiarid areas). Native vegetation or noninvasive species should be preferred.

A wide variety of artificial wetlands have been constructed and other designs can adapt to local needs (Figure 3.14). For instance, in some projects,

(a)

(b)

(c)

(d)

FIGURE 3.14 (See color insert.)
An artificial wetland in Mexico. (a) Sieving operation (foreground) and tanks allowing for flow regulation. (b) The reed plantation is irrigated from below ground; the high evapotranspiration rates of the reed help keep up with the wastewater flow of 2000 people. (c) Vents of the slotted underground PVC pipes allow for unclogging them with pressurized air. (d) Effluent water can be further treated and used in other ponds. (© Photos by Marina Islas-Espinoza.)

watercress has helped remediate aquacultural discharges otherwise bound to eutrophicate the fish ponds or contaminate downriver. The systematic reuse of aquaculture waste by plants in artificial ecosystems is called aquaponics. Traditional variants thereof can be found in Bangladesh, Cambodia, Inle Lake in Myanmar, or Lake Titicaca in Bolivia and Peru. Unlike polders, which gained land over water at a high energetic cost, some ancient designs, which combined agriculture and the surrounding natural wetland, are still in operation. This is the case of the Chinampas of Mexico City, which are a network of small-plot islands emerging from the lacustre environment. As they sink under their own weight, their owners dig more nutrients from the sediment and apply them to the top of the island to keep it above water level.

References

Adams J, Bartram J, Chartier Y, Sims J (2009) *Water, sanitation and hygiene standards for schools in low-cost settings.* Geneva: WHO.

Akinjiola OP, Balachandran UB (2012) Concentrated solar thermal (CST) system for fuelwood replacement and household water sanitation in developing countries. *Journal of Sustainable Development* 5:25–34.

Allen L, Christian-Smith J, Palaniappan M (2010) *Overview of greywater reuse: The potential of greywater systems to aid water management.* Oakland, CA: Pacific Institute.

Azare B, Okwute S, Kela S (2007) Molluscicidal activity of crude water leaf extracts of *Alternanthera sesselis* on *Bulinus* (phy) *globosus. African Journal of Biotechnology* 6:441–444.

Clivus Multrum (2008) Composting dry toilet. www.clivusmultrum.com.au (accessed April 15, 2012).

Commons.wikimedia.org (2007a) Pakistan pump water system at my farm lands. Licensed under the Creative Commons Attribution 2.0 Generic license. http://commons.wikimedia.org/wiki/File:Pakistan_Pump_water_system_at_my_farm_lands.jpg (accessed June 18, 2013).

Commons.wikimedia.org (2007b) QanatFiraun. Licensed under the Creative Commons Attribution-Share Alike 3.0 Unported license. http://commons.wikimedia.org/wiki/File:QanatFiraun.JPG (accessed June 18, 2013).

Commons.wikimedia.org (2009) Big dew condenser in Corsica. Licensed as public domain. https://en.m.wikipedia.org/wiki/File:Big_Dew_Condenser_in_Corsica.jpg#filelinks (accessed June 18, 2013).

Dawney B, Pearce JM (2012) Optimizing the solar water disinfection (SODIS) method by decreasing turbidity with NaCl. *Journal of Water, Sanitation and Hygiene for Development* 2:87–94.

De Luca MJ (2012) Appropriate technology and adoption of water conservation practices: Case study of greywater reuse in Guelph. Master's dissertation. University of Guelph, Ontario, Canada.

Diwan I, Rodrik D (1989) *Patents, appropriate technology, and North-South trade.* Washington DC: World Bank.

En.wikipedia.org (2009) Radiative condenser. This file is in the public domain. http://en.wikipedia.org/wiki/File:Radiative_condenser_(section).jpg (accessed July 10, 2013).

Es.wikipedia.org (2009) Qanat-3. This image is in the public domain. http://es.wikipedia.org/wiki/Archivo:Qanat-3_es.svg (accessed June 22, 2012).

Espinoza CM, O'Donnell MK (2011). Evaluation of the Kanchan™ Arsenic Filter under various water quality conditions of the Nawalparasi District, Nepal. Masters dissertation. Massachusetts Institute of Technology. Cambridge.

Foxon KM, Brouckaert CJ, Remigi E, Buckley CA (2004) The anaerobic baffled reactor: An appropriate technology for onsite sanitation. *Water SA*, 30:44–50 (special edition).

Garfi M, Fruci A (2007) Tecnologías apropiadas para desinfección de agua. Cuadernos Internacionales de Tecnología para el Desarrollo No. 6.

Greenlee LF, Lawler DF, Freeman BD, Marrot B, Moulin P (2009) Reverse osmosis desalination: Water sources, technology, and today's challenges. *Water Research* 43:2317–2348.

Hashmi F, Pearce JM (2011) Viability of small-scale arsenic-contaminated water purification technologies for sustainable development in Pakistan. *Sustainable Development* 19:223–234.

Hussam A, Munir AKM (2007) A simple and effective arsenic filter based on composite iron matrix: Development and deployment studies for groundwater of Bangladesh. *Journal of Environmental Science and Health Part A* 42:1869–1878.

Kalbermatten JM, Julius DS, Gunnerson CG (1980) *Appropriate technology water supply and sanitation technical and economic options*. Washington DC: World Bank.

Khanna R, Rathore R, Sharma C (2008) Solar still an appropriate technology for potable water need of remote villages of desert state of India—Rajasthan. *Desalination* 220:645–653.

Martínez-Castillo A (2009) Filtración en múltiples etapas como tecnología apropiada para la potabilización de agua en una comunidad rural de Honduras. Análisis FIME El Naranjal. Master en Ingeniería del agua.

Massoud MA, Tarhini A, Nasr JA (2009) Decentralized approaches to wastewater treatment and management: Applicability in developing countries. *Journal of Environmental Management* 90:652–659.

Mohamed IA (2011) Economic perspective of indigenous knowledge systems. Technology transfer and rural water use in Darfur. *Middle East Studies Online Journal* 2:161–180.

Montgomery MA, Bartram J, Elimelech M (2009) Increasing functional sustainability of water and sanitation supplies in rural Sub-Saharan Africa. *Environmental Engineering Science* 26:1017–1023.

Morgan P (2008) The Arborloo. Licensed under the Creative Commons Attribution-Share Alike 3.0 Unported, 2.5 Generic, 2.0 Generic and 1.0 Generic license. http://en.wikipedia.org/wiki/File:Arborloo-en.svg (accessed January 26, 2012).

Mwabi JK, Mamba BB, Momba MN (2012) Removal of *Escherichia coli* and faecal coliforms from surface water and groundwater by household water treatment devices/systems: A sustainable solution for improving water quality in rural communities of the Southern African Development Community Region. *International Journal of Environmental Research and Public Health* 9:139–170.

Ongley LK, Chambreau SD, Guenthner A, Godaire TP, Vij A, Cheema H (2012) Validating the Digital Arsenator® and a composite iron matrix arsenic filter at the arsenic field event, Boron, CA. Boron: Chemists Without Borders and the Mojave Desert Section of the American Chemical Society.

Rosato M, Rojas F, Schemenauer RS (2010) *Not just beneficiaries: Fostering participation and local management capacity in the Tojquia fog-collection project, Guatemala.* FogQuest: Kamloops.

Sanders H, Fitts J (2011) Assessing the sustainability of rural water supply programs: A case study of Pawaga, Tanzania. Masters dissertation. Duke University.

Sas L (2004) Appropriate technology for the development of the "Third World." *Totem: The University of Western Ontario Journal of Anthropology* 12:73–78.

Singh A, Tuladha B, Shrestha R, Karki K, Nakarmi P, Kansakar LK, Karmacharya AP, et al. (2010) An investigational research on the synergy of water pasteurization and improved cook stove for providing safe drinking water and improved indoor air. In *Regional Conference on Appropriate Water Supply, Sanitation, and Hygiene (WASH) Solutions for Informal Settlements and Marginalized Communities*, Sabitri T, Shrestha HK, Rai YM, Rajiv C (eds), 33–43. Kathmandu: Nepal Engineering College, Imperial College London and Preston University, Pakistan.

Tiwari S-SK, Schmidt W-P, Darby J, Kariuki Z, Jenkins MW (2009) Intermittent slow sand filtration for preventing diarrhoea among children in Kenyan households using unimproved water sources: Randomized controlled trial. *Tropical Medicine and International Health* 14:1374–1382.

Tuladhar S, Shakya BM (2010) A study on the performance of Sono filter in reducing different drinking water quality parameters of ground water: A case study in Ramgram municipality of Nawalparasi district, Nepal. In *Regional Conference on Appropriate Water Supply, Sanitation, and Hygiene (WASH) Solutions for Informal Settlements and Marginalized Communities*, Sabitri T, Shrestha HK, Rai YM, Rajiv C (eds), 297–310. Kathmandu: Nepal Engineering College, Imperial College London and Preston University, Pakistan.

Usna P (1972) Power without pollution. Wind driven water pump 05/1972. Licensed in the public domain. http://www.dijitalimaj.com/alamyDetail.aspx?img=%7B2CDCF7B7-EEBA-45EB-BE25-DF651D09F350%7D single (accessed March 26, 2013).

WHO (2006) *Guidelines for the safe use of wastewater, excreta and greywater.* Geneva: World Health Organization.

Williams DE (2007) *Sustainable design: Ecology, architecture, and planning.* Hoboken, NJ: John Wiley & Sons.

WS Atkins International Ltd (2001) *Rapid assessment of household level arsenic removal technologies Phase II Report.* Dhaka: BAMWSP, DFID, WaterAid Bangladesh.

4

Soil

Marina Islas-Espinoza, Jeremy Burgess, and Margaret Wexler

CONTENTS

Nature of Soil

Possibly the most widely known fact about the Earth is that two-thirds of its surface is covered by water, and one-third by land. From a human perspective, the "land" is covered by a variety of things. In the case of a city dweller, by tarmac, buildings, parks, concrete, and gardens; to a miner, by material that has to be removed in order to reach mineral deposits; and to a farmer, by a thin layer that if she is lucky will provide a living in the form of crops. It is this last perception that gives a clue about the important nature of soil.

To a scientist, the soil is the upper layer of the regolith, the debris that covers the bedrock of the land. In rare instances, the regolith is so shallow that it has been converted entirely into soil, so that the soil rests on the bedrock. But what defines soil? Merely grinding rocks into ever smaller particles does not produce soil; it produces ground up rocks.

Two additional elements are required to produce soil. The first is the presence of some organic material. Soil is commonly regarded as a medium in which plants grow; indeed, it can be purchased as a bagged commodity specifically for this purpose. But it is more accurately regarded as a habitat for a very wide diversity of living creatures, from microorganisms such as bacteria and fungi, to microscopic animals, mollusks, and even ground-dwelling mammals. Most obviously, it is the habitat for terrestrial plants, which are the ultimate source of food for all other groups of living creatures, including humans. When a finely divided mineral layer covers the land, it is a matter of conjecture whether it is soil. Is the ground in the center of a dry desert truly to be called soil? (The answer is—it is called desert soil.)

It may be thought that if plants are growing on the ground layer, that layer must be called soil. This is partially correct; the ground at that point is on the way to becoming soil (the plant roots will decay eventually). But it should be remembered that a fortuitous sufficiency of dissolved nutrients can result in plant growth in a completely inert mineral medium. This is the basis for hydroponic culture in commercial greenhouses. A natural habitat of this type, where plants can apparently grow against the odds, is coastal shingle. Shingle is not soil; it is an unconsolidated and shifting mass of eroded pebbles. However, if shingle builds sufficiently to avoid constant movement by wave action, soil can start to form at loci within it, due to opportunistic plant growth. Lichens and mosses are early formers of soil on rocks (see the section "Soil Biological Crusts").

The second essential for full soil development is weathering. This includes cycles of heating and cooling, rainfall, and exposure to the atmosphere; processes that result in chemical changes to the mineral component of the soil and provide the soil inhabitants with the conditions and nutrients necessary for life.

The importance of this thin complex habitat (on average less than 2 m thick) can scarcely be overstated. It provides the food on which most creatures in the world depend. In addition, for humans, it provides building materials, fiber for clothing, and raw materials for a variety of industries. It is slow to form and easy to damage by poor agricultural practice and pollution. The basis of sustainability must therefore encompass preserving the unique properties and value of the soil wherever possible. Our demands upon it are constantly rising.

Soil and the Nutrient Cycles

The soil is thus a complex and very variable medium. How does the idea of sustainability relate to soil? Soils have come into being over various time scales; the majority of which are long and predate any possibility of human effect, for good or bad. In pristine environments—a failing brand these days—soils are by and large robustly stable, despite their complexity and apparent fragility. It is as though they are a self-sustaining phenomenon.

Consider for a moment a pendulum clock. It tells the time day in day out, by means of the movement of a swinging weight whose motion is determined by gravity. At first glance, it appears to be a self-sustaining way of measuring time. But as we all know, clocks are not self-sustaining; they have to be wound up (or these days, the battery has to be charged or replaced). In other words, the working of the clock depends on an input of energy. The energy input corresponds exactly to how the clock was designed; it was intended to be wound up. Other forms of energy will not serve to keep the clock going; if you shake a pendulum clock that has stopped, it will not cause it to restart.

The soil receives one principal sort of energy, the same energy that the entire planet relies upon: that of the sun. A smaller energy source is the planet's residual heat, most dramatically resulting in earthquakes and volcanoes. For our current global discussion, this source can be discounted. The sun's energy has many effects, which we generally call the weather. That's to say it heats things up, blows things around, causes water to evaporate, drives photosynthesis, and by its seasonal patterns of availability may shatter rocks and cause landslides and erosion. These days, it may even change the shape of landforms as ocean levels rise.

The continuous input of energy to a system can be a problem. If you were to wind a clock every day or every hour or every minute, you would sooner or later cause the main spring to break. Despite the continuous input of solar energy, soils in pristine environments do not break. This is because the soil system continually utilizes the energy to a useful end. The energy is used to recycle soil components. In the following sections, the nature of these cycles is described.

Before entering into details, there are a few more generalities about soil that should be kept in mind. Soil is comprised of all three phases of matter (solid, liquid, and gaseous) and consists of mineral particles, organic compounds, water, air, and living organisms. The diversity of soils is derived from a variation in the relative proportions of organic and inorganic matter and biomass, and also from the nature of the underlying bedrock, the local climate, and the extent of weathering. Just as the composition of the atmosphere has been molded over time by the presence of living organisms on the planet, so too, do the properties of soil depend both on lifeless starting materials (the bedrock and regolith), and the presence of life. On a lifeless Earth, the seas would still contain water, but the atmosphere would be unrecognizably different and there would be no soils. The biodiversity of the soil is greater than that of aqueous environments or the atmosphere. It is this—together with mineral diversity—on which we depend for our food, shelter, and raw materials.

The ultimate fate of all living creatures is to become deposited material, whether in soils on land or as underwater sediments. However, this apparently simple inevitability contains within it many subtle cycles. Small organisms may spend time suspended in the atmosphere or trapped in ice before reaching their final destination. Likewise, minerals within aqueous environments—derived from erosion—serve as nutrients for life, and in so doing may be redeposited on land. Soil itself can circulate, both as particles in the atmosphere (wind erosion) or as silts in flowing water. The importance of this type of cycle may be illustrated by Amazonia; a region that depends on airborne dust from the Sahara to replenish the minerals in its soils.

When considering soil cycles, it is worth reflecting on one of the meanings of sustainability, namely, the ability to continue. The word *cycle* implies a system in which individual components move between states, eventually closing a circle and thus keeping the system going. Like the pendulum clock, this requires energy; the soil system is kept going by the sun.

Biogeochemical Cycles

The term *biogeochemical* relates to the connection between biology, geology, and the chemistry of life. Living organisms that live and die on this planet are the biological part; the Earth into which they decompose comprises the geological part; and the reactions by which organic and inorganic materials are degraded for reuse is the chemical part. The biogeochemical cycles transform and transport to the Earth the substances that are required for all living organisms. The cycling of water, carbon, nitrogen, phosphorus, and sulfur are strongly interrelated. We will describe briefly each of these major cycles.

Water Cycle

The water cycle is fundamental to life and has a profound effect on temperature and climate. Living organisms acquire water in a variety of ways. Land plants absorb it from the soil through their roots; some may be adapted to absorb water directly from the atmosphere. Animals either drink water or eat other organisms that themselves consist largely of water.

The water within living things is returned to the atmosphere, where it forms clouds. The most significant group of living things in control of atmospheric humidity is land plants; the majority of water taken up by the roots of a plant passes rapidly out of the leaves. But all living creatures give back the water they have absorbed by some means; for example, in their breath, urine, perspiration, and after death. Atmospheric humidity is also very importantly determined by evaporation from surface waters. The cycle is closed when clouds form, and rain or snow falls, replenishing groundwater, freshwater, and oceans.

The water cycle appears to be relatively simple to grasp; water vapor moves into the atmosphere by evaporation or transpiration, even perspiration; it returns from the atmosphere in the form of liquid water in rain, snow, and fog. However, it is not a purely physical process; it depends on living creatures. Current life forms are adapted to live in current conditions. They react to the climate in which they live. If there were no living component in the water cycle, climate change would have little significance beyond changing the rate of its flux; the rate falling with lowered temperatures, rising with higher. But since the cycle depends on living organisms, climate change is rendered more hazardous, because the balance between evaporation and deposition of water would change not merely in terms of the rate of flux, but qualitatively as the organisms either took with gusto to the new conditions, or were destroyed by them.

Carbon Cycle

Understanding the biological mechanisms that regulate carbon exchange between land, oceans, and the atmosphere is central to our understanding of global climate change and the role of human activities that generate greenhouse gas emissions (Bardgett et al. 2008). Over 99% of the Earth's carbon is stored in rocks and sediments. This stable form of predominantly inorganic carbon contributes very little to carbon cycling. Organic carbon, found in the polysaccharides, proteins, lipids, and nucleic acids within living organisms, is the predominant form of carbon that is cycled. The soil is the largest organic carbon store in the world (1500 gigatons). This is derived mainly from humus, generated from dead and decaying biota, mostly from the cell walls of plants and microorganisms. Boreal forests cover about 11% of Earth's land surface and contain around 16% of the total soil organic carbon (Clemmensen et al. 2013).

The carbon cycle is closely linked to oxygen cycling. The fundamental process is photosynthesis. Photosynthetic organisms, both on land and in oceans, metabolize carbon dioxide in order to generate energy and carbohydrate; in so doing, they release oxygen. Animals do not utilize carbon dioxide; in order to produce energy they consume plants or other organisms, which themselves feed on plants. (They also use the oxygen produced by photosynthesis.) Carbon dioxide is released back into the atmosphere by the respiration of nonphotosynthetic organisms, which includes plants during darkness, and by all after death and decomposition. Under some conditions, decomposing plant materials may form relatively permanent and inert stores of carbon. For example, in the acidic and anaerobic conditions of a bog, peat deposits will form and remain stable for centuries. Coal and oil, which we regard as fuels, are also, until burned, large stores of stable carbon. Carbon is also stored in a potentially labile manner in the form of methane under the oceans and in permafrost. Additional carbon dioxide is released when organic matter (biomass) and fossil fuels are burned. The carbon dioxide returns to the atmosphere, where it can be taken in again by plants and stored in the wood of trees and bushes, or to the oceans as dissolved carbon dioxide from where it is taken up by aquatic plants, algae, and cyanobacteria.

In the oceans it may be more or less permanently sequestered as carbonates in the shells of sea creatures including diatoms, as after death they sink into the depths to form sedimentary layers.

Phosphorus Cycle

Most phosphorus in the biosphere is in a form of phosphate (PO_4^-), which cycles between living organisms, water, soil, and rocks. Unlike other biogeochemical cycles, the phosphorus cycle does not include a gas phase. Plants absorb inorganic or free organic phosphate from the soil. When animals consume plants or other organisms, they acquire phosphorus, which is incorporated into nucleic acids, adenosine triphosphate (ATP), phospholipids, and tissues such as bone. Organic phosphate is excreted from animals and released by decomposing plants and animals, forming free organic phosphorus (Figure 4.1). When free organic phosphorus returns to the soil, plants can absorb it again or it becomes part of the sediment layers that form rocks. The rocks erode by the action of water, and phosphate ions are returned to water and soil. Phosphates from fertilizers and other anthropogenic activity as well as from microorganisms feed into water bodies. These soluble phosphates may be utilized by aquatic organisms.

FIGURE 4.1 (See color insert.)
(a) Nutrient cycling in the pronghorn (*Antilocapra americana*) habitat. 1: Water and nutrients (e.g., phosphate) in the soil are taken up by plant roots. 2: Plants are eaten by pronghorns. 3: Organic and inorganic compounds are returned to the soil. Plant and animal necromass (b, c) as well as feces (d) are decomposed and mineralized in the soil. Plants take up these minerals and the cycle is closed. (© Photos by Marina Islas-Espinoza.)

The phosphorus cycle is unusual in the extent to which it is dependent on its use and conservation by humans. The requirement for high yields of food crops in particular means that the flux rate of the phosphorus cycle is far higher in developed countries now than it was in former agrarian times. Phosphate is a valuable commodity, extensively mined for the benefit of modern industrial farming. This has consequences for water pollution.

Nitrogen Cycle

Nitrogen is present in the nucleic acids and proteins of all living organisms and in the photosynthetic pigment, chlorophyll. Although it comprises 78% of the Earth's atmosphere, plants and animals cannot utilize gaseous nitrogen and it must be converted to other nitrogenous species prior to use. The only living organisms capable of carrying out this conversion are nitrogen fixing bacteria and archaea, known as diazotrophs.

The nitrogen cycle includes nitrogen fixation, nitrification, denitrification, and ammonification. Free-living and symbiotic bacteria fix gaseous nitrogen into ammonia, which accounts for around 65% of the total fixed nitrogen (Brady and Weil 2002). Symbiotic bacteria such as rhizobia carry out this transformation inside root nodules of legumes. Around 3% of fixed nitrogen in the form of nitrates and nitrites is generated by lightning. The remainder is generated industrially using the Haber-Bosch process, which combines gaseous nitrogen and hydrogen under high temperature and pressure to generate ammonia, and finally through the combustion of fossil fuels. During nitrification, bacteria oxidize ammonium into nitrates and nitrites. Other microorganisms reduce nitrate to nitrogen gas through a series of reductions known as denitrification. One of the intermediates of this series of reductions is nitrous oxide, a potent greenhouse gas. An alternative type of denitrification may occur under anoxic conditions where highly specialized bacteria generate gaseous nitrogen from nitrate and ammonium, in a process known as anammox (Kuenen 2008). Plants absorb nitrates or ammonium ions and convert them to organic compounds. Animals obtain nitrogen by consuming plants or other organisms. Therefore, the waste products of animals (e.g., urea from urine, peptides from feces) all contain nitrogen. Some nitrogen is released back into the ecosystem through excrement and dead plants or animals, which is subsequently converted to ammonium ions through microbial activity; this is known as ammonification.

Just as is the case with the phosphorus cycle, the flux rate through the nitrogen cycle is very agriculture dependent. Nonleguminous annual crops are heavily dosed with nitrogen fertilizers, again resulting in water pollution hazards as well as nitrous oxide emissions.

Sulfur Cycle

Most of the Earth's sulfur found in rocks and sediments is in the form of gypsum ($CaSO_4$) or sulfide minerals (e.g., pyrite). However, oceans comprise the largest sulfur reservoir, predominantly in the form of sulfate (SO_4^{2-}). The major biological transformations of the sulfur cycle are carried out by chemolithotrophic bacteria, although photosynthetic bacteria and archaea also contribute.

Sulfur is present in all living organisms, mostly in sulfur-containing amino acids. It is liberated from organic matter by microbial degradation and converted to sulfides, predominantly the volatile gas hydrogen sulfide (H_2S). Under aerobic conditions this is oxidized to sulfur or sulfate (SO_4^{2-}), abiotically or by aerobic bacteria. Sulfate may be reduced back to sulfide by sulfate reducing bacteria (a process known as dissimilatory sulfate reduction). Plants or microorganisms reduce sulfates to sulfides, before incorporating them back into amino acids (assimilatory sulfate reduction). Animals ingest this reduced form of sulfur, completing the cycle. Organic sulfur compounds may undergo further oxidation or reduction reactions; the most important of these being the microbial interconversion of dimethyl sulfide (DMS) to dimethyl sulfoxide (DMSO).

Organic sulfur compounds provide a particularly interesting example of the way living organisms can influence their own environment. Dimethylsulfonioproprionate is an osmoprotectant compound widely produced by marine organisms, partly in response to overheating. It biodegrades to DMS that escapes into the atmosphere. On reaching the troposphere, DMS is oxidized and forms sulfate aerosols. These act as condensation nuclei for cloud formation. The clouds reduce the amount of solar radiation reaching the surface of the ocean, and the marine organisms bask in their success as agents, albeit locally, of beneficial climate change. The role of DMS in influencing climate was first described by Charlson et al. (1987) and is now referred to as the CLAW hypothesis. It is an example of a negative feedback system in which a disturbance from an equilibrium state results in a response that restores the original conditions.

Other abiotic reactions play an important role in the global cycling of sulfur. Sulfur dioxide enters the Earth's atmosphere following the burning of fossil fuels and from volcanoes. Acidic gases, such as sulfuric acid and sulfur dioxide, react with water vapor molecules to form acid droplets, which precipitate and fall as acid rain, returning sulfur to the soils and oceans.

Methane, Clathrates, Sediments, and Permafrost

The soil in the ocean is covered by water. However, in essence the sediments are part of the Earth's surface with similar environmental roles in the biogeochemical cycles as a result of movement, transport, and deposition of matter in the seafloor. The wind, rain, flow, and gravity forces also determine the accumulation or lack of nutrients in these aquatic habitats.

In addition to carbon dioxide, the other major form of biologically produced carbon is methane. This is formed in anaerobic environments through the actions of methanogenic archaea, from carbon dioxide and hydrogen, or from acetate, produced by bacteria following fermentation. Methane, a potent greenhouse gas, may be released to the atmosphere, reoxidized to carbon dioxide by methanotrophic bacteria and archaea, or is sequestered in continental shelf areas, marine sediments, and underneath permafrost regions by an icy solid called methane clathrate (hydrate) (Bhattacharyya et al. 2012).

Estimates of the global inventory of methane clathrate may exceed 10^{19} g of carbon, which is comparable to estimates of potentially recoverable coal, oil, and natural gas. This methane reservoir on the seafloor could be released in response to climate change, such as increased temperature or lower pressure (Buffett and Archer 2004).

Biodiversity

Soil microorganisms are essential for the maintenance of all life on Earth, and are involved in the major biogeochemical cycles of elements and transformations of organic and inorganic substrates. They are responsible for the anaerobic and aerobic decomposition of organic molecules and the degradation of a wide range of compounds including recalcitrant natural polymers, xenobiotics, and toxic compounds. These diverse activities of mixed populations of soil microorganisms are fundamental to ecological processes. Diazotrophs, most of which are soil or root nodule dwelling bacteria, are the only organisms capable of fixing atmospheric nitrogen, which can then be used by plants and microorganisms. The Earth's soils contain an estimated 10^{30} bacteria, more than the estimated number of stars in the universe (Jansson and Prosser 2013). More than 99% of these microorganisms have not been cultured under laboratory conditions (Amann et al. 1995).

Soils that contain large amounts of decomposing roots and their symbiotic fungi store a large share of the world's carbon (up to 70% of the soil carbon). These organisms (mainly ectomycorrhizal fungi), colonize roots and gain nourishment from the plants while helping their hosts to absorb water and nutrients from the soil (Perkins 2013).

Soil Biological Crusts

A biological crust is a living community of lichens, nitrogen-fixing cyanobacteria, algae, liverworts, and mosses growing on the soil surface and binding it together. This helps stabilize the soil surface (Figure 4.2), protecting it from erosion and increasing water infiltration. Soil surface disturbances such as livestock grazing or human trampling can adversely affect biological crusts. In arid and semiarid ecosystems biological crusts can provide a significant amount of nitrogen for plant growth. In general, the relative importance of biological crusts increases as annual precipitation and potential plant cover decrease (USDA Natural Resources Conservation Service 2001).

FIGURE 4.2 (See color insert.)
Soil biological crust growing after a fire. The major components are cyanobacteria, green algae, fungi, mosses, liverworts, and lichens. (© Photo by Marina Islas-Espinoza.)

It should be noted that crustose organisms may also be found on a variety of surfaces other than developed soils. For example, tree bark is a habitat for many plants (mosses and lichens), as are manmade surfaces (walls, buildings, gravestones). Exposed bare rock may also host a variety of living crusts, notably lichens (a symbiotic association between fungi and photosynthetic algae or cyanobacteria). Such crusts can represent the very beginnings of soil formation in rocky habitats. Atmospheric dust trapped by the crust may be a sufficient condition for the arrival of higher plant life, and this in turn results in the eventual formation of soil as the plant dies. In some rock types, even a slight degree of penetration of faults in the rock by growing plant roots (or lichen rhizoids) may accelerate the process of weathering of the rock, again contributing to the development of new soil.

Quality of Soils

We have seen that soils are variable in their origins, their makeup, and in the functions they perform or the uses to which we put them. A "good" soil is one that performs its functions well. A natural example might be the soils within a tropical rainforest. Its function is to sequester the materials of the forest (fallen leaves, dead animals) and to recycle them as quickly and efficiently (without loss) as possible. However, good soil most usually means soil that is able to repeatedly produce good crops. Examples of good soil areas include the Liverpool Plains in New South Wales, Australia; the Fens of Eastern England; and reclaimed lands in Holland. These are all areas of intensive crop production. How sustainable is the quality of such soils? The Liverpool Plains have been an important agricultural area since the middle of the 19th century. The Dutch polders vary in age, with some dating from the 11th century, with the most recent being less than 100 years old. The Fenlands of England were finally drained to produce the farmland Fens in the late 18th century.

The Fenlands of Eastern England provide a good illustration of the difficulties of maintaining a good soil. Characteristically, fen soil has a high soil organic matter (SOM) content, in the form of peat. To maintain it, natural soil cycles are overridden, in that water has to be continually pumped out of the fenland basin. Over two centuries, this has resulted in both the shrinkage of the peat layer and its partial loss due to erosion, so that the underlying clays approach the surface more closely as time passes. In the most pessimistic view, the end result of this process will be the loss of function of some of the best cropland in Britain.

Is there any way of detecting or quantifying loss of soil quality over a shorter timespan than a couple of hundred years? Research in Canada (Carter 2002) seeks to show that analysis of SOM content of soils may be one such method, enabling scientists to detect subtle changes in soils that would escape the notice of a day-to-day grower. This is hopeful for sustainability. By careful husbanding of organic materials arising from agriculture, such as crop residues or animal manures, the loss of organic material from soils can be slowed or arrested. All too commonly at present, industrial farming relies heavily on the inputs of chemical fertilizers to maintain crop yields. The danger is also masked by the development of higher yielding strains of crop plants. But this is an unsustainable approach, because higher yielding varieties often require yet higher fertilizer inputs in order to achieve their potential. This serves the short-term goal of producing food at reasonable prices, but it entirely neglects the necessity to maintain the quality of the soil itself. This is as clear a definition of unsustainable practice as can be imagined.

Desertification, land degradation, and drought affect over 1.5 billion people in more than 110 countries, and each year the planet loses 24 billion tons of topsoil (European Commission 2012). Soil degradation, caused by contamination, erosion, loss of organic matter, compaction, salinization, soil sealing, landslides, and climate change, has severe consequences for humans and the environment. These factors may interact and result in irreversible damage to soils.

Increasing concentrations of atmospheric carbon dioxide can affect microbial activity and water content, and stimulate the production of important greenhouse gases such as nitrous oxide (N_2O) from upland soils and methane (CH_4) emissions from artificial and constructed wetlands (van Groenigen et al. 2011). Agricultural soils treated with nitrogen fertilizer are the main source of human-induced nitrous oxide emissions, which result from incomplete microbial denitrification. Such anthropogenic sources generate almost the same amount of nitrous oxide as soils under natural vegetation. Nitrous oxide has over 300 times the potential for global warming compared with carbon dioxide. Artificial and constructed wetlands such as rice paddies contribute a third to a half of all global emissions of methane. Some methane is eliminated from the atmosphere through oxidation by methanotrophic bacteria, however, this is generally low. Such changes in nitrous oxide and methane fluxes have the potential to greatly alter how terrestrial ecosystems affect our climate (van Groenigen et al. 2011).

The Conservation of Soils

Soil is a habitat. Living organisms that dwell within the soil both contribute to its value and exploit its virtues. They depend upon its mineral and carbon content, its structure, and its ability to acquire and recycle water. Conservation of soils must have regard to this dynamic relationship between the soil and its inhabitants.

The use of phosphate fertilizers provides a case study in this context. Phosphorus is an essential element for life, and phosphate ore is a nonrenewable resource. Mining of phosphate ore has increased markedly since the beginning of the Green Revolution in the 1940s and correlates with world population growth. Worldwide, 90% of phosphate ore is used for mineral fertilizers and animal feed additives (Kalmykova et al. 2012).

Phosphate ore contains insoluble calcium phosphate, mostly as tricalcium phosphate. Prior to its use as a fertilizer, this must be converted to a soluble form; monocalcium phosphate, by addition of acid, usually phosphoric acid. Around 95% of the world's known phosphate reserves are controlled by only five countries: the United States, China, Jordan, South Africa, and most significantly, Morocco, where 85% of all known reserves are located (Cordell and White 2011).

There is ongoing debate as to the length of time the world's usable phosphate reserves will last, ranging from hundreds of years to decades. Meanwhile, the demand for fertilizer is likely to expand further due to population growth, increasing food demand (including a shift to more meat), and the growing use of crops for biofuels.

On a global scale, considerable losses of phosphorus are generated in agriculture due to its leaching into water bodies. Phosphorus depletion can be minimized by practicing efficient farming methods. It can be recycled by the reuse of animal manure and crop residues. There is scope for recovering phosphorus from various sources in urban areas. Sewage sludge and incineration residues contain large amounts of phosphorus that could be recycled. Activated sludge cake is commonly sold as phosphate fertilizer.

Struvite (magnesium ammonium phosphate) may be recovered from human waste and used (following treatment) as an alternative phosphate fertilizer. This may involve the separation of urine, a waste product high in phosphorus. Other sources of nonmineral phosphate currently include bone meal and guano. Landfill sites are also potential sites of phosphate recovery. Separate collection and digestion of both household and commercial food waste can provide phosphorus for agriculture. Overall, a sustainable approach to managing phosphorus is needed, which will involve multiple strategies that consider the roles of phosphate in terms of resource depletion, agriculture, and pollution.

The Problem of Erosion

Soils are the main terrestrial reservoir of nutrients and organic carbon; however, these may be significantly altered by mobilization and deposition of particles by wind, rain, and gravity, giving soil a dynamic status, spatially and temporally. In addition, agriculture, mining, construction, and other anthropocentric activities can significantly alter nutrient mobilization and carbon cycling via changes in erosion rates (Figure 4.3). In particular, erosion can result in significant lateral fluxes of nitrogen and phosphorus. Conversely, the translocation and burial of soil could potentially lead to long-term carbon storage, due to a reduction in the aerobic decomposition of organic carbon (Quinton et al. 2010).

FIGURE 4.3
Road construction in a forest impacts the structure of the soil provoking loss of nutrients, affecting the local biogeochemical cycles, and ultimately resulting in a decrease in biodiversity. (© Photo by Marina Islas-Espinoza.)

Wind erosion events tend to increase with decreasing rainfall, whether across climatic regions or from year to year. In arid and semiarid regions the level of protection afforded to the soil surface by vascular plants or their residues is drastically reduced. Wind and water erosion are not necessarily problematic for soils with significant plant cover. During dry periods when much of the soil cover has been removed, biological crusts may provide organic matter and vital protection against erosion. Land management activities have a strong influence on surface erosion rates. Poor management strategies that reduce biological crusts, plant cover, or soil aggregate stability are likely to be more significant as soil texture becomes coarser. Burning also reduces aggregate stability of soils and damages biological crusts. Agricultural strategies should aim to evenly distribute livestock, in order to avoid concentrated zones of impact. Simple management strategies are applicable here: the provision of multiple water sources and feeding locations, and minimizing or removing livestock during dry seasons or periods of prolonged drought (Eldridge and Leys 2003).

The Pollution of Soils

Pollution is a word that always implies something undesirable. In terms of the study of soils, it may have multiple meanings. Generally, pollution expresses the presence within a soil of components that are deleterious to its function, and which in extreme cases may render it useless; that is to say, not capable of its prosaic function to humanity of growing food crops as well as its general function of supporting natural ecosystems. Organic compounds such as crude oil (hydrocarbons) and biological waste products, together with inorganic compounds, including heavy metals, explosives, pharmaceuticals, radionuclides, and gases all contribute to the contamination of soils. They may originate in ongoing processes such as mining, construction, industrial processes, and accidents; or legacy pollution may remain when a site previously used for some industrial process is redeveloped.

Tackling Soil Contamination

A fundamental principle of sustainability is to use ecological processes to achieve desired goals. Bioremediation—the use of microorganisms and plants to clean up pollutants in the environment—is precisely such a process.

In many of these situations bioremediation techniques are an effective response. The basis of bioremediation was given in Chapter 2.

Phytoremediation

The use of plants in the bioremediation of water, soil, or air is known as phytoremediation. Just as in the case of microorganisms, plants have been in the remediation business for a very long time. Indeed, they represent the most important carbon capture system on the planet, and its by-product is the oxygen we breathe. They themselves grow by taking up chemicals from the soil, which include some that locally may reach pollutant proportions, such as nitrates and phosphates. They also require small quantities of metals for metabolic purposes, and can tolerate and accumulate others. Plants are by no means as versatile as microorganisms in their ability to degrade organic materials, although current genetic research may widen the spectrum of possibility. The advantage of plants in bioremediation is that they can be deployed exactly where required, stay put, and can be harvested or removed at any time, if appropriate. Furthermore, when pollution occurs within built-up environments, green plants may represent an additional bonus amenity value quite different from the sight of an industrial plant performing a similar function. Phytoremediation is most efficiently used when treating low levels of pollution that are present in the upper layers of the soil.

Plants may be used in a variety of ways to reduce contamination of soil or water (Vidali 2001). Phytoextraction is the process whereby plants are used as an intermediary means of exporting toxic materials away from a site of contamination. The plant accumulates toxins within its own tissue; these can include heavy metals such as cadmium, nickel, copper, and zinc, as well as toxic elements such as arsenic. Later harvesting of the crop and its removal from the site results in lowered levels of contamination remaining *in situ*. The plant material, now itself contaminated, must be treated *ex situ*, for example, by incineration. In some cases, metals may be recovered. Plants that can accumulate high concentrations of heavy metals are known as hyperaccumulators. Around 400 plant species from Asteraceace, Brassicaceae, Caryophyllaceae, Poaceae, Violaceae, and Fabaceae families tolerate high levels of heavy metals in the soil. Some Brassicaceae may accumulate up to 100 times more metal than nonaccumulating species. One example is *Thlaspi caerulescens* (alpine pennycress), which accumulates high concentrations of cadmium, zinc, and nickel (Milner and Kochian 2008). The disadvantage of using alpine pennycress for bioremediation is that it grows slowly. Faster growing herbaceous plants, such as grasses (e.g., *Vetiveria* species) are more suitable and have been used to decontaminate former mine sites.

A variation of phytoextraction is known as phytostabilization. In this case, no attempt is made to remove contamination from the site. Rather, the plants absorb the toxic material, thus preventing its leaching away from the site; they are then simply left to continue their growth *in situ*. This process is in common view throughout the world on disused slag heaps next to abandoned mines of all sorts. Certain trees have a remarkable ability to tolerate and immobilize high levels of toxic metallic elements. Species of willow trees (*Salix*), for example, have been shown to help remediate mineral-oil-contaminated soil as well as cadmium-contaminated soil (Vervaeke et al. 2003). The coppiced willow can be used in various ways such as a fuel source. Many naturally occurring grasses (e.g., red fescue) are found growing on metal-contaminated coal slags.

In wetland soils or sediments, phytoremediation takes the form of rhizofiltration. The roots of plants are the site of mineral absorption; if water contaminated with nitrate or phosphate (both already considered under wastewater treatment) is passed through a bed of reeds, for example, the influent will stimulate growth of the reeds and the effluent will be partially decontaminated as a result of flowing past submerged roots. The same concept can be used in static water such as lagoons, ponds, or created wetlands; rapid plant growth removes many pollutants and can immobilize others within the plant tissue. Sunflowers were successfully used in this way to remove radioactive caesium and strontium from contaminated pond water at the Chernobyl nuclear power plant in the Ukraine (Dushenkov et al. 1997).

Phytotransformation is a further remediation process, which depends on the, albeit limited, ability of plants to absorb contaminants, including organic compounds, and metabolize them into harmless chemicals or gases. And plant roots also exude materials that can stimulate the growth of their more versatile coworkers in the remediation business: soil microorganisms.

Finally, while it is quite proper to consider the deliberate use of plants in the context of remediation, it should not be forgotten that they are our allies in any context. The case of the mine slag heap has been mentioned in terms of toxic metals, for example. Planting a slag heap with trees is not merely a chemical benefit. Trees stabilize such sites; possibly forestalling disastrous accidents such as was seen in Aberfan in Wales in 1966. They may, just possibly, even beautify them. Natural forests stabilize mountainsides and reduce flooding; large stands of trees influence patterns of rainfall and represent habitats of the highest value by whatever means measured. Even in city parks, trees reduce airborne particulates and so remediate the city-living experience. We should certainly exploit the character of plants in whatever way possible to attain sustainability goals. But one goal remains paramount: the sustaining of natural plant communities themselves.

The Case for Composting

Composting is a method of bioremediation of contaminated soil or sludge that results in useful end products—fertilizers or soil conditioners. Its main objectives are to break down organic matter to reduce its mass, destroy pathogens and parasites, and reduce odor. The soil or sludge is mixed with a bulking agent such as wood chips in order to create an open texture. The mixture is air cured, and at the end of the process the bulking agent is recovered. There are several ways of composting. In the most common, the bulked up material is spread in rows 1 to 2 m high—windrows—and turned several times a week in order to mix the material and maintain aeration. Alternatively, the material may be formed into heaps and a blower used to provide aeration, a method called aerated static piles. Both systems typically take 1 to 2 months to produce the end product. Closed composting systems are also used, some of these including the Dano biostabilizer allow ferrous metal recycling (Scragg 2005).

Composting is an aerobic degradation process, which consists of several temperature-related phases, each driven by different groups of microorganisms that develop within the bulked material. The efficiency of the overall process may be improved if samples of remediated soil that already contain acclimated microorganisms are added to the bulked material at the start.

Degradation proceeds via an initial mesophilic phase dominated by mesophilic bacteria and fungi, which raise the temperature of the material to around 40°C as a result of their metabolic activity (Beffa et al. 1995). During the following thermophilic phase, bacteria including thermotolerant actinomycetes and spore-forming *Bacillus* break down organic matter including carbohydrates and proteins. In this phase also, thermophilic fungi such as *Aspergillus* degrade more recalcitrant compounds such as cellulose and lignin. Temperatures regularly reach 60°C and may increase up to 80°C. It is during this phase that most decomposition and biomass formation takes place. The final cooling and maturation phase recruits mesophilic microorganisms, which play a further role in mineralization as well as nitrogen, sulfur, and carbon nutrient cycling. This is the phase during which higher level biota such as protozoa, beetles, and worms are recruited—familiar to gardeners who make compost from food and garden waste to fertilize their gardens.

Composting is low cost and can be done *in situ*, however, it does take a comparatively long time and uses a lot of land space. Furthermore, pathogen removal is not always complete, especially if temperatures exceeding 55°C are not maintained for at least 3 days (Bitton 2010).

Biodegradation of Xenobiotics

Xenobiotics are compounds that do not occur naturally; they are synthetic products of the modern chemical industry. Their novelty poses a particular problem for bioremediation, since microorganisms will not have encountered them before. Common examples include pesticides, polychlorinated biphenyls, dyes, chlorinated solvents, and petroleum-derived plastics. Many of these compounds degrade extremely slowly; for example, polystyrene takes hundreds of years to degrade and as it does, releases toxic styrene into the environment. Often xenobiotic degradation is incomplete, resulting in additional toxic products, or can only proceed by cometabolism. Most plastics take decades or more to degrade and they are a major source of pollution especially in the marine environment. The use of plastics of this nature may fairly be said to be unsustainable.

For some uses, alternatives are becoming available. Plastics such as polyhydroxyalkanoates (PHAs)—carbon storage compounds synthesized by numerous bacterial species—are biodegradable. PHAs include numerous different hydroxyacid polymers, many of which have properties resembling petroleum-based plastics. They are completely degraded in 3 to 9 months by a large and diverse array of bacteria and fungi present in soils (Park et al. 2012). Current research efforts are trying to establish cheaper ways to synthesize these PHAs.

Another xenobiotic source of particular concern is the degradation of pharmaceuticals and personal care products. Synthetic oestrogens from the oral contraceptive pill are continually being discharged into the environment via the sewage system in addition to natural oestrogens. Although degraded by many bacterial species and consortia during the wastewater treatment process, low concentrations may remain and enter receiving waters. Even at very low concentrations oestrogens can disrupt endocrine function in a variety of aquatic animals. As populations increase, additional treatments beyond that of wastewater treatment may be necessary to prevent this causing harm to wildlife. Physical methods, such as membrane filtration and chemical treatments such as oxidation by ozone, are costly and not always effective. Combined technologies such as membrane bioreactors, which integrate biological degradation of waste products with membrane filtration, may prove to be effective solutions.

Engineering and Limitations

Both *in situ* and *ex situ* bioremediation techniques are used to clean polluted soil. The choice of method may depend in part on the nature of the pollution event itself. Bioremediation is always a comparatively slow process, so that if there is a serious threat to the immediate environment, it may be preferable simply to remove the polluted soil to a less sensitive site for treatment.

An example of an *in situ* method is bioventing, which involves introducing oxygen into the soil in order to encourage aerobic biodegradation by indigenous microorganisms. Water and air or hydrogen peroxide (readily converted to oxygen) is pumped into contaminated soil to stimulate growth of aerobic microorganisms. Mineral nutrients such as nitrogen or phosphorus may also be added. *In situ* methods are low cost and fulfill sustainability criteria well. However, they depend for their effectiveness on an amenable soil structure. Clay soils, with a high proportion of very fine particulate matter, and conversely, gravelly or rocky soils, are all less susceptible to successful treatment. Very heavy contamination may also result in an unsatisfactory outcome.

Contaminated soils that have been moved to an appropriate *ex situ* treatment site may be dealt with in a variety of ways (Vidali 2001).

If the volume of contaminated soil is relatively small then slurry-based bioreactors can be used. Contaminated soil is added to vessels containing water and fertilizer. Degradation is enhanced by controlling the nutrient status of the slurry, as well as its pH, temperature, and oxygen content. In addition, mixed microbial populations preadapted to degradation of specific contaminants may be added. Bioreactor designs are often based on those used for wastewater treatment.

Solid-based techniques such as land farming, soil biopiles, and composting require large tracts of land. Land farming has been used for many years, particularly for treatment of oil-contaminated soils. Contaminated soil is spread out over existing land and mixed on a regular basis to enable aeration. As with most bioremediation methods, land farming relies on the indigenous microbial population degrading the contaminants. The nutrient status of the soil being treated can be controlled by the addition of fertilizers, including natural manures. Land farming is low cost but slow. Furthermore, it presents the risk that organic pollutants may leach into groundwater. To prevent this, contaminated soil may alternatively be spread onto a filter pad or a liner, enabling water and leachates to be collected from below. Land farming is mostly suited to treat soils with adequate moisture content, with around neutral pH, and at temperatures above 10°C.

If soil contains contaminants with a high proportion of volatile constituents, spreading over a wide area is inappropriate, since toxic gases may easily escape. In this situation, biopiles are a preferable approach. The material is formed into piles, which may be several meters high. A liner is used to enable leachate collection and recirculation, and fertilizers are also usually added. A system of piping is placed in the pile, enabling air to be pumped into the soil. In some instances, biofilters may be used to trap gaseous emissions.

Oil spills may represent a particularly difficult case. They do not only occur in the marine environment. There is much contamination of groundwater, soils, and deserts. This may result from accidental events or natural seepage, particularly in desert regions overlying large oil deposits, and it may involve large amounts of contaminants.

Bioremediation of deserts is particularly problematic due to the dry conditions, the lack of dispersal, and adhesion of oil to rocks and sand. There are also fewer oil-degrading species present. Plants, including members of the Compositae, have been found to promote the growth of oil-degrading bacteria in lightly polluted desert areas and soils.

One way in which oily sludges and oil-contaminated soils with normal levels of fertility may be subjected to bioremediation is by composting.

A Different Agriculture?

There is much about current agricultural practice that can be criticized in the context of sustainability. Very high energy inputs in terms of machinery and fuel use, for example; or heavy use of insecticides and herbicides, which historically and currently have profound effects on biodiversity; the increasing abandonment of good practice in the form of crop rotation and periodic fallows, leading to loss of soil structure and fertility; and the replacement of that fertility by energy-intensive chemical fertilizers instead of organic amendments. Organic and fair-trade entrepreneurship has made important inroads in urban markets in developed countries and their Third World suppliers. These approaches may presage a better fate for soils around the world.

Organic farming attempts to eliminate all artificial inputs, whether in terms of plant fertilizers or pest control. This style can seem retrograde in developed countries; however, it is admirably sustainable in terms of land use and soil protection.

Some of the bad effects of industrialized agriculture can also be addressed at a biological level. The use of biological agents to control pests, for example, is widely adopted even by the most industrially efficient sectors such as glasshouse-grown vegetables (tomatoes, capsicums). It would also be more "efficient" to use biological pest control and draught animals in place of diesel-engine machines, where possible.

Another example of an "alternative agriculture" is permaculture, a system of growing plants in dense diverse communities, which exploits the natural synergistic relationships between plant groups to provide an enhanced environment for all group members, while at the same time minimizing inputs and the effects of pests.

Finally, the soil encompasses many types of microenvironment. One of these is landfill sites for rubbish. The soil used merely as a waste dump is clearly a foolish and unsustainable idea. There is clear potential for such sites to be managed to capture their waste products, such as methane, for use in energy generation. This topic is covered elsewhere (see Chapter 10).

In summary, the soil is a resource of inestimable importance to the future of humankind. It is complex and once exploited, a surprisingly fragile environment, the most biodiverse habitat on the planet, and the one on which not only humans depend absolutely for their well-being. Our duty is to sustain the quality of the soils we have. Some of the problems and some solutions have been the subject of this chapter. If humans succeed in this goal, even with selfish motive, all life will benefit.

References

Amann RI, Ludwig W, Schleifer KH (1995) Phylogenetic identification and *in situ* detection of individual microbial cells without cultivation. *Microbiological Reviews* 59:143–169.

Bardgett RD, Freeman C, Ostle NJ (2008) Microbial contributions to climate change through carbon cycle feedbacks. *The ISME Journal* 2:805–814.

Beffa T, Blanc M, Marilley L, Lott Fischer J, Lyon P-F, Aragno M (1995) Taxonomic and metabolic microbial diversity during composting. In *The science of composting*, de Bertoldi M, Sequi P, Lemmes B, Papi T (eds), 149–161. Glasgow: Blackies Academic and Professional.

Bhattacharyya S, Cameron-Smith P, Bergmann D, Reagan M, Elliott S, Moridis G (2012) Tropospheric impact of methane emissions from clathrates in the Arctic Region. *Atmospheric Chemistry and Physics* 12:26477–26502.

Bitton G (2010) *Wastewater microbiology*. Hoboken, NJ: John Wiley & Sons.

Brady NC, Weil RR (2002) *The nature and properties of soils*. Upper Saddle River, NJ: Prentice Hall.

Buffett B, Archer D (2004) Global inventory of methane clathrate: Sensitivity to changes in the deep ocean. *Earth and Planetary Science Letters* 227:185–199.

Carter MR (2002) Soil quality for sustainable land management. *Journal of Agronomy* 94:38–47.

Charlson RJ, Lovelock JE, Andreae MO, Warren SG (1987) Oceanic phytoplankton, atmospheric sulphur, cloud albedo and climate. *Nature* 326:655–661.

Clemmensen KE, Bahr A, Ovaskainen O, Dahlberg A, Ekblad A, Wallander H, Stenlid J, Finlay RD, Wardle DA, Lindahl BD (2013) Roots and associated fungi drive long-term carbon sequestration in boreal forest. *Science* 339:1615–1618.

Cordell D, White S (2011) Peak phosphorus: Clarifying the key issues of a vigorous debate about long-term phosphorus security. *Sustainability* 3:2027–2049.

Dushenkov S, Vasudev D, Kapulnik Y, Gleba D, Fleisher D, Ting KC, Ensley B (1997) Removal of uranium from water using terrestrial plants. *Environmental Science and Technology* 31:3468–3474.

Eldridge DJ, Leys JF (2003) Exploring some relationships between biological soil crusts, soil aggregation and wind erosion. *Journal of Arid Environments* 53:457–466.

European Commission (2012) The implementation of the soil thematic strategy and ongoing activities. Report from the Commission to the European Parliament, the Council, the European Economic and Social Committee and the Committee of the Regions. 13.2.2012 COM(2012) 46 final. Brussels: European Commission.

Jansson JK, Prosser J (2013) The life beneath our feet. *Nature* 494:40–41.

Johnson BD, Hallberg KB (2005) Acid mine drainage remediation options: A review. *Science of the Total Environment* 338:3–14.

Kalmykova Y, Harder R, Borgestedt H, Svanäng I (2012) Pathways and management of phosphorus in urban areas. *Journal of Industrial Ecology* 16:928–939.

Kuenen JG (2008) Anammox bacteria: From discovery to application. *Nature Reviews Microbiology* 6:320–326.

Milner MJ, Kochian LV (2008) Investigating heavy-metal hyperaccumulation using *Thlaspi caerulescens* as a model system. *Annals of Botany* 102:3–13.

Park SJ, Kim TW, Kim MK, Lee SY, Lim SC (2012) Advanced bacterial polyhydroxy-alkanoates: Towards a versatile and sustainable platform for unnatural tailor-made polyesters. *Biotechnology Advances* 30:1196–1206.

Perkins S (2013) Fungi and roots store a surprisingly large share of the world's carbon. *Nature News.* doi:10.1038/Nature: 12698.

Quinton JN, Govers G, Oost KV, Bardgett RD (2010) The impact of agricultural soil erosion on biogeochemical cycling. *Nature Geoscience* 3:311–314.

Scragg A (2005) *Environmental biotechnology.* Oxford: Oxford University Press.

USDA Natural Resources Conservation Service (2001) Rangeland soil quality—Physical and biological soil crusts. Soil Quality Information Sheet. Rangeland Sheet 7.

van Groenigen KJ, Osenberg CW, Hungate BA (2011) Increased soil emissions of potent greenhouse gases under increased atmospheric CO_2. *Nature* 475:214–216.

Vervaeke P, Luyssaert S, Mertens J, Meers E, Tack FM, Lust N (2003) Phytoremediation prospects of willow stands on contaminated sediment: A field trial. *Environmental Pollution* 126:275–282.

Vidali M (2001) Bioremediation. An overview. *Pure and Applied Chemistry* 73:1163–1172.

5

Land-Use Impacts on Ecosystem Services

Agnieszka Ewa Latawiec and Bernardo B.N. Strassburg

CONTENTS

Land as a Key for Sustainability

Land is a source of sustenance, resources and "is literally the base upon which all human societies are built" (Owens and Cowell 2011). In addition to being a material base, land is charged with diverse social and cultural dimensions (Owens and Cowell 2011), while terrestrial ecosystems provide a number of ecosystem services (Daily 1997). The concept of *ecosystem services* is increasingly an influential way to consider conservation and is becoming a central tool to express humanity's need for the rest of living nature. Notwithstanding a number of definitions, the most commonly cited one is from the Millennium Ecosystem Assessment (MEA), which defines ecosystem services as "the benefits people obtain from ecosystems" (MEA 2005). Terrestrial ecosystems provide regulating functions, for instance, for climate change; provisioning services for food, fuel, and water resources; supporting services for biodiversity and nutrient recycling; and finally, cultural, such as recreation (Figure 5.1). The capability of terrestrial ecosystems to provide these services is determined by changes in land use and land cover, socioeconomic factors, biodiversity, atmospheric composition, and climate (Metzger et al. 2006).

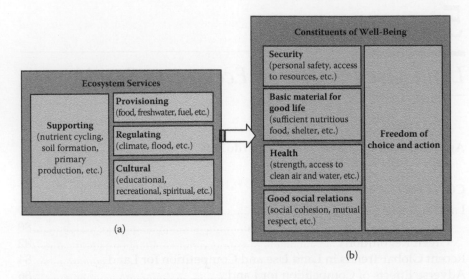

FIGURE 5.1
Four categories of ecosystem services and how these ecosystem services influence human well-being. (b) This box illustrates the constituents of human well-being and how ecosystem services (a), grouped into four categories within the Millennium Economic Assessment (MEA), affect human well-being (b). (Modified from the Millennium Economic Assessment [MEA], 2005, *Ecosystems and human well-being: Current state and trends*, vol. 1, New York: Island Press.)

In this chapter we will specifically discuss land-cover and land-use change, their recent trends and their impacts on ecosystem services provided by terrestrial ecosystems. Land-use activities, either by converting natural landscapes or by changing management practices on existing lands, have indeed transformed a large proportion of the planet's land surface (Foley et al. 2005). These human-driven changes on the terrestrial surface influence the structure and function of terrestrial ecosystems, with far-reaching consequences for human well-being (Turner et al. 2007). The extent and significance prompted the emergence of land-change science (Turner et al. 2007), which seeks to understand the dynamics of land cover and land use within a coupled human–environment system. This interdisciplinary field has emerged as a fundamental component of global environmental change and sustainability research and it addresses concepts, models, and implementations associated with environmental and societal problems, including the intersection of the two.

Land-Use and Land-Cover Change

Land uses are human-defined functions (agricultural, mining, industrial, residential, or natural use) and land cover is the resulting surface (e.g., forest, prairie, barren, inland water body). Although land-use and land-cover change have enabled mankind to appropriate an increasing share of the planet's resources to provide food, fuel, water, and shelter, they also potentially undermine the capacity of ecosystems to provide a wide range of ecosystem services.

Several decades of research have revealed the environmental impacts of land use throughout the globe. There is extensive literature on impacts of land-use and land-cover change (Foley et al. 2005; Lambin and Geist 2006) and here we will just highlight a few. For instance, deforestation, agricultural misman-agement, overgrazing, fuelwood consumption, and urbanization lead to soil degradation caused by water erosion, wind erosion, chemical, and physical degradation (Foley et al. 2005). Changes in nutrient cycling such as nitrogen, phosphorus, and potassium depletion by runoff or leaching in terrestrial eco-systems affect food production through reduced fertility and declining crop yields. In addition, the loss of native habitats affects agricultural production by degrading the services of bees and other pollinators (Ricketts et al. 2004). The global hydrologic cycle has been transformed to provide freshwater for irrigation, industrial, and domestic consumption, while water quality and coastal and freshwater ecosystems have been affected through sediment load and nutrient inputs from fertilizers and atmospheric pollutants (Tilman et al. 2001). Indeed, approximately 40% of the land surface is dedicated to agricul-ture (Turner et al. 2007) and agriculture accounts for nearly 85% of annual water withdrawals globally (Foley et al. 2005). Water withdrawals are almost universally unsustainable and lead to declining water tables, whereas many large rivers, especially in semiarid regions, have greatly reduced flows and some routinely dry up (Foley et al. 2005). Moreover, land-use and land-cover change lead to salinization and sodication, identified to significantly con-tribute to land degradation and loss of agricultural production (Food and Agriculture Organization of the United Nations [FAO] 2009). Land-use and land-cover change also cause declines in biodiversity through the loss, modi-fication, and fragmentation of habitats, degradation of soil and water, and overexploitation of native species. Deforestation has been identified as the single major cause of species extinctions worldwide (Baillie et al. 2004). Recent projections suggest that up to 27% of forest species could disappear before the end of the century due to deforestation alone, a level 10,000 times higher than the historic background extinction rate (Strassburg et al. 2012). Land-use practices have played a role in changing the global carbon cycle and affecting both the regional and the global climates (through changes in surface energy and water balance). Indeed, agriculture and forestry combined contribute to 31% of anthropogenic greenhouse gas emissions (Intergovernmental Panel on Climate Change 2007), more than any other sector (Figure 5.2).

Land-cover change affects the water cycle and regional climates through its effects on net solar radiation and latent heat (Foley et al. 2005; Lambin and Geist 2006). Moreover, increased aerosol particle loading in the atmosphere may be affecting the water cycle on account of changes in precipitation driven by the number and size of cloud condensation nuclei in the atmo-sphere (Lambin and Geist 2006). Urban "heat islands" are an extreme case of how land use modifies regional climate because the reduced vegetation cover, impervious surface area, and morphology of buildings in cityscapes store heat, lower evaporative cooling, and warm the surface air (Bonan 2002).

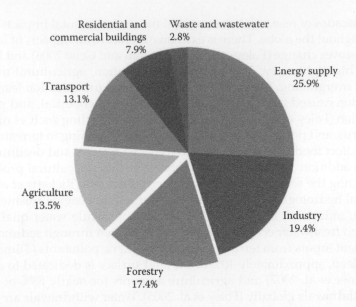

FIGURE 5.2
Greenhouse gas emissions per sector in 2004 (in CO_2 equivalents). (Modified from the Intergovernmental Panel on Climate Change [IPCC], 2007, Fourth assessment report: Climate change 2007, http://www.ipcc.ch/pdf/assessment-report/ar4/syr/ar4_syr.pdf.)

To reflect both the amount of area used by humans and the intensity of land use, an aggregated indicator of human appropriation of net primary production (HANPP) may be used. Although many definitions exist, HANPP, in essence, demonstrates how much of the trophic energy would be available for wildlife in the absence of human activities. It is considered an extremely valuable indicator of the intensity of socioeconomic "colonization of ecosystems" (Fischer-Kowalski and Haberl 1997) and of the "human domination of ecosystems" at the global scale (Vitousek et al. 1997). Empirical studies increasingly demonstrate that HANPP may indicate human pressures on ecosystems because it is directly associated with the provision of ecosystem services, such as the provision of biomass through forestry and agriculture. Indeed, the net amount of biomass produced each year by plants (NPP) may be a central parameter of ecosystem functioning impacted by human-induced changes in ecosystems. Because HANPP has been often interpreted as an indicator for ecological limits to growth, studies of global HANPP have been also considered in the context of sustainable development. Several metrics, in addition to net primary productivity, allow for comparisons of landscapes and their trade-offs of ecosystem services (Figure 5.3).

Green Revolution

Between 1940 and the late 1970s, the Green Revolution helped the agriculture to overcome physical and biotic constraints, such as insects, diseases, and

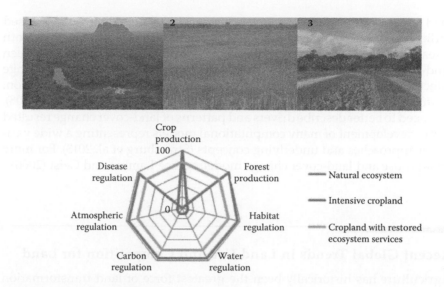

FIGURE 5.3 (See color insert.)
A conceptual framework for comparing land use and trade-offs of ecosystem services in three hypothetical landscapes: (1) a natural ecosystem, (2) an intensively managed cropland, and (3) a cropland with restored ecosystem services. The provisioning of multiple ecosystem services under different land-use regimes is presented. The condition of each ecosystem service is demonstrated along each axis. The natural ecosystem provides many ecosystem services but not food production. The intensively managed cropland produces food (maybe in the short term) at the cost of diminishing other ecosystem services. A cropland that is managed to maintain other ecosystem services (cropland with restored ecosystem services) may provide a broader portfolio of ecosystem services. (Figure modified from Foley et al., 2005, *Science* 309:570. Photos 1, 2, and 3 courtesy of Trond Larsen ©/Conservation International, Bernardo Strassburg/International Institute for Sustainability, Maureen Silos/the Caribbean Institute, respectively.)

weeds, with unprecedented increases in production. This negated dreary Malthusian forecasts that the Earth would not be able to support its growing human population. The Green Revolution was based on a range of scientific research and management solutions, and the combined effects of factors (such as development of improved, high-yielding cultivars; soil cultivation techniques; chemical fertilization, pest control via synthetic pesticides; and expansion of irrigation infrastructure) doubled the global food production in the past 50 years (Tilman et al. 2001). It greatly reduced food shortages and is believed to have saved millions of people from starvation. However, the Green Revolution's distinguishing feature was its dependence on pesticides and chemical fertilizers. As a consequence, the Green Revolution also led to soil, water, and air contamination; loss of soil and water quality; side effects on nontargeted species; loss of biodiversity and agrobiodiversity; and deterioration of human health, in particular for small landholders.

A better understanding of land-cover change and its impacts on soil degradation (Trimble and Crosson 2000), biodiversity loss (Baillie et al. 2004; International Union for Conservation of Nature 2011), climate change, and food security (IPCC 2007), among other local and global effects discussed earlier, is considered paramount for sustainable land management by both researchers and decision makers (Verburg et al. 2002). The linkages between land-cover change and policies are bidirectional, with land-cover change affecting and being affected by decisions, such as infrastructure expansion, tariffs, taxes, subsidies, and the creation of protected areas (Reid et al. 2008). The need to better describe drivers and patterns of land-cover change resulted in the development of many computational models representing a wide variety of approaches and underlying concepts (Strassburg et al. 2013). For more on land-use and land-cover change modeling, see Lambin and Geist (2006).

Recent Global Trends in Land Use and Competition for Land

Agriculture has historically been the greatest force of land transformation (Ramankutty et al. 2007), with population growth and per capita consumption driving global land-use change (Tilman et al. 2001). Global cropland area expanded from 3 million to 4 million km^2 in 1700 to 15 million to 18 million km^2 in 1990, mostly at the expense of forests (Goldewijk and Ramankutty 2004). Similarly, Gibbs et al. (2010) shows that throughout the tropics, between 1980 and 2000, more than 80% of new agricultural land resulted from deforestation.

As much as 50% of the Earth's ice-free land surface has now been transformed and virtually all land has been affected to some extent by land-use and land-cover change. Croplands and pastures are one of the largest terrestrial biomes on the planet. In 2011, about 12% (more than 1.5 billion ha) of the globe's land surface (13.0 billion ha, excluding "inland water") were used for crop production (arable land and land under permanent crops) while 3.3 billion ha were occupied by permanent meadows and pastures. Together croplands and pastures rival in extent the global forest area (around 4 billion hectares) (FAO 2013). Currently, arable land occupies some 28% of the prime (very suitable) and good (suitable and moderately suitable) land (Alexandratos and Bruinsma 2012).

According to projections (Alexandratos and Bruinsma 2012), the future demand for agricultural land will continue to grow over the next few decades at least, fueled by demand for commodities such as food, fodder, and timber, and driven by population and per capita consumption growth. These multiple demands do not only put pressure on scarce land resources, but they also lead to competition for land (Smith et al. 2010) and adverse effects associated with this competition (Lambin and Meyfroidt 2011). According to projections (Smith et al. 2010), the competition for land is expected to continue over at least the next four decades, driven both by a growing world population, changing dietary preferences, and the pursuit of higher standards of living

(Smith et al. 2010). In addition to growing demand for agricultural products, competition for land will escalate in the future on account of environmental degradation that reduces the pool of available land for production (see earlier and Smith et al. 2010). Furthermore, in a world seeking for solutions to energy demand and substitutes for high-priced petroleum products, GHG-emitting fossil fuels, and energy supplies originating from politically unstable countries (Naylor et al. 2007), biofuels offer an attractive solution (Tilman et al. 2009). Yet, crops destined for biofuels also compete with staple crops for land. In addition, they dictate an abrupt increase in demand for agricultural commodities traditionally used for food and feed, placing upward pressure on crop prices and altering the fundamental economic dynamics that have governed global agricultural markets for the past century (Naylor et al. 2007). As a consequence, this may lead to deforestation elsewhere to meet the displaced demand, having therefore adverse effects far beyond the land already taken by biofuels (Tilman et al. 2009). Furthermore, conservation and restoration may be also perceived as an additional component in competition for land (Smith et al. 2010). Figure 5.4 shows the continuing decline of arable land (in use) per person (Alexandratos and Bruinsma 2012), which is often cited as an indicator of imminent land-associated problems. Indeed, concerns have recently been raised (International Assessment of Agricultural Knowledge, Science and Technology for Development 2009; Godfray et al. 2010) that agriculture run as business-as-usual might no longer be able to produce the food needed to sustain a still growing world population at levels required to lead a healthy and active life. Solutions proposed to diminish the pressure on scarce land resources are discussed in the next chapter of this book.

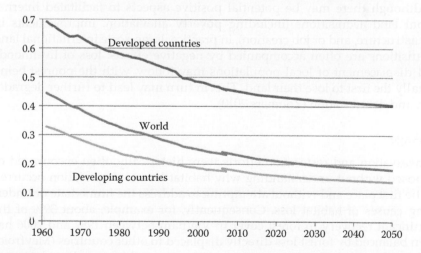

FIGURE 5.4
Arable area per capita (hectares in use per person). (Alexandratos N, Bruinsma J, 2012, World agriculture towards 2030/2050, The 2012 Revision, ESA Working Paper No. 12-03, Rome: FAO. With permission.)

Adverse Effects of Competition for Land

Land Grab

A convergence of global crises (financial, environmental) and competition for land driven by the need for energy and food in recent years have contributed to a dramatic reevaluation of and rush to control land, especially land located in developing countries. Transnational and national actors from various business sectors (including the oil, mining, forestry, food, chemical, and bioenergy industries) are acquiring large chunks of land on which to build, maintain, or extend large-scale agroindustrial enterprises. National governments in "finance-rich, resource-poor" countries are looking to "finance-poor, resource-rich" countries to secure their future food and energy needs, while many national governments in finance-poor, resource-rich countries are searching for possible land investors. As a result, there is an ongoing rise in the number of cross-border large-scale land deals. Some now refer to this phenomenon as a new "global land grab." Indeed, global rush for new land has recently sparked a vivid discussion after the release of a number of papers and reports (e.g., World Bank 2011) that criticized these kinds of transactions. In 2009, over 50 Mha of farmland in Africa were subject to such negotiations or transactions (Friis and Reenberg 2010), with the food and biofuel production grown in these areas destined for exports. In its report, the World Bank claims there are now about 45 million ha covered by recent large-scale land acquisitions, 70% of which are in Africa (Mozambique, Democratic Republic of Congo, and Zambia, among others).

Although there may be potential positive aspects to facilitated international land acquisitions (including poverty alleviation, improvements in infrastructure, and or job creation), in practice, large-scale international land acquisitions are often accompanied by negative effects: loss of livelihoods and displacement of local populations may follow, with the poorest being usually the first to lose their land. This in turn may lead to further degradation and deforestation (Zoomers 2010).

Leakage

Conservation and restoration efforts were historically often encouraged or imposed without understanding why habitat loss or degradation occurred in the first place and without attempting to address the immediate or underlying causes of habitat loss. Consequently, for example, about 39% of the significant recovery in forest cover in Vietnam between 1987 and 2006 has been balanced by forest loss directly displaced to other countries (Meyfroidt and Lambin 2009). Furthermore, increasing demand for wood products and new forest conservation programs in China and Finland increased pressure on forests through wood imports in neighboring Russia (Mayer et al. 2005). In Brazil, it has been demonstrated that such an indirect land-use change may overcome carbon saving from biofuels (Lapola et al. 2010). There has also often been a mismatch between social and ecological goals of conservation and restoration; either restoration has aimed to fulfill social or economic needs without reference to its wider ecological impacts or it has had a narrow conservation aim without taking into account people's needs. The success of restoration projects could therefore be undermined by making restoration another competitive use of land, possibly leading to "leakage" (Lambin and Meyfroidt 2011). Leakage can be associated with conservation measures, where restrictions to land-use change in one area (e.g., establishment of a protected area) without measures to address demand for agricultural land lead to deforestation elsewhere (Lambin and Meyfroidt 2011). Indeed, leakage can be transboundary, meaning that although demand increase occurs in one part of the world, pressure to provide commodities may be shifted elsewhere, given economical benefits for commodity-providing countries (Lambin and Meyfroidt 2011) and the globalization of agricultural markets. Such leakage, especially from developed to tropical countries may be detrimental to the environment because environmental protection tends to be weaker in developing countries, and their ecosystems are usually more productive, biodiverse, and carbon rich. Leakage may occur either via migrations or by increasing imports of agricultural or wood products to satisfy demand, thus shifting pressure on natural ecosystems elsewhere. Possible leakages can stem, for example, from a badly designed Reducing Emissions from Deforestation and Forest Degradation (REDD) scheme. REDD is an initiative under the United Nations Framework Convention on Climate Change (UNFCCC), which seeks to offer financial incentives to reduce greenhouse gas emissions related to deforestation and degradation in developing countries. Implementation of a narrowly focused REDD mechanism that does not take into account social and economic aspects of increasing demand for agricultural products driving deforestation in the first place, could result in unintended perverse land-cover change and carbon leakage. Potentially harmful side effects for some biodiversity areas have been reported (Strassburg et al. 2009). Addressing drivers of deforestation and degradation within the agricultural sector is crucial for successful REDD interventions, as discussed in the following chapter. As deforestation practices may be an outcome of deeply embedded social, economic, and cultural histories, attempting to alter these drivers may be challenging. Yet, conservation or restoration projects that ignore these immediate or underlying causes of forest loss will likely be affected by leakage.

Rebound Effect

Although an increase in productivity carries the potential to abate competition for agricultural products and to spare wilderness (see the following chapter), if complementary measures are not implemented it can lead to rebound, a classic economic effect where increased productivity leads to an increase in demand for its input, land in this case (Parrotta et al. 2012). In Brazil, for example, a remarkable increase in productivity of soybean made soybean farming more attractive, transforming it into a leading cause of deforestation. The moratorium on soy production that followed virtually eliminated direct deforestation for soybean production, although indirect deforestation, where soybean expands onto pasturelands and pushes ranchers into the forest (an instance of leakage), still remains a challenge. The abolition of the soy moratorium also remains a possible future challenge.

Conclusions

If the global population stabilized at about 9 billion people, the next 50 years might be the last episode of extensive global agriculturally driven land-use change. Over the next decades, agriculture has the potential to have irreversible environmental impacts. An ever-increasing demand for agricultural products will likely lead to further competition for finite natural resources. Increasing demand for agricultural products, urbanization, industrial uses, and biofuel crops will have an impact on already scarce global land resources. Changing land availability because of environmental degradation and the need to preserve resources through conservation and reforestation also contribute to competition for land. In this chapter we discussed drivers of competition for land, and the following chapter will present selected solutions to attenuate this competition. Noteworthy, many issues associated with (un)sustainable land use and land-use change are inherently complex and involve a variety of environmental, social, and economic aspects. Land is used by people both in material and nonmaterial ways, and a vast number of them are still craving to develop. Convincing countries to undergo a different route than most of the developed countries have chosen in the past may prove problematic not only on account of ethics but also due to the sheer number of people involved and their expectations of higher standards of living than the land may support. On the other hand, there is an uncontested need to protect land resources on account of ecosystem services (Figures 5.5 and 5.6). Part of the difficulty with land-use and land-cover change is that they happen, as discussed earlier, through a multiplicity of direct, indirect, sometimes cumulative, and often uncertain effects. They operate at different scales, have transboundary and global effects; and have economic, legal, social, and political dimensions. Conflicts may arise between multiple rights and jurisdictions, which environmental science alone cannot resolve. Notwithstanding complex and challenging and sometimes gloomy prognoses of future food and resources catastrophes, and discussions of fast approaching tipping points of unsustainability, we next review some tangible approaches and concepts to reconcile multiple and growing demands. This may offer us a more optimistic road to the future of the natural environment.

FIGURE 5.5
Atlantic Rainforest, Brazil. Rainforests provide a wide range of global and local ecosystem services. (Photo by Agnieszka Latawiec, International Institute for Sustainability.)

FIGURE 5.6 (See color insert.)
Natural, unconverted habitats are home to a variety of species. Losing these habitats often leads to species loss, following a basic ecological relationship known as the species–area curve. The cheetah is an example of a threatened species according to the Red List compiled by the International Union for Conservation of Nature (IUCN). (Photo courtesy of Bernardo Strassburg, International Institute for Sustainability.)

References

Alexandratos N, Bruinsma J (2012) World agriculture towards 2030/2050. The 2012 Revision. ESA Working Paper No. 12-03. Rome: FAO.

Baillie JEM, Hilton-Taylor C, Stuart SN et al. (2004) *A global species assessment*. Gland: IUCN.

Bonan GB (2002) *Ecological climatology*. Cambridge: Cambridge University Press.

Daily GC (1997) *Nature's services: Societal dependence on natural ecosystems*. Washington DC: Island Press.

Fischer-Kowalski M, Haberl H (1997) Tons, joules and money: Modes of production and their sustainability problems. *Society and Natural Resources* 10:61–85.

Foley JA, DeFries R, Asner GP, Barford C, Bonan G, Carpenter SR, Chapin SF, et al. (2005) Global consequences of land use. *Science* 309:570.

Food and Agriculture Organization of the United Nations (FAO) (2009) Advances in the assessment and monitoring of salinization and status of biosaline agriculture. Report of an expert consultation held in Dubai, United Arab Emirates, November 26–29, 2007. World Soil Resources Reports 104. Rome: FAO.

Food and Agriculture Organization of the United Nations (FAO) (2013) FAOSTAT. http://faostat.fao.org/ (accessed February 15, 2013).

Friis C, Reenberg A (2010) Land grab in Africa: Emerging land system drivers on a teleconnected world. GLP report No.1. Copenhagen: GLP-IPO.

Gibbs HK, Ruesch AS, Achard MK, Clayton MK, Holmgren P, Ramankutty N, Foley A (2010) Tropical forests were the primary sources of new agricultural land in the 1980s and 1990s. *PNAS* 107:16732–16737.

Godfray HCJ, Beddington JR, Crute IR, Haddad L, Lawrence D, Muir JF, Pretty J, Robindon S, Thomas SM, Toulmin C (2010) Food security: The challenge of feeding 9 billion people. *Science* 327:812–818.

Goldewijk KK, Ramankutty N (2004) Land cover change over the last three centuries due to human activities: The availability of new global data sets. *GeoJournal* 61:335–344.

Intergovernmental Panel on Climate Change (IPCC) (2007) Fourth assessment report: Climate change 2007. http://www.ipcc.ch/pdf/assessment-report/ar4/syr/ar4_syr.pdf (accessed January 20, 2013).

International Assessment of Agricultural Knowledge, Science and Technology for Development (IAASTD) (2009) Agriculture at Crossroads. Global Report. Washington DC: Island Press.

International Union for Conservation of Nature (IUCN) (2011) Global figures for 2011.1 IUCN Red List of Threatened Species. http://www.iucnssg.org/tl_files/Publications/June%202011%20IUCN%20Red%20List%20update_EN.pdf (accessed July 15, 2013).

Lambin EF, Geist H (2006) *Land-use and land-cover change: Local processes and global impacts*. Berlin: Springer-Verlag.

Lambin EF, Meyfroidt P (2011) Global land use change, economic globalization, and the looming land scarcity. *PNAS* 108:3465–3472.

Lapola DM, Schaldach R, Alcamo J, Bondeau A, Koch J, Koelking CA, Priess JA (2010) Indirect land-use changes can overcome carbon savings from biofuels in Brazil. *PNAS* 107:3388–3393.

Mayer AL, Kauppi PE, Angelstam PK, Zhang Y, Tikka PM (2005) Ecology. Importing timber, exporting ecological impact. *Science* 308:359–360.

Metzger MJ, Rounsevell MDA, Acosta-Michlik L, Leemans R, Schroter D (2006) The vulnerability of ecosystem services to land use change. *Agriculture, Ecosystems and Environment* 114:69–85.

Meyfroidt P, Lambin EF (2009) Forest transition in Vietnam and displacement of deforestation abroad. *PNAS* 106:16139–16144.

Millennium Economic Assessment (MEA) (2005) *Ecosystems and human well-being: Current state and trends*, vol. 1. New York: Island Press.

Naylor RL, Liska A, Burke MB, Falcon WP, Gaskell JC (2007) The ripple effect: Biofuels, food security and the environment. *Environment* 49:30–43.

Owens S, Cowell R (2011) *Land and limits: Interpreting sustainability in the planning process*, 2nd ed. New York: Routledge.

Parrotta JA, Wildburger C, Mansourian S (2012) Understanding relationships between biodiversity, carbon, forests and people: The key to achieving REDD+ objectives. A global assessment report prepared by the Global Forest Expert Panel on biodiversity, forest management and REDD+. IUFRO World Series Vol. 31. Vienna: IUFRO.

Ramankutty N, Gibbs HK, Achard F, DeFries R, Foley JA, Houghton RA (2007) Challenges to estimating carbon emissions from tropical deforestation. *Global Change Biology* 13:51–66.

Reid R, Gichohi H, Said M, Nkedianye D, Ogutu J, Kshatriya M, Kristjanson P, et al. (2008) Fragmentation of a peri-urban savanna, Athi-Kaputiei plains, Kenya. In *Fragmentation in semi-arid and arid landscapes: Consequences for human and natural systems*, Galvin KA, Reid RS, Behnke Jr RH, Hobbs NT (eds), 195–224. Berlin: Springer.

Ricketts TH (2004) Tropical forest fragments enhance pollinator activity in nearby coffee crops. *Conservation Biology* 18:1262–1271.

Smith P, Gregory PJ, van Vuuren D, Obersteiner M, Havlik P, Rounsevell M, Woods J, Stehfest E, Bellarby J (2010) Competition for land. *Philosophical Transactions of the Royal Society B* 365:2941–2957.

Strassburg B, Latawiec AE, Creed A, Nguyen N, Sunnenberg G, Miles L, Lovett A, et al. (2013) Biophysical suitability, economic pressure and land-cover change: A global probabilistic approach and insights for REDD+. *Sustainability Science* DOI 10.1007/s11625-013-0209-5.

Strassburg B, Rodrigues ASL, Gusti M, Balmford A, Fritz S, Obersteiner M, Turner RK, Brooks TM (2012) Impacts of incentives to reduce emissions from deforestation on global species extinctions. *Nature Climate Change* 2:350–355.

Strassburg B, Turner RK, Fisher B, Schaeffer R, Lovett A (2009) Reducing emissions from deforestation: The "combined incentives" mechanism and empirical simulations. *Global Environmental Change* 19:265–278.

Tilman D, Fargione J, Wolff B, D'Antonio C, Dobson A, Howarth R, Schindler D, Schlesinger WH, Simberloff D, Swackhamer D (2001) Forecasting agriculturally driven global environmental change. *Science* 292:281–284.

Tilman D, Socolow R, Foley JA, Hill J, Larson E, Lynd L, Pacala S, Reilly J, Searchinger T, Somerville C, Williams R (2009) Beneficial biofuels: The food, energy, and environment trilemma. *Science* 325:270–271.

Trimble SW, Crosson P (2000) Land-use: U.S. soil erosion rates—Myth and reality. *Science* 289:248–250.

Turner BL, Lambin EF, Reenberg A (2007) The emergence of land change science for global environmental change and sustainability. *PNAS* 104:20666–20671.

Verburg PH, Soepboer W, Veldkamp A, Limpiada R, Espaldon V, Mastura SSA (2002) Modeling the spatial dynamics of regional land-use: The CLUE-S model. *Environmental Management* 30:391–405.

Vitousek PM, Mooney HA, Lubchenco J, Melillo JM (1997) Human domination of Earth's ecosystems. *Science* 277:494–499.

World Bank (2011). *Rising global interest in farmland: Can it yield sustainable and equitable benefits?* Washington DC: World Bank.

Zoomers A (2010) Globalisation and the foreignisation of space: Seven processes driving the current global land grab. *Journal of Peasant Studies* 37:429–447.

6

Conciliating Ecosystem Services and Human Needs through Improved Land Use

Agnieszka Ewa Latawiec and Bernardo B.N. Strassburg

CONTENTS

Having discussed the drivers of land-use and land-cover change, possible conflicts over land, and their adverse effects in the previous chapter, here we introduce selected solutions to mitigate competition for land and to aid meeting future demands without compromising either development or nature. Before moving to specific examples, we contextualize land management within a sustainable development framework and present complexities associated with sustainable land management. Because many of the solutions to abate competition for land claim to be "sustainable," we briefly discuss the intricacies of sustainability as a concept.

Sustainability

Over the last decades a number of definitions of *sustainability* have been proposed (Dresner 2008). Almost every article or book on sustainability bemoans that the concept of sustainability is broad and lacks consensus, usually followed by the authors' own preferred definitions. The term *sustainable development* emerged in the 1980 World Conservation Strategy of the International Union for Conservation of Nature and Natural Resources, and was defined as "the integration of conservation and development to ensure that modifications to the planet do indeed secure the survival and well-being of all people" (Dresner 2008). However, probably the most well-known definition is that of the World Commission on the Environment and Development (WCED), the Brundtland Report, *Our Common Future*, which defines sustainable development as "development that meets the needs of current generations without compromising the ability of the future generations to meet their needs and aspirations" (WCED 1987). Broadly speaking, sustainability is the capacity of any system or process to maintain itself indefinitely and thus sustainable development indicate human, social, and economic systems that are able to maintain themselves indefinitely in harmony with the biophysical systems of the planet (Hak et al. 2007). Although intuitively simple, the dynamic concept of sustainability poses challenges in practical implementations.

Interpretation of sustainable development has paramount influence on land use. Attempts to operationalize sustainable development frameworks for land management have inexorably led to perplexities to be kept in mind when considering sustainable solutions: what contributes to the well-being of current and future generations, whose good is to be taken into account, does maximizing the good necessarily define the right course of action (Owens and Cowell 2011), as well as space and time dimensions of sustainability (sustainable where and for how long) (Bell and Morse 2008). Because sustainable development is not theorized like gravitation it may be considered along with other broadly consensual concepts, for instance, liberty, whose argumentation must be defensible (Owens and Cowell 2011).

In practice, whenever we start determining what is sustainable, we will likely need to make certain fundamental choices of ethical and political nature, over which people frequently disagree (Owens and Cowell 2011). As a consequence, it might be easier to pursue sustainable solutions knowing what is unsustainable (as discussed in Chapter 5) or as Owens and Cowell (2002) state: developing in the way that outcomes are "less environmentally damaging than they might otherwise have been." Also, in the recent development of sustainability science, sustainability implies an attempt to bridge the natural and social sciences, seeking creative solutions to complex challenges (Komiyama et al. 2011). Failing to understand these fundamental issues and challenges of sustainability has often fueled unrealistic expectations, while purposely understating these complexities may allow for dialog between different parties with different goals and visions of sustainability (Owens and Cowell 2011).

Because sustainability involves certain boundaries within which it operates, the concept of "space and sustainability" has recently been discoursed (Morse et al. 2011). In the terms of policy, sustainability has tended to focus on administrative boundaries because this is the space in which policy makers operate. However, there are other viable spaces such as the landscape in the ecological sense. Notwithstanding the fact that neither scale is perfect for exploring relationships between components of sustainable development (Morse et al. 2011) (in particular environmental quality and social deprivation may cut across both administrative and landscape boundaries), the "landscape approach" to land management is increasingly being brought into discussions on future land management.

The Landscape Approach

The landscape approach embraces all aspects of land management (ecological, economic, and social). In that, both scientific and policy approaches encompass natural and agricultural ecosystems, along with the links between them, for example, interdependencies and synergies involved in land-cover change in the whole landscape. This holistic view to land management includes conservation and restoration, addresses the ecological value of environment, as well as the necessity to reconcile any conservation activity with human needs (DeFries and Rosenzweig 2010). In other words, the goals of the landscape approach are to reconcile demand-driven increases in agricultural production with protection and restoration of natural ecosystems, maximizing global and local ecosystem services in a socially fair way. An integrated landscape approach can increase synergies among multiple local, regional, and global societal objectives. Although reconciling competing objectives is never easy, such a holistic approach provides a platform for addressing diverse goals such as food production, biodiversity conservation, climate mitigation, the provision of ecosystem services, and delivery of development. Initiatives that promote a landscape approach include CGIAR's Going from Red to Green and the Ecoagriculture Partners' Landscapes for Food, People and Nature.

Yield Gap and Sustainable Intensification

There is substantial evidence showing that current agricultural productivity is below potential yield (Food and Agriculture Organization of the United Nations [FAO] and International Institute for Applied Systems Analysis [IIASA] 2010; Licker et al. 2010; Foresight 2011; Mueller et al. 2012). Licker et al. (2010), for example, in their spatial analysis present yield gap patterns for 18 crops around the year 2000. In general, low yield gaps are concentrated in regions with relatively high-input agriculture, mostly in developed countries. They conclude that although biophysical factors like climate are key drivers of global crop yield patterns, land management practices are crucial for yields.

Sustainable intensification of agricultural production on current agricultural lands has been suggested as a key solution to the conflict between expanding agricultural production and conserving natural ecosystems globally (Godfray et al. 2010; Licker et al. 2010; Smith et al. 2010; Foresight 2011).

Indeed, Mueller et al. (2012) have demonstrated that meeting the food security and sustainability challenges of the coming decades is possible but will require considerable changes in nutrient and water management. Their global-scale assessment demonstrates that large production increases (45% to 70% for most crops) are possible when closing yield gaps to 100% of attainable yields through change in management practices, while opportunities to reduce the environmental impact of agriculture by eliminating nutrient overuse may still allow an approximately 30% increase in production of major cereals (maize, wheat, and rice). The concept of sustainable intensification, increasingly endorsed by researchers and policy makers (Godfray et al. 2010; Foresight 2011) in essence means "producing more food from the same area of land, while reducing the environmental impacts" (Royal Society 2009) through resource conservation and improvements in land management (Tilman et al. 2002; Herrero et al. 2010). Impacts on the environment can potentially be reduced by changing agronomic practices, adopting integrated pest management methods, implementing agroforestry, and the integrated management of waste in livestock production (Godfray et al. 2010). Strategies to improve yields while simultaneously improving environmental conditions or preventing degradation include zero or reduced tillage, mulches, and cover crops. Precision agriculture that involves a series of technologies that allow the application of nutrients, water, and pesticides only to the places and at the times they are required also optimizes the use of inputs contributing to sustainable intensification (Day et al. 2008).

With respect to pasturelands, the introduction of mixed systems such as silvopastoral (cattle and trees) or crop–livestock systems, and also rotational systems for pasture management, have been shown to result in positive effects both to farmers and the environment (Herrero et al. 2010). Contrary to the Green Revolution, whose success has depended on pesticides and chemical fertilizers, these advances are modern adaptations of techniques that have been used in the past. This, in most cases, means simple changes in farm management. For instance, a significant increase in productivity can be obtained via the adoption of appropriate grass species for given edaphoclimatic conditions, the prevention of pasture degradation by not allowing animals to graze on short grass, by dividing a pasture into modules for rotational grazing, and the incorporation leguminous grasses to fix atmospheric nitrogen. Pretty and coauthors (2006) have also shown a 79% increase in crop yields in 57 poor countries after the application of resource-conserving practices, such as integrated pest management (utilizing ecosystem resilience and diversity for pest control), conservation tillage, agropasture, and agroforestry. At the same time, these initiatives have improved critical environmental services provided by soils including water-use efficiency and carbon sequestration.

The transition of extensive, low productivity pastoralism to multiple-use silvopastoral or crop–livestock systems and the like may result in a range of socioeconomic benefits. In addition to maintaining agricultural productivity and supplementary farm outputs, they may enhance the supply of diverse market products and contribute to risk reduction due to the provision of alternative products and higher incomes. Especially, poorer people rely on forests as a safety net to avoid or mitigate poverty, or as a means out of poverty. Sustainable intensification of pasturelands can also result in positive returns to landowners, smallholders, processors, traders, and ultimately governments through increased tax returns and multiplier effects on the economy (Strassburg et al. 2012). For example, the transition to well-managed sustainable intensive rotational systems has been shown to improve a range of environmental and economic aspects of agriculture through reduction of land degradation and reversal of soil erosion (Drewry 2006).

To realize the full potential of sustainable productivity increase, complementary policies such as territorial planning, improved law enforcement, monitoring, and tenure security must be put in place to avoid undesirable outcomes, such as leakage and rebound effects, as discussed in Chapter 5. Furthermore, agricultural land can be managed in ways specifically designed to reduce negative impacts on biodiversity (Godfray et al. 2010). The concepts of "land sparing" or "land sharing" have been proposed to best reconcile agriculture and biodiversity. In short, land sparing suggests saving biodiversity through the intensification of agricultural lands, which could lead to lower demand for new land clearance and as a consequence larger areas dedicated to nature conservation (Balmford et al. 2012). Land sharing on the other hand proposes a concept of coexistence of biodiversity and agriculture on the same area (Perfecto and Vandermeer 2010).

REDD+

The scheme of Reducing Emissions from Deforestation and Forest Degradation (REDD+), a climate mitigation strategy included in the United Nations Framework Convention on Climate Change (UNFCCC), is fundamentally based on the premises that developing countries would receive financial incentives for reducing deforestation, degradation, and increasing carbon stocks in terrestrial ecosystems. It reflects the recognition of the role that land plays in the global climate system. It can be understood as an international payment for ecosystem services (PES) scheme for better land management, which may be a powerful tool to address climate change and other deforestation-related issues such as biodiversity loss (Strassburg et al. 2012). PES projects aim to provide financial incentives to landowners or managers for implementing conservation actions that they would not otherwise adopt. The original concept of reducing emissions from deforestation only proposed incentives to reduce deforestation, yet soon it became apparent that interventions for reducing deforestation without complementary policies addressing the agricultural drivers of forest loss and demand for land may have limited effectiveness in climate change mitigation (Strassburg et al. 2009). If national REDD+ policies are to be effective, they must be part of a comprehensive and integrated international system in order to avoid leakage.

As posited by the landscape approach to land management, ecological and policy linkages have to be encompassed between changing forested and agricultural landscapes. Figure 6.1 presents what synergies and feedbacks exist between REDD+ and its social and economic impacts within landscapes, and Box 6.1 discusses the role of tenure in forest management.

FIGURE 6.1
Economic and social impact of REDD+ management actions on different stakeholders within a landscape. (Reprinted from Parrotta JA et al., 2012, Understanding relationships between biodiversity, carbon, forests and people: The key to achieving REDD+ objectives, Vienna: IUFRO.)

BOX 6.1   THE ROLE OF TENURE AND PROPERTY RIGHTS IN CONTEXT OF FOREST MANAGEMENT

Decisions on forest management have been shaped by the ways in which forest use is governed, that is, the structure of rules and institutions that shape human interaction, and detailed tenure and property rights arrangements (Parrotta et al. 2012). These factors are vital to understanding the context within which particular interventions might impact social and economic well-being. Tenure security, which indicates the extent to which the rights of land or forest owners are recognized, may provide incentives for sustainable management. On the other hand, inadequate recognition of such rights excludes the poor from decision making and may deny them access to potential benefits from market-based interventions, such as PES. Formally recognized management rights or titles are often essential to receive direct benefits from participation in PES contracts, such as REDD+. For REDD+ implementation to be effective and sustainable, tenure and property rights, including rights of access, use, and ownership, need to be clarified. Clear tenure arrangements are also critical for forest certification and restoration. Without property and use rights, local populations may not have an incentive to invest in restoration. Communities may fear that planting by the government may signify land appropriation, they may fear that they will no longer be able to harvest the restored land, and they may also fear that without adequate mechanisms to transfer benefits they may not receive due payment for providing an ecosystem service. Tenure security is also a key consideration when addressing drivers of deforestation in the agricultural context. When tenure is secure, sustainable agricultural intensification can benefit small farmers, while also creating the possibility of land sparing for forest conservation. Conversely, weak tenure security may facilitate "land grabbing" and other irregularities related to land transactions, which typically result in the expropriation of lands from the most vulnerable groups. Another important aspect in the context of forest management is the prevailing type of property regime. Communal property systems typically recognize multiple interests and usually require some form of collective action to be effective. Although maintaining collaboration is not always easy and frequently involves the collaboration of multiple stakeholders, often from different sectors (e.g., states and communities working together in systems of joint forest management), participatory regimes are potentially mechanisms for promoting mutually beneficial outcomes.

Examples of Approaches to Sustainable Land Management

Sustainable Intensification of Pasturelands in Brazil

Cattle ranching is the prime driver of deforestation in the Amazon and degradation of Cerrado, the Brazilian savannas. The conversion of natural ecosystems into cattle farms over the last decades has created a dramatic forest loss known as the "arc of deforestation" along the southern and eastern edges of the Amazon forest. Forest biomass has been perceived by the farmers as an input in agricultural production (slash-and-burn practices transiently supply nutrients to both croplands and cattle pastures). The competition between agriculture and forestry is stimulated by their relative marginal benefits and the low price of the land in the Amazon. Most deforestation for cattle production yields low productivity or unproductive pastures but is a source of hundreds of tons of CO_2 per hectare. Overall, tropical deforestation released approximately 1.5 billion metric tons of carbon annually throughout the 1990s, which accounted for almost 20% of all atmospheric greenhouse gas emissions (Gullison et al. 2007). Expansion into the forest also entails destruction of one of the most pristine and biodiverse ecosystems on Earth.

Currently, 75% of all agricultural land uses in Brazil are dedicated to cattle ranching, generally in extensive, low productivity systems. Considering climatic and edaphoclimatic conditions, the productivity of Brazilian pasturelands can be sustainably increased to meet future demand for agricultural commodities (sugarcane, maize, soybean, and timber) while sparing land for nature (Strassburg et al. 2011). However, as mentioned earlier, it has become increasingly clear that socioeconomic concerns cannot be overlooked, if sustainable intensification coupled with land sparing is to be successful.

In line with an integrated approach to land, a new concept of "Land-Neutral Agriculture Expansion" (LNAE, Strassburg et al. 2012) may allow farmers to demonstrate that their agricultural expansion has not caused direct or indirect (such as displacement) impacts over natural environments. It may also contribute to the implementation of large-scale restoration within landscapes without exacerbating the conflict for land. In a context of incentives related to avoided deforestation, this mechanism would allow claims to PES. In the absence of such a mechanism, it can still be used to demonstrate commitment to sustainability goals, be it in order to gain access to specific markets or to meet their or their partners' sustainability commitments. Indeed, there is a recent increase in environmental awareness from private actors, such as the Consumer Goods Forum, mostly fueled by increased awareness in final consumers that have pledged to remove from their supply chains products related to deforestation. The Consumer Goods Forum is an association that brings together over 400 retailers and manufacturers from 70 countries, with combined sales of US$3.1 trillion and nearly 10 million people employed.

The ability to access markets that represent a substantial fraction of global agricultural trade, by pursuing a sustainable agriculture production without deforestation, would bring an important competitive advantage to goods produced sustainably in developing countries.

The LNAE mechanism consists of a series of coordinated steps to link concerted efforts of expanding agriculture into a certain area and mitigating or compensating the displacement of the original production in the area. Such efforts can be understood as a closed system with zero land leakage, which could merit a very robust claim to avoiding deforestation equal to the averted land leakage.

Although many opportunities exist for sustainable intensification and restoration (Menz et al. 2013), they require a rigorous management, administrative control, and permanent adjustments based on careful monitoring. There are also initial costs that will need to be incurred. Even though the investment can be recovered in a relatively short period of time (3 to 4 years), there may be problems with initial financing (Strassburg et al. 2012). In addition, the complexity of introducing more intensive agricultural systems may demand specialized knowledge and technical assistance. Also, unless risk barriers are reduced or eliminated, a shift toward a more productive agriculture may be heavily constrained.

Developing a Greener Agricultural Sector in Suriname

Sustainable development and the development of a sustainable agricultural sector have been increasingly recognized worldwide as a strategy to achieve long-term environmental, social, and economic benefits. Although complex and often requiring active participation by a range of stakeholders, sustainable development can be found in governmental agendas across both developed and developing countries. There are various alternatives to arrive at the development of a sustainable agricultural sector. In Figure 6.2, a selection of means is presented by which the development of the sustainable agricultural sector can be promoted and stimulated, which have been proposed for Suriname (Latawiec et al. 2012).

Suriname is in an extraordinary position to benefit from incentives to conserve forest carbon and biodiversity, and simultaneously meet an increasing demand for agricultural commodities through agricultural intensification and better land management strategies. First, most of Suriname's forests, large extents of which have not been deforested over the last decades, present top levels of both carbon and biodiversity, and Suriname may therefore benefit from REDD+ funds that may reach up to US$40 billion per year globally. Second, productivity levels encourage the demand for land from the rice sector, the main agricultural crop in Suriname. If rice productivity stagnates at current levels (approximately 4.2 tonnes per hectare), high production targets would mean that the rice production area in Suriname would need to

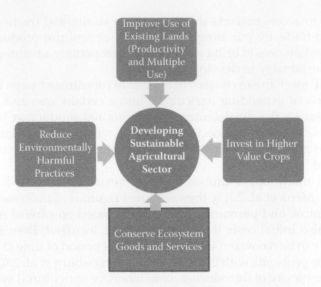

FIGURE 6.2
Selected ways to developing a more sustainable sector in Suriname, producing more agricultural outputs, and saving the tropical forest of the Surinamese Amazon. (Reprinted from Latawiec AE et al., 2012, Developing sustainable agricultural sector in Suriname, www.iis-rio.org.)

increase by more than 20,000 hectares by 2022 (Latawiec et al. 2012). On the other hand, an accelerated productivity increase combined with modest increases in production targets could liberate 10,000 hectares from rice production. If productivity increase keeps pace with production targets, 15,000 hectares could be available for other crops after meeting production targets from the rice sector. This area is three times as large as the area currently occupied by vegetables and fruit crops in Suriname. Economic returns from these crops are on average 10 times higher than returns from rice production. It has been shown that conflicts over land can be avoided as long as rice productivity does not stagnate at current levels, suggesting that Suriname already has enough land cleared for agriculture to meet ambitious targets from the rice sector and increase the area dedicated to higher value crops without deforestation.

Organic farming offers another opportunity to contribute to greening and simultaneously adding value to the Surinamese agricultural sector (Figure 6.3). Organic farming may provide a wide range of economic, environmental, and social benefits, and over the past two decades, global markets for certified organic products have grown rapidly with sales expected to continue to grow over the next years. By developing a framework to stimulate organic farming and by working with smallholder farmers, Suriname may benefit from an increased value of its national agriculture, create both alternative and higher incomes (also by investing in high cash products, such as açai), offer an alternative path for rural people, create new job opportunities,

FIGURE 6.3 (See color insert.)
"Safe food" market in Suriname. (Photo courtesy of Maureen Silos, the Caribbean Institute.)

achieve food security both in terms of provision and healthier products, among many other benefits. Recent initiatives toward sustainable agriculture, such as Safe Farming, have shown that there is a national interest in and a market for more sustainable agricultural products in Suriname. In recent years, there has been an increasing interest of the Surinamese farmers to produce healthier and more environmentally friendly food. This has been driven by increasing concern over the overuse of pesticides and risks associated with consumption of contaminated agricultural products. The Caribbean Institute (http://www.caribbean-institute.org/), for example, has been leading a successful initiative to assist farmers in their transformation toward a more environmentally friendly agriculture, called Safe Food. Upon entering the domestic market these products have been received with enthusiasm and, more importantly, have not necessarily been more expensive. These projects and existing infrastructure, such as the Centre for Agricultural Research (CELOS), may provide a starting point for the development of a national organic farming framework. Furthermore, the global market for organic products is likely to continue to expand, with global trade moving toward higher-quality products, demanding higher social and environmental standards (see also the aforementioned Consumer Goods Forum). With respect to exports of organic products, Suriname still needs to develop high technical and legal expertise. To this end, liaison should be sought with regional, as well as worldwide organizations, which could provide the necessary technical and policy-relevant know-how (Figure 6.4).

FIGURE 6.4
Training "Plant Doctors" in Suriname. Recognizing the problems with agricultural products is paramount for avoiding excessive use of agrochemicals and a key step toward organic farming. (Photo courtesy of Maureen Silos, the Caribbean Institute.)

Although a range of opportunities to develop the organic farming market in Suriname exists, there are some constraints to overcome, including management skills for integrated land management. However challenging, through the development of an organic farming market in Suriname, there is a potential to establish a more sustainable, higher-income agricultural sector.

Large-Scale Restorations and Landscape Approach

To maximize the benefits of ecological restoration it should be accomplished at the large scale. This allows significant increases in landscape connectivity and improvements in the flow of ecosystem services to society. Furthermore, balancing ecological and social needs is always difficult, but is most likely to succeed when working on a large enough area. Different landscape units with different land uses by local people are likely to be more resilient but require negotiation and trade-offs among different demands. Several large-scale restoration initiatives have arisen across the world, such as the Greenbelt Movement in Kenya, the Payment for Ecosystem Services Program in Costa Rica (Sánchez-Azofeifa et al. 2007), the Terai Arc Landscape Project in India and Nepal, and the Atlantic Forest Restoration Pact in Brazil (Calmon et al. 2011). The Atlantic Forest Restoration Pact, for example, aims to restore 1 million hectares by 2020 and 15 million by 2050. Consequently, it is expected that millions of hectares currently occupied by agricultural land use will be converted to natural ecosystems in the near future.

The Atlantic Forest Restoration Pact is an example of a large-scale restoration initiative that has factored in land competition. There are 30.5 million hectares of planted pastureland in the Atlantic Forest supporting 36 million heads of cattle. This corresponds to a stocking rate of 0.82 animal units per hectare (IBGE 2003; PROBIO 2009), a very low efficiency by international standards (FAOSTAT 2012). Doubling this productivity over the next three decades could liberate 15.3 million hectares for forest restoration. However, when agricultural fields are converted to forests, food security and leakages of demand into other areas should be taken into account. Sustainable intensification has therefore been suggested as a solution to meet demands for food and spare land for the Atlantic Rainforest restoration. It has been shown (Strassburg et al. 2011) that increasing pasture productivity in the state of Espirito Santo that is currently at 29% of its sustainable carrying capacity will enable the expansion of agriculture and silvopastoral systems over an additional 674,000 hectares. This would keep meat production constant while sparing land for nature by increasing by 50% native forest cover in the state in 20 years. However, in order to realize the land-sparing potential from increased cattle ranching productivity, complementary policies such as territorial planning, improved law enforcement, monitoring, and tenure security must be put in place to avoid leakage and rebound effects.

Conclusions

Changes in land use have led to human appropriation of an increasing share of the planet's resources, potentially undermining the capacity of ecosystems to sustain food production, maintain freshwater and forest resources, and regulate climate and air quality. Especially in tropical countries where the agricultural sector continues to expand mainly at the cost of the forest, and land potentially available for agriculture is further degraded, there is a need to reconcile development and conservation. Sustainable intensification of agricultural production generates an opportunity to plan and implement the whole landscape approach, which combines increased productivity of agriculture with conservation and restoration of natural environments and takes into account not merely developmental aspects but also environmental, social, and economic ones. In this chapter we have discussed selected alternatives to contribute to sustainable development and sustainable land management. The next chapter will specifically discuss the potential of biochar to improve soil quality and agricultural productivity.

References

Balmford A, Green R, Phalan B (2012) What conservationists need to know about farming. *Proceedings of the Royal Society B*, 279:2714–2724.

Bell S, Morse S (2008) *Sustainability indicators. Measuring the immeasurable?* London: Earthscan.

Calmon M, Brancalion PHS, Paese A, Aronson J, Castro P, da Silva SC, Rodrigues RR (2011) Emerging threats and opportunities for large-scale ecological restoration in the Atlantic Forest of Brazil. *Restoration Ecology* 19:154–158.

Day W, Audsley E, Frost AR (2008) An engineering approach to modelling, decision support and control for sustainable systems. *Philosophical Transactions of the Royal Society B* 363:527–541.

DeFries R, Rosenzweig C (2010) Toward a whole-landscape approach for sustainable land use in the tropics. *PNAS* 107:19627–19632.

Dresner S (2008) *The principles of sustainability*. London: Earthscan.

Drewry JJ (2006) Natural recovery of soil physical properties from treading damage of pastoral soils in New Zealand and Australia: A review. *Agriculture, Ecosystems and Environment* 114:159–169.

Food and Agriculture Organization of the United Nations (FAO) (2012) FAOSTAT. http://faostat.fao.org/ (accessed December 11, 2012).

Food and Agriculture Organization of the United Nations (FAO) and International Institute for Applied Systems Analysis (IIASA) (2010). Global agro-ecological zones (GAEZ v3.0). Laxenburg and Rome: IIASA and FAO.

Foresight (2011) The future of food and farming. Challenges and choices for global sustainability. Final project report. London: The Government Office for Science.

Godfray HCJ, Beddington JR, Crute IR, Haddad L, Lawrence D, Muir JF, Pretty J, Robindon S, Thomas SM, Toulmin C (2010) Food security: The challenge of feeding 9 billion people. *Science* 327:812–818.

Gullison RE, Frumhoff PC, Canadell JG, Field CB, Nepstad DC, Hayhoe K, Avissar R, et al. (2007) Environment: Tropical forests and climate policy. *Science* 316:985–986.

Hak T, Moldan B, Dahl A (2007) *Sustainability indicators: A scientific assessment.* Washington DC: Island Press.

Herrero M, Thornton PK, Notenbaert AM, Wood S, Msangi S, Freeman HA, Bossio D, et al. (2010) Smart investments in sustainable food production: Revisiting mixed crop livestock systems. *Science* 327:822–825.

IBGE (2003) Pesquisa pecuária municipal 2002. http://www.ibge.gov.br/home/estatistica/economia/ppm/2002/default.shtm (accessed January 20, 2013).

Komiyama H, Takeuchi K, Shiroyama H, Mino T (2011) *Sustainability science: A multidisciplinary approach*. Tokyo: United Nations University Press.

Latawiec AE, Strassburg B, Rodrigues AM, Matt E (2012) Developing sustainable agricultural sector in Suriname. Report available at www.iis-rio.org.

Licker R, Johnston M, Foley JA, Barford C, Kucharik CJ, Monfreda C, Ramankutty N (2010) Mind the gap: How do climate and agricultural management explain the "yield gap" of croplands around the world? *Global Ecology and Biogeography* 19:769–782.

Menz MHM, Dixon KW, Hobbs RJ (2013) Hurdles and opportunities for landscape-scale restoration. *Science* 339:526–527.

Morse S, Vogiatzakis I, Griffiths G (2011) Space and sustainability: Potential for landscape as a spatial unit for assessing sustainability. *Sustainable Development* 19:30–48.

Mueller ND, Gerber JS, Johnston M, Ray DK, Ramankutty N, Foley JA (2012) Closing yield gaps through nutrient and water management. *Nature* 490:254–257.

Owens S, Cowell R (2002) *Land and limits: Interpreting sustainability in the planning process*. New York: Routledge.

Owens S, Cowell R (2011) *Land and limits: Interpreting sustainability in the planning process*, 2nd ed. New York: Routledge.

Parrotta JA, Wildburger C, Mansourian S (2012) Understanding relationships between biodiversity, carbon, forests and people: The key to achieving REDD+ objectives—A global assessment report prepared by the Global Forest Expert Panel on Biodiversity, Forest Management and REDD+. IUFRO World Series Vol. 31. Vienna: IUFRO.

Perfecto I, Vandermeer J (2010). The agroecological matrix as alternative to the land-sparing/agriculture intensification model. *PNAS* 107:5786–5791.

Pretty JN, Noble AD, Bossio D, Dixon REH, Penning de Vries WT, Morison JIL (2006) Resource-conserving agriculture increases yields in developing countries. *Environmental Science and Technology* 40:1114–1119.

PROBIO (2009) Land use and land cover classification of Brazilian biomes. Ministry of Environment, Brazil. www.mma.gov.br/probio (accessed February 15, 2012).

Royal Society (2009) *Reaping the benefits: Science and the sustainable intensification of global agriculture*. London: The Royal Society.

Sánchez-Azofeifa GA, Pfaff A, Robalino JA, Boomhower JP (2007) Costa Rica's payment for Environmental Services Program: Intention, implementation, and impact. *Conservation Biology* 21:1165–1173.

Smith P, Gregory PJ, van Vuuren D, Obersteiner M, Havlík P, Rounsevell M, Woods J, Stehfest E, Bellarby J (2010) Competition for land. *Philosophical Transactions of the Royal Society B–Biological Sciences* 365:2941–2957.

Strassburg B, Latawiec A, Cronemberger F (2011) Programa estadual de conservação e recuperação da cobertura florestal—Espírito Santo. Análise integrada do uso da terra no Estado do Espírito Santo. The state program for conservation and recuperation of forest. Integrated analysis of the land use in the state of Espirito Santo. Report available at www.iis-rio.org.

Strassburg B, Micol L, Ramos F, Seroa da Motta R, Latawiec A, Lisauskas F (2012) Increasing agricultural output while avoiding deforestation: A case study for Mato Grosso, Brazil. Report available at www.iis-rio.org.

Strassburg B, Turner RK, Fisher B, Schaeffer R, Lovett A (2009) Reducing emissions from deforestation: The "combined incentives" mechanism and empirical simulations. *Global Environmental Change* 19:265–278.

Tilman D, Cassman KG, Matson PA, Naylor R, Polasky S (2002) Agricultural sustainability and intensive production practices. *Nature* 418:671–677.

World Commission on Environment and Development (WCED) (1987) *Our common future*. Oxford: Oxford University Press.

7

Sustaining Soils and Mitigating Climate Change Using Biochar

Lewis Peake, Alessia Freddo, and Brian J. Reid

CONTENTS

Introduction

Biochar is a product of a biomass-heating process in an oxygen-limited environment, yielding little or no CO_2 (pyrolysis). This process also produces syngas and bio-oil that can be used in heat and power generation. The yields of each component (syngas, bio-oil, and biochar) are dependent upon the temperature of pyrolysis, the residence time of the process, and the type of feedstock used.

Biochar holds the potential to reduce atmospheric CO_2 concentrations by sequestering carbon (C) from the atmosphere, into biomass, and "locking up"

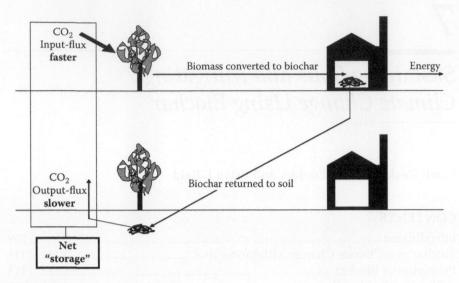

FIGURE 7.1
Net C storage in a biomass-to-biochar cycle.

this C when this biomass is converted into biochar (Figure 7.1). Biochar is recalcitrant and physically stable to the extent that once applied to soil it becomes a persistent component within the soil matrix.

The modern use and scientific study of biochar barely exceeds a decade, but interestingly, it has an ancient pedigree in the form of Amazonian dark earths (ADEs), which have attracted a great deal of retrospective research. Spanish Conquistadors' reports of extensive settlements on black soils in 16th century Amazonia were initially dismissed as fables. Subsequent investigations by archaeologists and soil scientists have partially confirmed these accounts and established the existence of ADEs, locally called *Terra Preta* (black earth in Portuguese). These anthrosols (manmade soils) are the result of humans adding charcoal and midden waste to the indigenous soil over many centuries, but it is not known whether this practice was anthropic (unintentionally formed) or anthropogenic (intentionally formed). Radiocarbon dating suggests that the process began at least as early as 450 BC and continued until perhaps 1500 AD. Before being transformed in this way, the soils were typically reddish or yellowish, acidic oxisols, ultisols, or entisols low in nutrients and organic matter. ADEs, even centuries since they were actively managed, by contrast are very dark to a depth of 1.5 to 2 m and contain relatively high levels of organic C and nutrients, especially phosphorus and calcium (Lehmann et al. 2003).

There is mounting evidence that biochar influences a wide range of soil properties in ways that predominantly have the potential to increase agricultural productivity. The nature and extent of such influences vary widely and depend upon soil type, agro-ecological factors, and the type and quantity of biochar used. The variables affected collectively have a direct bearing on physical, chemical, and biological soil characteristics. Yet unlike most other soil amendments, such as fertilizer, manure, compost, or lime, the effects of biochar are not yet well understood, either in terms of the precise mechanisms involved or their longevity.

Embracing all of these aspects, the European Commission (Verhaijen et al. 2010) recently defined biochar as

> charcoal (biomass that has been pyrolyzed in a zero or low oxygen environment) for which, owing to its inherent properties, scientific consensus exists that application to soil at a specific site is expected to sustainably sequester C and concurrently improve soil functions (under current and future management), while avoiding short- and long-term detrimental effects to the wider environment as well as human and animal health.

In this chapter, we first present biochar with respect to its potential to reduce levels of atmospheric CO_2 and thereafter give an account of the mechanisms through which biochar can deliver soil improvements and increase crop yields.

Biochar as a Climate Change Mitigation Tool

The total C present in the planet is, to all intents and purposes, constant (Houghton 2007). However, the amounts of C present in the various environmental compartments, such as the atmosphere, biosphere, pedosphere, hydrosphere, and lithosphere can and do change (Macías and Arbestain 2010). Natural cycles and anthropogenic activities are the main drivers of change. When compared to the amount of C in other compartments, the total amount of C present in the atmosphere is relatively small (800 Pg C) (Macías and Arbestain 2010). In contrast, fossil fuel (5000 Pg C) (Archer et al. 2009) and soil C reservoirs (3200 Pg C) (Macías and Arbestain 2010) are much larger. As a consequence of the burning of fossil fuels and to a lesser extent changes in land use and soil cultivation practices, atmospheric CO_2 concentrations have increased by 37.5% since the preindustrial era (CO_2 levels have risen from about 280 to 385 ppmv) (IPCC 2007).

Several studies have shown the necessity to keep the cumulative anthropogenic greenhouse gas emissions below a maximum upper limit (Broecker 2007; Matthews and Caldeira 2008; Solomon et al. 2009). Hansen et al. (2008) proposed a maximum concentration threshold of atmospheric CO_2 of 350 ppm, versus the present 385 ppm. Thus, if dangerous changes in the climate are to be avoided, future anthropogenic emissions must approach zero (Hansen et al. 2008). Consequently, global action is necessary to reduce atmospheric CO_2 concentration. Adoption of "sustainable" or "low-C" or "C-neutral" or indeed "C-negative" approaches to global energy provision are key to a strategy to curb CO_2 emission to the atmosphere.

The use of biomass as feedstocks from which to produce energy is not a new concept. However, originality exists where these resources are used to provide energy and at the same time the opportunity to sequester C from the atmosphere. The pyrolysis of biomass serves to provide energy (via bio-oil and syngas that are subsequently used to run steam turbines) and the purposefully produced material "biochar." The conceptual foundations of biochar as an atmospheric CO_2 removal mechanism lie in the photosynthetic processes that produce the biomass to be used for biochar production (Figure 7.1). As biomass grows it removes atmospheric CO_2. The production of biochar converts comparatively labile C present in the biomass into recalcitrant C compounds that resist mineralization. In this way the rate of return of C to the atmosphere is greatly inhibited. It is the difference between the (relatively fast) rate of atmospheric CO_2 sequestration into biomass compared to the subsequent (relatively very slow) rate at which biochar C is mineralized that gives rise to net storage of C; and by this token the opportunity to produce heat and power by C-*negative* means.

Several studies have attempted to predict the extent to which biochar can reduce atmospheric CO_2 levels. For example, Lehmann and Rondon (2006) estimate that biochar may be able to sequester 5.5 to 9.5 Gt C per year, or about 20 to 35 Gt CO_2 per year by 2100. Lenton and Vaughan (2009) suggest that the capture of CO_2 by plants destined to provide bioenergy and subsequent C capture and storage, combined with afforestation and biochar production, may have the potential to remove 100 ppm of CO_2 from the atmosphere. Woolf et al. (2010) suggest that biochar can potentially offset a maximum of 12% of current anthropogenic CO_2-C equivalent emissions to the atmosphere (i.e., 1.8 Pg emissions can be avoided out of the 15.4 Pg of CO_2-C equivalent emitted annually), decreasing significantly the emissions of CO_2 by preventing decay of biomass inputs. Moreover, it has been suggested that biochar presence in soil might initiate a positive feedback wherein soil physical and chemical properties are improved and plant yields increased as a result; this feedback further enhancing the amount of CO_2 removed from the atmosphere (Woolf et al. 2010). Additional positive feedback might also be realized where biochar suppresses the emissions of other GHGs, such as nitrous oxide and methane (both significant agricultural pollutants and far more harmful in their radiative forcing impact than CO_2). Further research is required to substantiate the circumstances under which such positive feedback are initiated and sustained.

Properties of Biochar

Biochar Physical Properties

The matrix of biochar has been determined by X-ray diffraction revealing an essentially amorphous structure with crystalline areas (Lehmann and Joseph 2009) consisting of random polycyclic aromatic (graphene) layers rimmed by functional groups (Zhu et al. 2005) and mineral compounds (Lehmann and Joseph 2009). Associated with the pyrolysis process above 330°C is the formation of polyaromatic sheets, which create turbostratic structures (Keiluweit et al. 2010) and increased porosity as temperatures increase. Studies have demonstrated that higher temperatures lead to a decrease in particle size (Downie et al. 2009) and the development of nanoporosity (<2 nm), which underpin the high surface area of biochar (Downie et al. 2009). Physical properties, of course, vary depending upon the biomass feedstock used and the thermochemical conditions of char formation.

Biochar Chemical Properties

In keeping with the European Commission definition of biochar (Verhaijen et al. 2010) as presented in the introduction, three groups of chemical attributes are worthy of consideration. First, if biochar is to achieve greater longevity as a means of C sequestration, then it must be stable and resistant to mineralization (back to CO_2). Second, if soil fertility improvements are to be realized then the levels and availabilities of key macro- and micronutrients are of significance. Finally, if biochar is to be adopted as a soil amendment it cannot represent a hazard to soil health.

Owing to different production conditions and a variety in feedstock materials used to produce biochar, chemical attributes vary considerably. At an elemental level, biochar properties can be ascribed to ratios of C, H, O, and N. Particularly, ratios of H/C and O/C are used to determine the degree of biochar aromaticity, that is, the lower the ratio, the greater the aromaticity (Kookana et al. 2011). H/C and O/C ratios have been reported to be higher in biochars produced at low temperatures, due to incomplete charring of the feedstock. H/C and O/C ratios decrease with increasing temperatures of production (Baldock and Smernik 2002). Thus, higher temperature chars are inherently more resistant to chemical attack and therefore are more recalcitrant.

The nutrient content in biochar also varies depending upon feedstock type and pyrolysis conditions used. Higher temperatures and faster heating rates strongly influence the retention of nutrients within the biochar formed: nitrogen (N) and sulfur (S) compounds, for example, volatize at 200°C and 375°C, respectively. As to potassium (K) and phosphorus (P), they become depleted when biochar is produced above 700°C and 800°C, respectively (DeLuca et al. 2009). Minerals such as magnesium (Mg), calcium (Ca), and manganese (Mn) volatilize at temperatures above 1000°C (Neary et al. 1999; DeLuca et al. 2009); pH, electrical conductivity (EC), and extractable NO_3^- tend to be higher with high temperatures (800°C), while low temperatures (350°C) result in greater extractable amounts of P, NH_4^+, and phenols. Feedstock type is responsible for different ratios of C/P and C/N; in particular, wood- and nut-based biochars show high C/P and C/N ratios, while manure-, crop- and food-waste biochars have lower ratios (Kookana et al. 2011).

In terms of risks to soil health, the three most likely toxicity drivers are (1) metals and metalloids, (2) polycyclic aromatic hydrocarbons, and (3) dioxins. Regarding metal and metalloid concentrations in biochar, Freddo et al. (2012) reported their concentrations to be broadly in keeping with levels observed in background soils and below concentrations ascribed to compost. Thus, biochar application, up to 100 t ha⁻¹, is unlikely to make any real difference to metal and metalloid concentrations in the receiving soil. Polycyclic aromatic hydrocarbons (PAHs) are formed during combustion and pyrolysis processes. Studies have shown that PAH concentrations are higher when biochar is produced at lower temperatures (300°C) (Freddo et al. 2012; Hale et al. 2012), and that concentrations vary with different feedstocks produced using the same pyrolysis temperature. Like metal and metalloid concentrations, the levels of PAHs are unlikely to be of concern from a soil health perspective. Dioxins, also produced during combustion processes, are extremely potent toxins. Hale et al. (2012) reviewed their levels in a range of biochars and concluded them to be present at levels that are not cause for particular concern. Thus, all three of these potential risk drivers have been reported to be below critical thresholds for concern. It must be stressed, however, that these studies considered "clean" feedstocks such as wood, bamboo, and straw. Should wastes such as household refuse or treated timber be diverted into biochar production, then levels of toxins might be expected to increase.

Influence of Physical and Chemical Properties on Biochar Stability

The complex structure of biochar affords its great stability in the environment (Schmidt and Noack 2000): the peculiar cross-linking and the steric protection of the refractory macromolecules present in biochar prevent hydrolytic enzymes from attacking the matrix itself (Derenne and Largeau 2001; Lehmann et al. 2009). Nevertheless, some studies show the decay of biochar due to metabolic processes (Shneour 1966; Baldock and Smernik 2002). Moreover, different biochar products have different decomposition potentials. These present different physical and chemical structures depending upon the feedstock and pyrolysis temperatures used (Lehmann and Joseph 2009). Biochar found in the Amazon region has suggested millennium-scale persistence with radiocarbon indicating ADE char to be 500 to 7000 years old (Neves et al. 2003). Liang et al. (2008) found no changes in the aromaticity determined by x-ray techniques in biochar particles coming from the same area. These results provide further evidence of biochar's potential for long-term C storage.

Influence of Biochar upon Soil Properties

Influence of Biochar and Soil Physical Properties

The physical properties of soil range from the electrostatic forces binding its microscopic particles to the structural cohesion, which helps it resist erosion. These properties include bulk density, porosity, aggregate stability, penetrability, tensile strength, and its hydrological characteristics, that is, the way in which it absorbs, retains, and releases water. All of this controls the ability of plant roots to penetrate the soil to obtain water, air, and nutrients, and has a direct impact on the chemistry and biology of soils. The factors that control these properties include particle size distribution (texture), that is, the relative proportions of clay, silt and sand, its clay mineralogy, and the quantity and quality of soil organic matter (SOM). Biochar is a low density porous material with a very large surface area. It is largely these characteristics that are responsible for its influence on soil physics.

Of all the physical effects biochar has on the soil, perhaps the most important is its potential to increase the availability of water to plants on contrasting soil types. The large surface area of biochar gives it a water-holding capacity comparable to clay, but its porosity provides it with the aeration that clay lacks. This means that the effect of biochar on some properties, like infiltration or hydraulic conductivity, varies according to soil texture (Tryon 1948). As a result biochar can counteract both the droughtiness prevalent in sandy soils (Uzoma 2011) and the waterlogging prevalent in heavy clay soils (Asai et al. 2009), and in this respect has been compared to SOM (Chan et al. 2007). Glaser et al. (2002) report ADEs with field water retention capacity 18% higher than surrounding soil without biochar. In various experiments around the world biochar-amended soils have shown increases in water-holding capacity from 11% to 481% with the higher values usually occurring on sandier soils (Iswaran et al. 1980; Dugan et al. 2010; Karhu et al. 2011; Southavong and Preston 2011; Uzoma 2011). Kammann et al. (2011) also reported greater water-use efficiency after applying biochar to a sandy soil.

The other physical effects of biochar often reflect its own physical properties, that is, its low density and high porosity, and include reduced bulk density (Laird et al. 2010), reduced tensile strength (Chan et al. 2007), and decreased soil penetration resistance (Busscher et al. 2010). Results are mixed, however, and will always reflect the type of biochar applied and the soil type being treated. Downie et al. (2009) report that biochar has been experimentally linked to improved soil structure or soil aeration in fine-textured soils. Biochar's influences on soil structure and aggregation (Liang et al. 2006) are subtle and are linked to its porosity, granularity, and surface charge (Major et al. 2009b). Piccolo et al. (1997) go on to suggest these effects could increase resistance to erosion. Teixeira and Martins (2003) contrast ADEs with similar soil but lacking biochar additions, as being more granular, workable, porous, structurally resilient, well drained, and having lower bulk densities, but it is difficult to isolate the effects of biochar from other factors (especially native SOM) in a historical context and over a large geographical area. Much more research is required before a full understanding of these important influences is possible.

Biochar and Soil Chemical Properties

Soil chemistry impacts directly on plant nutrition—or toxicity—at the most fundamental level. Biochar influences the chemistry of soil in ways that are highly dependent on the biochar's biomass feedstock and production process. In terms of plant nutrition this includes direct fertilizing effects, usually temporary, that involve the immediate addition of compounds in mineral form, and subsequent indirect effects, that are often longer term, such as changes to hydrogen ion concentration (pH) or cation exchange capacity (CEC), which can increase the availability of nutrients to plants and reduce losses by leaching.

Although biochar is not normally described as a fertilizer, it would be wrong to say that it contains no nutrients (other than C). As noted earlier, all biochars contain various nutritive elements (sometimes in considerable amounts) but not all are in plant-available forms. For example, the two nutrients applied most widely by farmers, nitrogen (N) and phosphorus (P), are frequently found in biochar at total levels comparable to those found in soil or much higher, especially in biochar of animal origin, but the available N is usually negligible while the amounts of available P vary considerably (Chan and Xu 2009). Cations, such as potassium, calcium, and magnesium, are frequently abundant in biochar, which explains its tendency to raise the base saturation of soil.

Like clay and SOM, biochar contributes a strong negative charge that raises the CEC by adsorbing positively charged ions (Major et al. 2009b). The intrinsic CEC of biochar is usually higher than that of mineral soil or SOM (Sohi et al. 2009). Laird et al. (2010) found up to 2% biochar raised CEC by up to 20% and pH by up to 1 pH unit. Chan et al. (2007) found a similar increase in pH that was halved in the presence of N fertilization. However, what type of biochar is added to what soil type is of critical importance. For example, the pH of biochar can vary from 4 to 12 (Lehmann 2007) while soil pH typically varies between 5 and 8, so inappropriate combinations of the two can lead to critical levels of micronutrient deficiencies (Kishimoto and Sugiura 1985). There is also evidence that chemical reactions that occur on the surface of the biochar long after it has been applied to soil can increase its nutrient-holding ability (Glaser et al. 2001) and pH (Cheng et al. 2008).

Biochar can influence nutrient transformations within the soil, such as increased nitrification and plant uptake of N, and increased availability and uptake of P (DeLuca et al. 2009). Glaser et al. (2002) found increased bioavailability of P, metal cations, and trace elements. Thies and Rillig (2009) report increased availability of N and P in the rooting zone. Van Zweiten et al. (2010) found significantly increased N uptake. Most of these effects can be partly traced back to physical characteristics like porosity, sorption capacity, surface area, and charge density, and to biological changes described in the next section.

Biochar and Soil Biological Properties

SOM (in various stages of decomposition) plays a vital role in drainage, aeration, plant nutrition, maintaining soil structure, and in providing C to the soil's biological life. Soil biota, primarily microorganisms, break down organic residues into plant-available nutrients and humus, a sponge-like matrix that attracts and retains moisture and nutrients, and releases humic substances that bind mineral particles together. The humus created by the biota also provides it with a habitat that protects from predation and desiccation, as well as being an energy-rich substrate. This partly living assemblage holds—in cellular form—most of the soil's C and nitrogen. C has been called the common currency of the soil system, and ecosystem functions are primarily driven via the energy generated by these transformations as SOM is decomposed (Kibblewhite et al. 2008).

Biochar is a form of thermally decomposed recalcitrant organic matter that provides an environment that is usually conducive to biota for a variety of physical and chemical reasons (Thies and Rillig 2009). Although biochar lacks the nutritional value of SOM, it provides a physical habitat that is more persistent than humus, and acts as a reservoir for water, air, and nutrients. Its large surface area attracts particles and facilitates chemical reactions and its porosity facilitates gaseous exchange. By encouraging microbial activity, biochar may initially speed up the decomposition of SOM (Yoshizawa et al. 2007), which may seem counterproductive in terms of C sequestration, but it is only by being broken down that SOM can contribute to soil productivity, and the net effect of applying biochar is a positive addition of C to the soil (Steiner et al. 2007). Furthermore, the long-term presence of biochar in *Terra Preta* soils has produced not only larger microbial biomass, but lower respiration and hence higher metabolic rates (Thies and Rillig 2009). In other words, over time C turnover decreases and the overall C stock is increased. Biochar's sorptive properties mean it has great affinity for organic compounds (Smernik 2009). Research using microscopy and fractionation has also revealed that biochar actively binds soil into microaggregates and these in turn protect it and SOM from oxidation (Brodowski et al. 2006). An underlying mechanism for this process is that aluminosilicate clay minerals in soil attach to the surface of the biochar, but this is just one of many complex organomineral interactions (Joseph et al. 2010). This growing body of evidence seems to confirm earlier suggestions that biochar, in addition to its own intrinsic properties, may also have a valuable role in helping to stabilize SOM and increase its longevity (Glaser et al. 2002).

Globally N is the single most limiting nutrient for primary production. In N-poor environments biochar has been shown to increase N_2-fixing bacteria and their associated mycorrhizal fungi (Rondon et al. 2007). Kim et al. (2007) found 25% greater microbial diversity and more N_2-fixing organisms in ADE soils than in equivalent, but unamended, soils nearby. In N-rich environments, however, biochar can supply the C to ameliorate a microbiologically unfavorable C/N ratio. Excess N tends to reduce *in situ* soil biodiversity and leach into adjacent surface water causing harmful eutrophication (Manning 2012). Biochar can "soak up" excess N both by NH_4^+ adsorption and microbial immobilization, creating a temporary reservoir of organic N (Steiner et al. 2007), which may subsequently become available in slow-release form (Steiner et al. 2009).

Anything that increases the availability of a range of nutrients, for example, through increasing pH, CEC, or base saturation, as biochar often does, as well as retaining moisture and particulate matter, and stabilizing humus, will tend to facilitate microbial activity. There is evidence that biochar has a positive effect on microbiological abundance and diversity, possibly through improved resource use, and no evidence so far of it inhibiting root growth. Certain changes brought about by biochar may simply displace some species in favor of others, for example, raising pH from acidic to neutral tends to favor bacterial species over fungi, and biochar has caused a decrease in some fungal symbionts possibly by obviating their role (Lehmann et al. 2011). Little is known about the long-term influence of biochar on soil fauna. The effect of biochar on soil ecology is complex, variable, and subject to temporal change. This is perhaps one of the least understood aspects of biochar, but one that is rightly attracting a great deal of research.

Influence of Biochar on Agricultural Productivity

Introduction

Changes to soil properties lead naturally to a discussion on soil productivity and other less direct ways in which the sustainability of agroecosystems may be affected. From an ecosystems services perspective, natural systems, frequently with human intervention, provide a range of benefits to people, which fall into the broad categories of supporting, provisioning, regulating, and cultural enhancement. Agriculture is a major beneficiary of such services, principally in terms of provisioning (in the form of primary production) and supporting, for example, soil conservation and the long-term maintenance of soil structure, water flow and water-holding capacity, nutrient cycling, and suppression of pests and diseases. But agriculture, if it is to avoid conflict with society at large, is also highly dependent on regulatory services, such as water purification, reduction of greenhouse gas emissions and C sequestration, and protection of human health. Agricultural operations are typically caught in a trade-off between provisioning and regulating (Kibblewhite et al. 2008). Biochar, which offers a range of enhancements to both of these contrasting types of ecosystems service, could provide part of a win-win solution.

Biochar and Crop Yields

Crop yield data for biochar is highly variable and usually attributable to multiple factors of the kind already presented. This said, in the vast majority of field or pot trials of biochar, yield increases far outnumber any observed negative or neutral effects. In the very rare examples of statistically significant yield decline, the cause has sometimes been due to inappropriate use that can and should easily be avoided (Kishimoto and Sugiura 1985). One review reports changes in biomass yield ranging from −71% to 324% (Sohi et al. 2009). Notable increases have been reported where biochar has been used in combination with fertilizer (Chan et al. 2007; Gathorne-Hardy et al. 2009; Peng et al. 2011). Steiner et al. (2007) found biochar with NPK fertilizer doubled rice and sorghum grain yields compared with fertilizer alone. A global meta-analysis (Jeffery et al. 2011) reported an average yield increase of 10%, spanning a wide range of biochar application rates, but it was observed that the optimum rate appeared to be approximately 100 t ha^{-1}. Common causal factors (where they could be identified) were thought to be water-holding capacity and liming effects, with nutrient availability also being important. Forest plots against categories of pH, soil texture, biochar feedstock, and crop type show no clear trends or contrasts. However, the analysis included only 16 studies, none of which were long-term and several of the crop/soil/biochar combinations were unique within the analysis and possibly beyond, for example, the one negative yield result was based on ryegrass grown with biochar from biosolids. Yet in another extreme example, algal biochar (also classed as biosolid) with low C and high ash increased sorghum yields 3200% (Bird 2012).

Significantly, productivity increases due to biochar tend to be greater on degraded or intrinsically infertile soils across a wide range of crop types, including cereals, legumes, and trees (Glaser et al. 2002; Kimetu et al. 2008; Major et al. 2009a; Haefele et al. 2011). These observations offer hope of added food security in regions of the world where poor quality soil and lack of access to agricultural resources cause an endemic cycle of poverty. Few studies have made direct comparisons of contrasting soil types, but some trends have emerged. Soils with higher pH and especially calcareous soils tend to show lower yield responses (Van Zwieten et al. 2010; Jeffery et al. 2011). Sandy soils tend to show higher yield responses than silty or clay soils (Yeboah et al. 2009; Haefele et al. 2011). The type of biochar used can be critical, for example, manure-based biochars tend to show higher yield responses (Chan et al. 2008; Jeffery et al. 2011; Uzoma 2011). Legumes tend to respond better to biochar than gramineae species (Lehmann and Rondon 2006). But these are all general observations with great variability, depending on the characteristics of the biochar, the response of the soil, and the requirements of the crop. There is a need for a greater understanding of not only the soil properties involved but the underlying mechanisms.

The Role of Biochar in Agricultural Resource Use Efficiency

In addition to direct increases in crop yield, biochar has the potential to bring about indirect increases in agricultural production and farm income. Water is a scarce resource for which agriculture must compete. Urbanization, soil sealing, climate change, and salinization of arable land (which requires leaching with additional irrigation) are all contributing to decreased availability of water. Improved water-holding capacity and water use efficiency has been repeatedly demonstrated as a characteristic and significant feature of biochar, which could help reduce water demand. By absorbing fluids and adsorbing particulate matter, biochar filters water passing through it and reduces leaching, leading to a greater efficiency of agrochemicals added to the soil. By improving drainage and aeration, biochar can also mitigate the harmful effects of waterlogging, such as acidification. The capacity of biochar to maintain soil C, stabilize SOM, and improve soil structure and cohesion has the potential to prevent erosion and counteract compaction. In the developed world these factors could contribute to farm incomes but in much of the developing world, where many soils are degraded, they could be critical to subsistence (Lehmann and Joseph 2009).

Converting biomass into biochar, especially if done close to its point of use, could be a highly efficient and valuable form of waste reuse. Biochar provides an inconspicuous service that accrues from its other benefits, namely, a reduced demand for fossil fuel, by improving the efficiency of fertilizers, reducing the demand for water, improving water quality (Chen et al. 2010; Beck et al. 2011), conserving soil and improving its workability, and consuming waste.

The Role of Biochar in Agricultural Good Practice and Environmental Risk Mitigation

By improving water quality of agricultural runoff and the surrounding watercourses, biochar can reduce offsite pollution and eutrophication, therefore having indirect effects on the costs of downstream water treatment. Biochar has a high capacity to absorb and neutralize harmful substances in the environment such as pesticides (Smernik 2009), herbicides (Spokas et al. 2009), PAHs (Beesley et al. 2010; Chen and Yuan 2011), and potentially toxic elements (PTEs) (Gomez-Eyles et al. 2011). More research is required to understand the long-term implications of this, but enough is known to consider deploying biochar in contaminated soil with a view to remediation (Beesley et al. 2011; Ennis et al. 2012). Biochar has also been shown to suppress GHG emissions such as nitrous oxide (N_2O) and methane (CH_4), both of which are disproportionally stimulated by agriculture (Rondon et al. 2005). While little research has been conducted into the direct influence of biochar on biodiversity, evidence suggests a positive relationship (Lehmann et al. 2011). Biodiversity, in particular species richness, is only one measure of ecological activity, but it is one that frequently contributes to a wide range of ecosystems services of direct and indirect benefit to farmers and society.

Conclusions

Current knowledge suggests that biochar can be applied to agricultural soils in order to boost crop yields and simultaneously sequester atmospheric C. Most research in this area has been, and continues to be, aimed at measuring one or both of these effects. Holding out the promise of ameliorating two of the major potential environmental catastrophes that face humanity—climate change and food scarcity—biochar is understandably attracting considerable attention.

References

Archer D, Eby M, Brovkin V, Ridgwell A, Cao L, Mikolajewicz U, Caldeira K, et al. (2009) Atmospheric lifetime of fossil fuel carbon dioxide. *The Annual Review of Earth and Planetary Sciences* 37:117–134.

Asai H, Samson BK, Stephan HM, Songyikhangsuthor K, Homma K, Kiyono Y, Inoue Y, Shiraiwa T, Horie T (2009) Biochar amendment techniques for upland rice production in Northern Laos. 1. Soil physical properties, leaf SPAD and grain yield. *Field Crops Research* 111:81–84.

Baldock JA, Smernik RJ (2002) Chemical composition and bioavailability of thermally altered *Pinus resinosa*. *Organic Geochemistry* 33:1093–1109.

Beck DA, Johnson GR, Spolek GA (2011) Amending greenroof soil with biochar to affect runoff water quantity and quality. *Environmental Pollution* 159:2111–2118.

Beesley L, Moreno-Jimenez E, Gomez-Eyles JL (2010) Effects of biochar and greenwaste compost amendments on mobility, bioavailability and toxicity of inorganic and organic contaminants in a multi-element polluted soil. *Environmental Pollution* 158:2282–2287.

Beesley L, Moreno-Jiménez E, Gomez-Eyles JL, Harris E, Robinson B, Sizmur T (2011) A review of biochars' potential role in the remediation, revegetation and restoration of contaminated soils. *Environmental Pollution* 159:3269–3282.

Bird MI (2012) Algal biochar: Effects and applications. *GCB Bioenergy* 4:61.

Brodowski S, John B, Flessa H, Amelung W (2006) Aggregate-occluded black carbon in soil. *European Journal of Soil Science* 57:539–546.

Broecker WS (2007) Climate change: CO_2 arithmetic. *Science* 315:1371.

Busscher WJ, Novak JM, Evans DE, Watts DW, Niandou MAS, Ahmedna M (2010) Influence of pecan biochar on physical properties of a Norfolk loamy sand. *Soil Science* 175:10–14.

Chan KY, Van Zwieten L, Meszaros I, Downie A, Joseph S (2007) Agronomic values of greenwaste biochar as a soil amendment. *Soil Research* 45:629–634.

Chan KY, Van Zwieten L, Meszaros I, Downie A, Joseph S (2008) Using poultry litter biochars as soil amendments. *Soil Research* 46:437–444.

Chan KY, Xu ZH (2009) Biochar: Nutrient properties and their enhancement. In *Biochar for environmental management: Science and technology*, Lehmann J and Joseph S (eds), 67–84. London: Earthscan.

Chen B, Yuan M (2011) Enhanced sorption of polycyclic aromatic hydrocarbons by soil amended with biochar. *Journal of Soils and Sediments* 11:62–71.

Chen Y, Shinogi Y, Taira M (2010) Influence of biochar use on sugarcane growth, soil parameters, and groundwater quality. *Soil Research* 48:526–530.

Cheng CH, Lehmann J, Engelhard MH (2008) Natural oxidation of black carbon in soils: Changes in molecular form and surface charge along a climosequence. *Geochimica et Cosmochimica Acta* 72:1598–1610.

DeLuca TH, MacKenzie MD, Gundale MJ (2009) Biochar effects on soil nutrient transformation. In *Biochar for environmental management: Science and technology*, Lehmann J and Joseph S (eds), 251–270. London: Earthscan.

Derenne S, Largeau C (2001) A review of some important families of refractory macromolecules: Composition, origin, and fate in soils and sediments. *Soil Science* 166:833–847.

Downie A, Crosky A, Munroe P (2009) Physical properties of biochar. In *Biochar for environmental management: Science and technology*, Lehmann J and Joseph S (eds), 13–32. London: Earthscan.

Dugan E, Verhoef A, Robinson S, Sohi S, Ahmedna M (2010) Bio-char from sawdust, maize stover and charcoal: Impact on water holding capacities (WHC) of three soils from Ghana. 19th World Congress of Soil Science, Soil Solutions for a Changing World, Brisbane, Australia.

Ennis CJ, Evans GA, Islam M, Ralebitso-Senior TK, Senior E (2012) Biochar: Carbon sequestration, land remediation and impacts on soil microbiology. *Critical Reviews in Environmental Science and Technology* 42:2311–2364.

Freddo A, Chao C, Reid BJ (2012) Environmental contextualisation of potential toxic elements and polycyclic aromatic hydrocarobons in biochar. *Environmental Pollution* 171:18–24.

Gathorne-Hardy A, Knight J, Woods J (2009) Biochar as a soil amendment positively interacts with nitrogen fertiliser to improve barley yields in the UK. *IOP Conference Series: Earth and Environmental Science* 6:372052.

Glaser B, Haumaier L, Guggenberger G, Zech W (2001). The Terra Preta phenomenon: A model for sustainable agriculture in the humid tropics. *Naturwissenschaften* 88:37–41.

Glaser B, Lehmann J, Zech W (2002) Ameliorating physical and chemical properties of highly weathered soils in the tropics with charcoal: A review. *Biology and Fertility of Soils* 35:219–230.

Gomez-Eyles JL, Sizmur T, Collins CD, Hodson ME (2011) Effects of biochar and the earthworm *Eisenia fetida* on the bioavailability of polycyclic aromatic hydrocarbons and potentially toxic elements. *Environmental Pollution* 159:616–622.

Haefele SM, Konboon Y, Wongboon W, Amarante S, Maarifat AA, Pfeiffer EM, Knoblauch C (2011) Effects and fate of biochar from rice residues in rice-based systems. *Field Crops Research* 121:430–440.

Hale SE, Lehmann J, Rutherford D, et al. (2012) Quantifying the total and bioavailable polycyclic aromatic hydrocarbons and dioxins in biochars. *Environmental Science & Technology* 46:2830–2838.

Hansen J, Sato M, Kharecha P, et al. (2008) Target atmospheric CO2: Where should humanity aim? *Journal of the Atmospheric Sciences* 2:217–231.

Houghton RA (2007) Balancing the global carbon budget. *Annual Review of Earth and Planetary Sciences* 35:313–347.

IPCC (2007) Climate Change 2007: The Physical Science Basis. Contribution of Working Group I to the Fourth Assessment Report of the Intergovernmental Panel on Climate Change. Solomon S, Qin D, Manning M, et al. (eds). Cambridge: Cambridge University Press.

Iswaran V, Jauhri K, Sen A (1980) Effect of charcoal, coal and peat on the yield of moong, soybean and pea. *Soil Biology and Biochemistry* 12:191–192.

Jeffery S, Verheijen FGA, Velde MVD, Bastos AC (2011) A quantitative review of the effects of biochar application to soils on crop productivity using meta-analysis. *Agriculture, Ecosystems and Environment* 144:175–187.

Joseph SD, Camps-Arbestain M, Lin Y, Munroe P, Chia CH, Hook J, van Zwieten L, Kimber S, Cowie A, Singh BP, Lehmann J, Foidl N, Smernik RJ, Amonette JE (2010) An investigation into the reactions of biochar in soil. *Soil Research* 48:501–515.

Kammann CI, Linsel S, Gößling JW, Koyro HW (2011) Influence of biochar on drought tolerance of *Chenopodium quinoa* Willd and on soil-plant relations. *Plant and Soil* 345:195–210.

Karhu K, Mattila T, Bergström I, Regina K (2011) Biochar addition to agricultural soil increased CH$_4$ uptake and water holding capacity: Results from a short-term pilot field study. *Agriculture, Ecosystems and Environment* 140:309–313.

Keiluweit M, Nico PS, Johnson MG, Kleber M (2010) Dynamic molecular structure of plant biomass-derived black carbon (biochar). *Environmental Science and Technology* 44:1247–1253.

Kibblewhite MG, Ritz K, Swift MJ (2008) Soil health in agricultural systems. *Philosophical Transactions of the Royal Society of London B Biological Sciences* 363:685–701.

Kim JS, Sparovek G, Longo RM, Melo W, Crowley D (2007) Bacterial diversity of Terra Preta and pristine forest soil from the Western Amazon. *Soil Biology and Biochemistry* 39:684–690.

Kimetu J, Lehmann J, Ngoze S, Mugendi D, Kinyangi J, Riha S, Verchot L, Recha J, and Pell A (2008) Reversibility of soil productivity decline with organic matter of differing quality along a degradation gradient. *Ecosystems* 11:726–739.

Kishimoto S, Sugiura G (1985) Charcoal as a soil conditioner. *International Achievements for the Future* 5:12–23.

Kookana RS, Sarmah A, Van Zwieten L, Krull E, Singh B (2011) Biochar application to soil: Agronomic and environmental benefits and unintended consequences. *Advances in Agronomy* 112:103–143.

Laird DA, Fleming P, Davis DD, Horton R, Wang B, Karlen DL (2010) Impact of biochar amendments on the quality of a typical Midwestern agricultural soil. *Geoderma* 158:443–449.

Lehmann J (2007) Bio-energy in the black. *Frontiers in Ecology and the Environment* 5:381–387.

Lehmann J, Joseph S (2009) Biochar for environmental management: An introduction. In *Biochar for environmental management: Science and technology*, Lehmann J and Joseph S (eds), 1–12. London: Earthscan.

Lehmann J, Kern DC, Glaser B, Woods WI (2003) *Amazonian dark earths: Origin, properties, management*. Dordrecht: Kluwer.

Lehmann J, Rillig MC, Thies J, Masiello CA, Hockaday WC, Crowley D (2011) Biochar effects on soil biota: A review. *Soil Biology and Biochemistry* 43:1812–1836.

Lehmann J, Rondon M (2006) Bio-char soil management on highly weathered soils in the humid tropics. In *Biological approaches to sustainable soil systems*, Uphoff N, Ball AS, Fernandes E et al. (eds), 517–530. Boca Raton, FL: CRC Press.

Lenton TM, Vaughan NE (2009) The radiative forcing potential of different climate geoengineering options. *Atmospheric Chemistry and Physics Discussion* 9:1–50.

Liang B, Lehmann J, Solomon D, Kinyangi J, Grossman J, O'Neill B, Skjemstad JO, et al. (2006) Black carbon increases cation exchange capacity in soils. *Soil Science Society of America Journal* 70:1719–1730.

Liang B, Lehmann J, Solomon D, et al. (2008) Stability of biomass-derived black carbon in soils. *Geochemica et Cosmochimica* 72:6069–6078.

Macías F, Arbestain MC (2010) Soil carbon sequestration in a changing global environment. Mitigation and Adaptation Strategies for Global Change 15:511–529.

Major J, Lehmann J, Rondon M, Goodale C (2009a) Fate of soil-applied black carbon: Downward migration, leaching and soil respiration. *Global Change Biology* 16:1366–1379.

Major J, Steiner C, Downie A, Lehmann J (2009b) Biochar effects on nutrient leaching. In *Biochar for environmental management: Science and technology*, Lehmann J and Joseph S (eds), 271–288. London: Earthscan.

Manning P (2012) The impact of nitrogen enrichment on ecosystems and their services. In *Soil ecology and ecosystem services*, Wall DH, Bardgett RD, Behan-Pelletier V, Herrick JE, Jones H, Ritz K, Six J, Strong DR, van der Putten WH (eds), 256–269. Oxford: Oxford University Press.

Matthews HD, Caldeira K (2008) Stabilizing climate requires near-zero emissions. *Geophysical Research Letters* 35, L04705.

Neary DG, Klopaket CC, DeBano LF, Folliot PF (1999) Fire effects of below-ground sustainability: A review and synthesis. *Forest Ecology and Management* 122:51–71.

Neves NG, Petersen JB, Bartone RN, Silva CAD (2003) Historical and socio-cultural origins of Amazonian dark earths. In *Amazonian dark earths: Origin, properties, management*, Lehmann J, Kern D, Glaser B, Woods W (eds), 29–50. Dordrecht: Kluwer.

Peng X, Ye LL, Wang CH, Zhou H, Sun B (2011) Temperature- and duration-dependent rice straw-derived biochar: Characteristics and its effects on soil properties of an Ultisol in southern China. *Soil and Tillage Research* 112:159–166.

Piccolo A, Pietramellara G, Mbagwu JSC (1997) Use of humic substances as soil conditioners to increase aggregate stability. *Geoderma* 75:267–277.

Rondon M, Lehmann J, Ramírez J, Hurtado M (2007) Biological nitrogen fixation by common beans (*Phaseolus vulgaris* L.) increases with biochar additions. *Biology and Fertility in Soils* 43:699–708.

Rondon M, Ramirez JA, Lehmann J (2005) Greenhouse gas emissions decrease with charcoal additions to tropical soils. 3rd USDA Symposium on Greenhouse Gases and Carbon Sequestration in Agriculture and Forestry, Baltimore.

Schmidt MWI, Noack AG (2000). Black carbon in soils and sediments: Analysis, distribution, implications and current challenges. *Global Biogeochemical Cycles* 14:777–794.

Shneour EA (1966) Oxidation of graphitic carbon in certain soils. *Science* 151: 991–992.

Smernik RJ (2009) Biochar and sorption of organic compounds. In *Biochar for environmental management: Science and technology*, Lehmann J and Joseph S (eds), 289–300. London: Earthscan.

Sohi S, Lopez-Capel E, Krull E, Bol R (2009) Biochar, climate change and soil: A review to guide future research. CSIRO Land and Water Science Report 05/09. Highett: CSIRO.

Solomon S, Plattner G, Knutti R, Friedlingstein P (2009) Irreversible climate change due to carbon dioxide emissions. *PNAS* 106:1704–1709.

Southavong S, Preston TR (2011) Growth of rice in acid soils amended with biochar from gasifier or TLUD stove, derived from rice husks, with or without biodigester effluent. Livestock Research for Rural Development 23, article 32. http://www.lrrd.org/lrrd23/2/siso23032.htm (accessed August 15, 2013).

Spokas KA, Koskinen WC, Baker JM, Reicosky DC (2009) Impacts of woodchip biochar additions on greenhouse gas production and sorption/degradation of two herbicides in a Minnesota soil. *Chemosphere* 77:574–581.

Steiner C, Garcia M, Zech W (2009) Effects of charcoal as slow release nutrient carrier on N-P-K dynamics and soil microbial population: Pot experiments with Ferralsol substrate. In *Amazonian dark earths: Wim Sombroek's vision*, Woods W, Teixeira W, Lehmann J, Steiner C, WinklerPrins A, Rebellato L (eds), 325–338. Netherlands: Springer.

Steiner C, Teixeira W, Lehmann J, Nehls T, de Macêdo J, Blum W, Zech W (2007) Long term effects of manure, charcoal and mineral fertilization on crop production and fertility on a highly weathered Central Amazonian upland soil. *Plant and Soil* 291:275–290.

Teixeira WG, Martins GC (2003) Soil physical characterization. In *Amazonian dark earths: Origins, properties, management*, Lehmann J, Kern D, Glaser B, Woods WI (eds), 271–286. Dordrecht: Kluwer Academic Publishers.

Thies JE, Rillig MC (2009) Characteristics of biochar: Biological properties. In *Biochar for environmental management: Science and technology*, Lehmann J, Joseph S (eds), 85–106. London: Earthscan.

Tryon EH (1948) Effect of charcoal on certain physical, chemical, and biological properties of forest soils. *Ecological Monographs* 18:81–115.

Uzoma KC (2011) Influence of biochar application on sandy soil hydraulic properties and nutrient retention. *Journal of Food, Agriculture and Environment* 9:1137–1143.

Van Zwieten L, Kimber S, Morris S, Chan KY, Downie A, Rust J, Joseph S, Cowie A (2010) Effects of biochar from slow pyrolysis of papermill waste on agronomic performance and soil fertility. *Plant and Soil* 327:235–246.

Verhaijen F, Jeffery S, Bastos AC, Valde Mvd, Diafas F (2010) Biochar application to soils. A critical scientific review of effects on soil properties, processes, and functions. EUR 24099 En. Luxemburg: Office for the Official Publications of European Communities.

Woolf D, Amonette JE, Street-Perrott FA, Lehmann J, Joseph S (2010) Sustainable biochar to mitigate global climate change. *Nature Communications* 1:1–9.

Yeboah E, Ofori P, Quansah G, Dugan E, Sohi S (2009) Improving soil productivity through biochar amendments to soils. *African Journal of Environmental Science and Technology* 3:034–041.

Yoshizawa S, Tanaka S, Ohata M (2007) Estimation of microbial community structure during composting rice bran with charcoal. Carbon 2007, Seattle.

Zhu D, Kwon S, Pignatello JJ (2005) Adsorption of single-ring organic compounds to wood charcoals prepared under different thermochemical conditions. *Environmental Science and Technology* 39:3990–3998.

8

Energy: Defining the Future

Dolores Durán and Eduardo A. Rincón-Mejía

CONTENTS

Energy: Definitions and Types

Nothing happens without energy. It allows all procedures from the simplest to the most complex to be carried out. The energy requirement of movement, like walking, is intuitive. Breathing, however, implies an energy transformation such as the oxidation of a fuel, like sugar, a process similar to combustion in an engine. When gasoline is burned with air it produces mostly carbon dioxide, water, and nitrogen:

$$C_8H_{18} + 12.5(O_2 + 3.76N_2) \rightarrow 8CO_2 + 9H_2O + 47N_2 \qquad (8.1)$$

During this reaction, the gasoline releases chemical energy and pollutants.

The first law of thermodynamics states that energy is not created or destroyed but transformed. In a car, if this energy was confined to the combustion chamber of an engine, it would just move the pistons and produce work. In reality, part of this energy is transformed into heat that will flow through the chamber walls serving no purpose. Also, the combustion products will be released into the atmosphere at a certain temperature and, hence, with remaining energy.

BOX 8.1   DEFINITIONS

Energy is the capacity for work. Work is an action of physical objects.

A heat flow is a transfer of energy from hotter space or bodies to cooler ones. Heat flows occur as energy is dissipated from work or another energy transformation.

When we are talking about the rate at which the energy is transferred, used, or transformed, we are referring to power.

Classical thermodynamics mostly studied isolated systems seldom encountered in real life, which is full of open and complex systems. For example, in the stars hydrogen reactions produce high-quality light, which fuels most planetary processes. Ecosystems and organisms exchange across their boundaries both energy and matter. Cities as well heavily rely on a steady stream of materials and energy across their boundaries.

BOX 8.2   ENERGY TYPES

In mechanics, an object possesses energy owing to its position (potential energy) or its motion (kinetic energy).

In chemistry, a fuel has energy that will be released by breaking up the bonds between atoms and molecules. Molecules, considering a closed tank filled with gas at a given pressure, are subject to three energy types:

- Intermolecular potential energy—Depends on the magnitude of intermolecular forces and the relative position of molecules
- Molecular kinetic energy—Associated with the displacement velocity of the molecules
- Intramolecular energy—Associated with the molecular and atomic structure and related forces

Entropy, Exergy, and Emergy

Entropy

The second law of thermodynamics explains the energy degradation process. It establishes that not all the heat can be transformed into work, and only part of it will be usefully exploited (Cengel 2011). As noted by Clausius (1867), only one part of the total wind energy spinning a windmill is transformed into mechanical energy, the other part being transformed into heat energy in the friction of the blades with the air or between the shaft and its bearing. Likewise, only a portion of the chemical energy of food is converted into mechanical energy used by human beings to perform an action or enters cell reproduction and synthesis, another part is excreted as waste products due to the inefficiency of metabolic processes. Clausius found that in fact physical, chemical, and electrical energy can be completely changed into heat. But the reverse cannot be fully accomplished without an inevitable loss of energy in the form of irretrievable heat. As said energy is not destroyed, it becomes irreversibly unavailable for work. The production of these irreversibilities or unavailable energy in the universe is measured by the abstract dimension that Clausius in the 1860s called entropy (from the Greek *entrope*, meaning "change").

Intuitively, some scientists refer to entropy as actual states such as disorder, waste, and the loss of time or information. More precisely, in thermodynamics the entropy indicates the irreversibility of a process, the natural decrease of energy quality as matter organizes adopting distributions that are no longer susceptible to become work. As an example, one can imagine a messy desk that work is required to sort. This action releases entropy, but when all has been ordered and can no longer be ordered anymore, it reaches equilibrium, the level of maximum entropy.

Thus, everything that happens in nature means an increase in entropy of that part of the world where it occurs. A living organism continually increases its entropy but is kept alive by extracting energy from farther afield in its environment. In short, when we use energy to carry out work we release entropy, that is, energy unable to be used in a useful process. The more transformations the energy undergoes, the more it is downgraded.

Exergy

Exergy measures the quality of energy, the distance from equilibrium, and the power to do something useful with this energy. Exergy expresses the limits of what can be done with energy, and entropy what has happened to the energy. Entropy reveals a hierarchy among the various forms of energy and an imbalance in their transformations. Exergy establishes criteria to determine the ideal performance of equipments and the direction of processes. That is, the maximum capacity to transform energy into work.

Engineers are interested in getting the most energy from a given system; however, as said, not all energy is equal. Some types can be used to power machines and set something in motion; others like heat cannot do any work unless it is transformed using a heat engine. So engineers must look for high exergy forms of energy (closer to stellar energy) and low entropy processes so as to maximize efficiency, that is, make the most of energy.

Emergy

Transformations between energy types require technology, infrastructure, financial resources, and trained personnel. Classical methods assess the efficiency of transformations based on energy and economy variables. But sustainability requires more integrated analyses whereby both environmental and socioeconomic costs are quantified. This is partly achieved by life-cycle analyses that deal with the materials and energy used to develop, build, and operate energy systems.

Only renewable energy systems can maintain a long-term dynamic balance between environmental conservation, social benefit, and economic profit. To this end, emergy complements life-cycle analyses with comparisons of alternative systems with different embodied energies. Emergy is useful energy of a certain type used directly and indirectly in a particular product or service (Odum 1988; Scienceman and El-Youssef 1993). It expresses the cost of units of energy from any source, relative to solar energy. This allows for comparisons of completely different systems in terms of emjoule (emergy joule), also called solar emjoule (sej).

The lesson of entropy is that the energy contained in the planet is exhaustible and easily degrades into less useful forms. This rules out fossil and nuclear fuels as sustainable energy sources. Moreover, exergy is not attainable due to technological inefficiencies, which include the release of heat and contaminant wastes in nonrenewable energy sources and some renewable applications (e.g., biomass combustion). Accordingly, energy systems have to be compared based on their emergy merits.

Zero-Carbon Energy Technologies?

For a technology, solar or otherwise, to be devoid of carbon emissions, all energy used in its raw material manufacture, transportation, installation, dismantling, and recycling should come from systems whose operation used energy from clean sources. Currently, the manufacture of monocrystalline silicon photovoltaic modules requires much energy to melt and purify the silicon ingot, cut, dope, and further manufacture the module before it starts converting radiant energy into power. Although at the end of its useful life the module may generate as much as 30 times more energy as it consumed in its development, continued manufacture with fossil fuels will not make photovoltaic modules free of greenhouse gas emissions. The same is true of all other technologies harnessing renewable sources.

Zero-carbon and carbon-negative solutions are possible, however, if their manufacture, operation, and recycling are offset by carbon sequestration activities. Furthermore, sustainable energy systems should compensate for socioeconomic inequalities and all pollutions incurred. Since the transition to sustainable energy systems requires fossil fuels under current manufacturing techniques, the tardier the transition the less energy will be available during this transition.

Transitioning to a Sustainable Energy Future

Energy is essential for economic, social, and global welfare, but most of it is produced and consumed in unsustainable ways (Yüksel 2008). Fossil fuels (oil, coal, and natural gas) are currently the main source, with more than 90% of the global commercial energy production. OPEC (Organization of the Petroleum Exporting Countries) forecasts further growth until 2030, both in developed and developing countries. Thereafter, energy poverty will become a crucial variable (OPEC 2011), whereas issues of gases and particulate matter emitted into the atmosphere will become more pressing as more polluting fossil fuels are tapped (e.g., tar sands, shale oil, and natural gas obtained via fracking). But innovations in the production, transmission, distribution, and consumption of energy cannot replenish nonrenewable reserves (Yüksel 2008).

Against this backdrop, renewable energy sources are potential candidates to meet global energy requirements in a sustainable manner. Renewable energy sources have many advantages when compared with fossil fuels (Demirbas 2000). The sun shines for everyone under most latitudes; is poised to last billions of years; and results in wind, rain, ocean currents, and photosynthesis, which withdraw CO_2 from the atmosphere. All these energy sources are likely sufficient to meet all energy needs for generations to come, and their zero-greenhouse effect potential is within our technological grasp.

The Risks of Nonrenewable Energy Sources

Oil Pcak

Well-to-wheel analyses of gasoline vehicles and many studies of electric power generation and transportation yield similar conclusions: less than 5% of the energy of these nonrenewable sources is useful in providing mechanical work (Ruiz 2009); the remaining 95% is thermal waste that aggravates global warming. It is also unjust, since the vast majority of countries are forced to import this energy, which is highly polluting, destroys ecosystems in oil spills, and is unable to meet the energy needs of humanity in the medium term.

Fossil fuel consumption at today's global rate is excessive and so has brought about atmospheric warming; ocean acidification; and derived ecological, economic, and health impacts. Water, air, and soil pollution—both chronic and acute—are linked to all the phases of fossil fuel production, transport, storage, and consumption.

None of the nonrenewable energy sources can ensure global energy supply in the long run, as far as current scientific knowledge can elucidate. All proven reserves of nonrenewable energy are negligible when compared to solar energy intercepted in just one year. Note that any nonrenewable resource sooner rather than later inescapably becomes scarce, especially if used irrationally, as has happened with petroleum. Thus, the globe faces a peak and posterior decline in oil production, predicted by Hubbert (1956). With increasing demand and decreasing supply, speculation, oil shocks, and reserves even more concentrated in some countries, economic recessions and armed conflicts over the control of reserves seem poised to become more frequent.

The largest share of global warming is attributed to the emission of greenhouse gases from fossil fuel consumption. With an insurance value for thermal stability at 350 parts per million (ppm by volume) CO_2 in the atmosphere (Hansen et al. 2008) and a current CO_2 concentration in excess of 390 ppm (Figure 8.1), we are in a dangerous zone of increased risk due to higher frequency and intensity of catastrophic floods, hurricanes, deadly heat waves

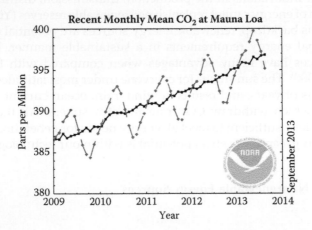

FIGURE 8.1
Keeling curve of the global atmospheric CO_2 concentration, measured at Mauna Loa, Hawaii. Seasonal variations are due to the photosynthesis cycle in the Northern Hemisphere, but the salient feature is the unabated increase in the average yearly value. (From Tans P, 2011, NOAA/ESRL, www.esrl.noaa.gov/gmd/ccgg/trends/.)

(deadly especially for the elderly and children), and infectious diseases. As the mass of the atmosphere is approximately 5.1×10^{18} kg, each additional parts per million of CO_2 in the atmosphere weighs nearly 7.75 billion tons, so more than 333 billion tons of CO_2 should be removed for the atmosphere to regain stability.

The alternative to expected damages from a more variable atmosphere with extreme events is a rapid transition to a global energy system based on solar energy (sunlight, wind, and biomass) as well as geothermal energy, coupled with halted deforestation and forest recovery, to bring back the atmospheric CO_2 concentration to 350 ppm initially and to preindustrial levels subsequently.

The Myth of a Safe Nuclear Bonanza

Some voices mention nuclear energy as part of the solution to energy and environmental problems. However, the reserves of uranium and thorium are very small compared to natural gas or coal, which in turn are insignificant compared to the sunlight. The proliferation of nuclear weapons is a concern that will maintain nuclear technology—both civil and military—in a few countries. This will in all likelihood enlarge socioeconomic gaps in the post-fossil-fuel era. Mining, processing, consuming, and disposing of nuclear fuels entail unsolved environmental problems. There are about 440 operating nuclear plants for electricity generation. Examples of recent accidents are the Swedish Forsmark plant (June 25, 2006) and the Japanese Kashiwazaki-Kariwa plant, the world's largest for power generation (July 16, 2007, when an earthquake damaged the nuclear reactor and released radioactive material to the environment). The Kashiwazaki-Kariwa plant had to be closed for over 21 months. The better known catastrophes to date are Three-Mile Island (1979), Chernobyl (1986), and Fukushima (2011) with local, regional, and far-reaching consequences for millions exposed around the world to radionuclides in the air, water, soil, and milk. Nuclear technology is costly, the construction of a nuclear plant is slow and lasts at least 5 years, and is exposed to military and terrorist attacks. In short, nuclear power is expensive, unacceptably risky, and insufficient to reduce the levels of greenhouse gas emissions. The supply of nuclear fuel in the long term and the necessary technology cause foreign dependency and correlate with military spending. The lack of transparency about safety conditions of the nuclear generators and confinement plants undermines normal democratic life. The only safe way to benefit from natural radioactivity at the planet's core is to exploit geothermal sources (see the section, "Geothermal Energy").

Technological Dependency

Technological independence in a population is correlated to the ability to exploit energy, to welfare, and to political and economic independence.

BOX 8.3 DEVELOPMENT BASED ON RENEWABLE ENERGY TECHNOLOGIES

Developing countries straddle the tropics and so are very well endowed with insolation. For example, Mexico receives two times more sunlight per square meter than Germany or Austria. And as Austria is some 25 times smaller than Mexico, insolation is 50 times higher in the latter. However, the Austrians have installed 300 W/inhabitant worth of thermal solar water heaters, 50 times more than Mexico. This means that the Austrians are 2500 times more efficient in making use of their solar energy resources than the Mexicans. This is not accounting for a lower absolute cost of one water-heater watt in Mexico. Reversing these figures would have a huge effect on energy and money savings as well as provide a stepping stone for further technological developments beyond the relatively simple water heater.

Wind-wise the situation is equally dramatic. The power provided by a windmill equipped with a turbine of a given size is proportional to the cube of the wind speed. So in an area like the Isthmus of Tehuantepec that separates the Atlantic and Pacific oceans, where the average wind speed is twice that of the European zone, wind power with same-sized machines would be eight times larger. At least four other similarly windy regions exist inland and onshore in Mexico. Wind turbines should be designed to operate advantageously with wind speeds at each site. Imported turbines designed for lower wind speeds may not withstand intense bursts, and make poor use of available wind speeds. These arguments can be extended to geothermal and other renewable energy sources.

The transition from oil to wind energy is likely to spread to many windy countries. Mexico, for instance, is five times larger than Germany and has 11,000 miles of coastline versus 1000 in Germany, many of which record strong winds. Overall, Mexico has a total installed electricity-generation capacity of 56,000 MW versus a wind energy capacity of 28,000 MW in Germany.

A clear difference separates advanced countries that develop technology from developing countries that mostly export mining and agricultural products and are technology importers. The development of systems using local energy sources creates a large opportunity for scientific and technological progress, and out of backwardness and (energy) poverty. It is not enough to import renewable energy sources. It is also essential, in order to spur development, to fund research and development on sustainable technologies in many countries, and to improve wages and working conditions for engineers and other professionals.

Social Inequality

Energy, similar to income, is very poorly distributed. Although there are a handful of super-rich, avid energy consumers, 2 billion people have insufficient food energy intakes and their energy consumption is limited to cooking with woodfuel or dung. Energy poverty reduces manpower efficiency due to a lack of tools to cultivate the land, and so is a contributing factor toward rural out-migration, land abandonment, and exacerbated urban issues. Millions lack electricity due to distance from major distribution lines or exceedingly dispersed settlement. It is not technically or economically feasible to provide them with electricity by conventional means, as it would require transformers, transmission lines, substations, and other equipment. Tens of thousands of dollars per home would often be required, which only affluent societies and settlers can afford.

However, renewable energy systems, such as photovoltaic panels and especially small wind turbines, are entirely feasible in remote communities in poor countries to emulate what in rich countries rural inhabitants achieve: living comfortably in autonomy with small energy systems based on solar, wind, and other renewable energy sources to power computers, dairy refrigerators, and other appliances. There, access to abundant and diverse technological and socioeconomic information via the Internet makes for more democratic societies.

An Overview of Renewable Energy Solutions

Conventional primary energy sources (e.g., petroleum, natural gas, coal, and uranium ore) are nonrenewable and nearing depletion or produce emissions

BOX 8.4   SUSTAINABLE ENERGY IN RURAL TROPICAL AREAS

Sustainable decentralized energy systems are a requirement for socioeconomic development in rural areas. Success stories highlight the necessary conditions. First, no off-the-shelf solution exists but rather a combination of self-taught technological savvy. Second, technology transfer is often ill-adapted to local conditions and may often create dependency. Put differently, costly high-tech solutions have to be substituted by homegrown adequate technology. However, external help with specific problem solving can foster technological skills.

The best candidate technologies for sustainable energy in tropical rural areas are solar cooking and water heating, combined with human-powered water pumps; electricity-wise, wind energy and bicycles adapted for human-powered electricity generation. Biomethane is most suitable for cattle ranchers.

in intolerable amounts. As to renewable sources (solar, wind, geothermal, hydro, and biomass sources), they are widespread and readily accessible. The most frequently exploited at present are as follows.

Solar Energy

The energy from the sun is the most abundant resource in the world. It supplies the climate and biosphere processes as well as other renewable energy sources (wind, ocean waves, bioenergy). Its availability depends on the local climate, but on average, the radiation that reaches the Earth's surface varies between 2.2 and 7 $kWh/m^2/day$, even though the clouds reflect part of the radiation back to space.

Photovoltaic systems transform solar energy into electricity. Photothermal collectors transform solar radiation into thermal energy available for industrial processes or electricity production. For instance, solar troughs reach 200°C and coupled to a thermal plant generate electricity (see Chapter 9).

Wind Energy

Wind energy has been the fastest growing renewable source in recent years. It is based on the transformation of the wind kinetic energy into work in a windmill. Small turbines operate at wind speeds as low as 3 m/s and big turbines can withstand wind speeds over 320 km/h.

The horizontal-axis turbines are the most common. Their main rotor shaft and electrical generator are atop a tower, and must be pointed into the wind. Most of them have a gearbox allowing increased power at higher rotation velocity. These turbines are installed in the large-scale wind farms. As to vertical axis turbines, their main rotor shaft is vertical and so does not need to be pointed into the wind to be effective. Also, this kind of turbine is able to operate with small wind velocities but have the disadvantages of low rotational speed, higher torque, and higher cost of the drive train.

The wind industry has grown exponentially, at rates close to 30% per year. The installed capacity has increased tenfold in the last 8 years. In June 2011, the installed capacity was just over 215,000 MWe (megawatt electrical) distributed among 80 countries, and generating 3% of the world's electricity. But this capacity is conservatively expected to reach more than 1.5 million MW by the end of 2020. In the countries with the largest installed capacity, China (52,000 MW), the United States (>42,000 MW), Germany (28,000 MW), and Spain (>21,000 MW) (WWEA 2011), wind energy can be predicted to reach or exceed the production of large hydroelectricity by 2018 (Rincón-Mejía 2010).

Bioenergy

Liquid biofuels, woodfuel, and solid waste incineration are allegedly renewable energy sources, but they are ridden with the following problems.

First, they compete for land with natural and agricultural uses (hence natural habitats are degraded and foodstuff prices increase). Second, their combustion egresses soot and greenhouse gases to the atmosphere. Finally, the production of some of these fuels releases, with the current technology, toxic byproducts.

More sustainable bioenergy sources come from recycling solid waste from agriculture and households, and sewage sludge from wastewater treatment plants, to produce biomethane. Sweden and Denmark already are the chief solid waste importers in the world and Germany the main biomethane producer (DENA 2013).

Geothermal Energy

Electricity generation uses steam produced from reservoirs of hot water found a couple of miles or more below the Earth's surface.

There are three types of geothermal power plants: dry steam, flash steam, and binary cycle. Dry steam power plants draw from underground resources of steam. The steam is piped directly from underground wells into the power plant, where it is directed into a turbine/generator unit. Flash steam power plants are the most common. They use geothermal reservoirs of water with temperatures greater than 182°C. This very hot water flows up through underground wells while water and condensed steam are injected back into the reservoir. Binary cycle power plants operate on water at lower temperatures (107°–182°C). These plants use the heat from the hot water to boil a working fluid, usually an organic compound with a low boiling point. The working fluid is vaporized in a heat exchanger and used to activate a turbine.

Geothermal energy, including heat pump microgeothermics, is a main energy source for 58 countries, and 39 of them could be 100% powered by geothermal energy. Solar and geothermal hybrids are being developed to reach high temperatures (see Chapter 9).

Hydropower

Hydropower is the most commonly used renewable energy source. It exploits the kinetic energy of water and the potential energy of dams. A very substantial drawback is that it involves flooding a large land area and, as a consequence, a large impact in the ecosystem and human settlements. Unproductive methane emissions from underwater vegetation decomposition contribute to the greenhouse effect. At present there are more than 45,000 big dams in the world.

Simple as it may seem, a transition toward a global energy system based on renewable energy sources, which include solar radiation and its secondary manifestations, hydropower, and geothermal energy, among others, is until now the only sustainable option for future energy demands.

References

Cengel YA, Boles M (2011) *Thermodynamics: An engineering approach.* New York: McGraw-Hill.

Clausius R (1867) *The mechanical theory of heat: With its applications to the steam engine and to physical properties of bodies.* London: John van Voorst.

Demirbas A (2000) Biomass resources for energy and chemical industry. *Energy Education, Science and Technology* 5:21–45.

DENA (2013) Evolución del Mercado. Ministerio Federal de Economía y tecnología. Agencia Alemana de Energía. http://www.renewables-made-in-germany. com/es/renewables-made-in-germany-pagina-de-inicio/biogas/biogas/ evolucion-del-mercado.html (accessed May 19, 2012).

Hansen J, Sato M, Kharecha P, Beerling D, Masson V, Delmotte M, Pagani R, Royer DL, Zacho JC (2008) Target atmospheric CO_2: Where should humanity aim? *The Open Atmospheric Science Journal* 2:217–231.

Hubbert MK (1956) Nuclear energy and the fossil fuels, presented before the Spring Meeting of the Southern Division of Production. American Petroleum Institute. Houston, Shell Development Company, Exploration and Production Research Division. San Antonio, TX.

Odum HT (1988) Self-organization, transformity, and information. *Science* 242:1132–1139.

Organization of the Petroleum Exporting Countries (OPEC) (2011) World oil outlook 2011. Technical Report OPEC. Vienna, Austria.

Rincón-Mejía EA (2010) Combining solar and wind in sustainable energy systems. In *Wind Energy International 2009–2010*, 383–385. Bonn: World Wind Energy Association.

Ruiz HV (2009) La electricidad solar térmica, tan lejos, tan cerca. Seville: Junta de Andalucía y Gas Natural.

Scienceman DM, El-Youssef BM (1993) The system of emergy units. Proceedings of the XXXVII Annual Meeting of the International Society for the Systems Sciences. New South Wales, Australia, July.

Tans P (2011) NOAA/ESRL. www.esrl.noaa.gov/gmd/ccgg/trends/ (accessed August 29, 2012).

Yüksel I (2008) Global warming and renewable energy sources for sustainable development in Turkey. *Renewable Energy* 33:802–812.

9

Solar Solutions

Iván Martínez-Cienfuegos and Eduardo A. Rincón-Mejía

CONTENTS

Solar Energy Transition

Soaring Solar Markets, Lingering Roadblocks

The oil and nuclear industries have started to decline worldwide: U.S. oil production peak occurred in the early 1970s, and Japan decided to put an end to its nuclear program after 2011. Meanwhile, solar technologies have attracted attention in China and the European Union (Patlitzianas et al. 2005; Pitz Pall et al. 2005; Hang et al. 2008; Lu 2010). The growth rates of these markets are indicative of their high profitability. Worldwide in 2004 photovoltaic (PV) solar energy production rose to 1256 MWp,[*] a 67% increase from 2003 (Flamant et al. 2006). Photothermal technology has exceeded 430 MW (Morse 2008). The PV industry doubles in size every 2 years. The installed capacity of grid-connected PV grew over 120% during 2010. No industry has grown

---

[*] MW$_P$: Megawatts at peak conditions is the power when solar irradiance is 1000 W/m^2.

so much in so little time, not even the cell phone, computer, or electronics industries.

The potential for synergies between energy savings and solar energy is still untapped. Passive solar heating in combination with energy-efficient building construction and practices can reduce the demand for space heating by up to 30%. Active solar energy can further reduce the fuel demand for water heating by 50% to 70% and for space heating by 40% to 60%. Daylighting can reduce the electricity demand for lighting by up to 50% (Faninger 2010).

The Magnitude of Solar Resources

In its annual orbit around the sun, the Earth intercepts 7300 times more energy than consumed by the planet in 2010 (3,900,000 as against 530 EJ) (Rincón-Mejía 2010). The total world demand for primary energy in 2008 was 12,272 million tonnes oil equivalent (International Energy Agency [IEA] 2010), 10.51% of the solar energy received in one day, 0.03% of that received in one year. Solar energy could meet any need of the present and future generations without impacting the environment, if coupled with the mitigation of greenhouse gas emissions. Solar energy is the most abundant (Table 9.1) and widely distributed (Figure 9.1) of all renewable energies. Due to its high exergy (see Chapter 8), it is also most suitable for the mass production of electricity. But more direct uses (heating, cooking, catalyzing) generate less entropy and are even more efficient.

Paradoxically, affluent countries have more solar installed capacity than sunnier tropical areas. This points first to the advantages of solar energy

TABLE 9.1

Gross and Sustainable Renewable Energy Sources Potentials

Energy Source	Gross Potential (EJ/yr)	Sustainable Potential (EJ/yr)	Global 2010 Consumption (EJ/yr)
Solar	3,900,000	3900	7.36
Geothermal	600,000	5000	9.43
Wind	111,000	1110	2.09
Marine currents and waves	27,500	27.5	0.05
Biomass	5490	302	0.57
Hydropower	166	23.2	0.04
Total	4,644,156	10,362.7	19.55

Source: Rincón-Mejía EA, 2010, Tecnologías solares de cero emisiones de carbono. In Memorias del Simposio de contaminación atmosférica y tecnologías de cero emisiones de carbon, Mexico: El Colegio Nacional, April 13–15.

Note: 1 EJ = 10^{18} J. Generally speaking there are two renewable energy sources: solar and geothermal. Wind, sea currents, waves, and biomass derive from solar energy. Additionally, there is tidal energy, dependent on lunar attraction. Large hydropower dams are not included among the sustainable energy sources as they entail massive socioeconomic and environmental costs; small hydropower, however, is potentially sustainable.

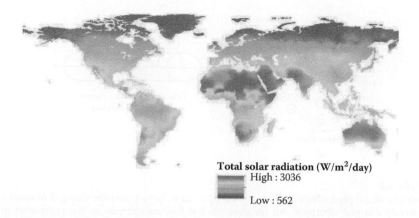

Total solar radiation (W/m²/day)
High : 3036

Low : 562

FIGURE 9.1 (See color insert.)
Average daily insolation estimated on the land surface (1961–1990). Median insolation is 1800 W/m²/day, in yellow. The insolation is strongly influenced by cloud cover in forested locations as well as by low solar elevation above the horizon at high latitudes. The intertropical zone has the highest insolation, especially in desert areas. Insolation is even higher on the larger ocean surface than on land. (Based on the Intergovernmental Panel on Climate Change [IPCC], 2005, Climate Research Unit [CRU] high resolution climate data, version 2.1. http://www.ipcc-data.org/observ/clim/cru_ts2_1.html.)

even at high latitudes and, second, to the oncoming transition as oil prices inevitably will rise in the near future due to increasing scarcity.

Photovoltaic Systems

A solar cell (or PV cell) is an electrical device that converts the energy of light directly into electricity by the photovoltaic effect. It is a form of photoelectric cell (in that its electrical characteristics—e.g., current, voltage, or resistance—vary when light is incident upon it) that when exposed to light can generate and support an electric current without being attached to any external voltage source (Kalogirou 2009). PV cells epitomize energy autonomy. They have been used in space vehicles, dwellings, and small and large generation systems interconnected to commercial power grids, among many others.

The photovoltaic effect was experimentally demonstrated by Becquerel in 1839, who built the world's first liquid state PV cell. In 1883, Fritts built the first solid state PV cell; he coated the semiconductor selenium with an extremely thin layer of gold to form the junctions. The device was only around 1% efficient. In 1888, Stoletov built the first photoelectric cell based on the outer photoelectric effect discovered by Hertz earlier in 1887 (Gevorkian 2007). The photoelectric effect was first observed by Hertz as metallic surfaces hit by ultraviolet light emitted charges. Successive observations by Hallwachs,

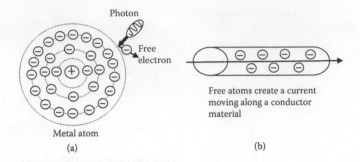

FIGURE 9.2
Schematic of the photoelectric effect. (a) A copper atom has a positively charged nucleus and negatively charged electrons. An incident photon hits an electron on the outermost orbit releasing it from the attraction of the atom. (b) An electric current is created when several free electrons move from the negative to the positive pole of an electric device bouncing between the outer layers of adjacent atoms.

Thomson, and Lenard paved the way for its final explanation by Einstein in 1905. This explanation was in terms of light quanta (particles), which were unheard of as light was thought of as waves. One such quantum, or photon, when sufficiently directed toward an electron on a metallic surface provides the electron with sufficient kinetic energy to free it from the energy attracting it to a proton in the metal atom (Serway and Jewett 2006). Photovoltaic applications were developed to channel free electrons so as to generate an electric current with high exergy, since incoming solar photons directly release a fraction of metal electrons to create an electric current (Figure 9.2).

Today, photovoltaic cells are made of composite materials, achieving conversion efficiencies exceeding 30%, with very competitive manufacturing costs and very broad application. This efficiency is an order of magnitude higher than the overall efficiency (3%) of fossil fuel energy sources taking into account the extraction, processing, distribution, and consumption (Ruiz 2005).

Novel Solar Technologies

PV technology, with 60 GW in operation and commercial plants in place since the 1980s, is at a turning point in development, having succeeded in matching the cost of conventional electrical power generation sources in a number of regions. With the first commercial plants in the United States, concentrated solar power (CSP) technology has been in use for over 30 years. Plants have improved over time and significant technological developments have led to cost reduction and higher efficiency. CSP is neither an experimental technology nor one undergoing testing, but rather a commercial solution that can be adapted to a variety of geographic locations.

Concentrated Solar Power

Among the various technologies available for converting solar energy into electricity, concentrated solar power (CSP) has the highest potential. These technologies use mirrors or lenses to concentrate solar radiation into thermal energy to heat a fluid and produce steam. The steam drives a turbine and generates power using thermodynamic cycles in the same way as conventional power plants. Other concepts are being explored and not all future CSP plants will necessarily use a steam cycle (International Renewable Energy Agency [IRENA] 2012). This is done reliably, manageably, and with the lowest cost of generation of all solar technologies (Quaschning and Blanco 2001; Becker et al. 2002). Manageability is the ability of a power plant to respond on demand during a specific period of time, that is, only when needed. CSP plants can be equipped with a heat storage system to generate electricity on overcast days or after sunset. This significantly increases the CSP capacity factor compared with PV and enables dispatchability, or the ability to adapt production to demand all day long, facilitating grid integration and economic competitiveness (IRENA 2012).

One of the first CSP plants was the Hispano-German GAST (Gas Cooled Solar Tower) at the Plataforma Solar de Almeria in Spain between 1981 and 1986, whose aim was to develop solar receivers for solar thermal power towers capable of supplying air at temperatures above 800°C and pressures in the vicinity of 9 bar. The low thermal efficiency of the prototypes prompted a shift from tubular receivers to volumetric receivers (Skocypec et al. 1989), that is, porous metal or ceramic media through which pressurized air flows. Said media are located inside a closed cavity wherein sunlight penetrates through a transparent quartz window. The receivers developed after completion of GAST displayed significant advances, represented by the Solgate prototype (European Commission 2005). The Solgate volumetric receiver concept tested in Almeria operated at 960°C with thermal efficiencies of 75 ± 6%. However, its design was complex (three receivers connected in series, with different geometries and materials) and so was the operation (under highly controlled conditions to ensure the physical integrity of the quartz windows at the entrance of the cavities of two receptors). Also, scalability was restricted by limitations in the manufacture of quartz windows of different sizes.

More recently, progress in the transformation of solar radiation in enthalpy[*] of a working fluid[†] and in material science (Osuna et al. 2000) were displayed by the first commercial solar thermal central towers. In June 2007, the Nevada Solar One plant, which reaches 75 MW peak and 64 MW nominal capacity, started operations. It covers an area of 121 hectares, with 184,000 parabolic mirrors.

---

[*] Enthalpy is a measure of the total energy of a thermodynamic system. It includes the internal energy, which is the energy required to create a system, and the amount of energy required to make room for it by displacing its environment and establishing its volume and pressure.
[†] In a thermodynamic system, the working fluid is a liquid or gas that absorbs or transmits energy.

This solar plant is located near Boulder City, Nevada, and it avoids the emission of 130,000 tons of CO_2 per year. Also, with 300 MW the San Lucar complex in Andalucía, Spain, will produce enough electricity to supply 180,000 homes, equivalent to the current demand in Seville. Its first plant was inaugurated on March 30, 2007. The complex will prevent the emission of 600,000 tons of CO_2 annually. Spain alone has over 50 solar plants planned, under construction, or already in operation (Richter et al. 2009), so their growth rate will be dramatic in the short term. These episodes are testament to the large-scale commercial CSP deployment worldwide.

These technologies have experienced surprising progress in the last decade (Figure 9.3), but there are still unresolved issues and room for improvement.

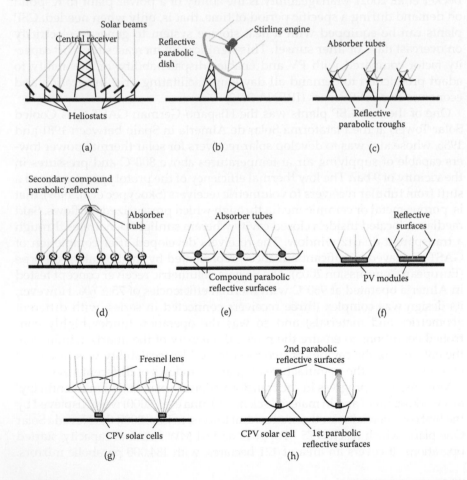

FIGURE 9.3 (See color insert.)
Solar power systems. (a) Central tower and heliostats. (b) Parabolic dish. (c) Parabolic trough. (d) Fresnel mirrors system. (e) Compound parabolic concentrator (CPC). (f) PV concentrators with plane mirrors. (g) Fresnel lenses system. (h) PV concentrators with plane mirrors and parabolic mirrors. (Modified from Purohit I, Purohit P, 2010, *Energy Policy* 38:3015–3029.)

Electricity production costs of solar thermal current concentration are still higher than those of conventional thermal power plants (IRENA 2012). To reduce these costs, it is necessary to improve the overall efficiency of conversion of solar energy into electricity via more efficient sunlight capture and conversion of thermal energy. Commercially available CSP relies on Rankine cycles, with overall yields hardly above 25%. But it is possible to obtain overall yields above 40% by designing a solar receiver to provide air at proper temperature and pressure for expansion in a gas turbine combined cycle. This requires a transformation from concentrated solar radiation into enthalpy of the working fluid with very high potential yields (>80%) and gains in efficiency in the upper (Brayton) cycle of a combined cycle (Figure 9.4).

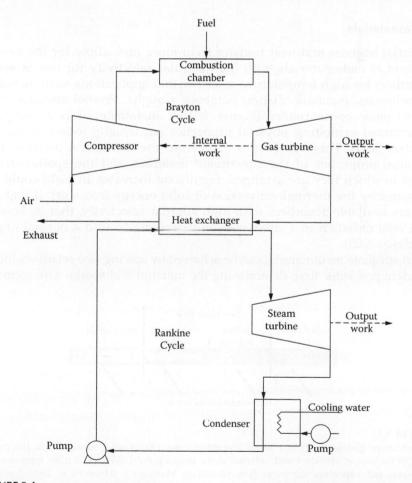

FIGURE 9.4
Combined cycle (Rankine and Brayton cycles). T and P increase in each cycle. (Modified from Çengel Y, Boles M, 2012, *Thermodynamics: An engineering approach,* New York: McGraw-Hill.)

Hybridizing Solar and Geothermal Energy

Direct steam generation (DSG) (Figure 9.5) in parabolic troughs in experiments and prototypes has demonstrated high feasibility and scalability to full-scale applications. This includes electricity generation in solar-geothermal hybrid plants as proposed by Almanza et al. (2002). Implementation steps have started to retrofit a preexisting 100 MWe geothermal electricity plant. From the underground fluid (60% brine, 40% vapor) only the energy from the vapor is used in the turbines. The parabolic troughs are intended to vaporize the brine thereby augmenting the amount of vapor and diminishing liquid waste. An additional 10% vapor will significantly augment the output of the turbines without increasing the salt concentration in the brine so much that they damage the absorbing tubes.

Metamaterials

Material sciences and heat transfer techniques now allow for the development of metamaterials with high angular selectivity for use as solar absorbers for high temperature solar thermal applications such as fields of heliostats, parabolic dishes, parabolic troughs, Fresnel systems, and photovoltaic concentrators (Figure 9.3). A metamaterial is a composite material exhibiting physical properties not usually found in nature (Walser 2001). These properties arise from the interaction between the physical properties of the constituent materials and the spatial structures in which they are arranged. Significant increases in yield could be obtained by the thermal conversion of solar energy into work if materials are available absorbers with high angular selectivity, that is, absorb and emit radiation in a small cone of directions around a normal angle (Badescu 2005).

An adequate metamaterial can be achieved by solving two relatively independent problems: first, determining the material and spatial arrangement

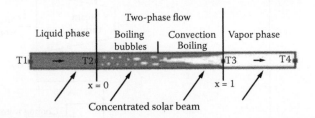

FIGURE 9.5
Direct steam generation (DSG) in an absorbing tube. Direct solar radiation on the outer wall of the tube is absorbed and radiated to the working fluid. Along the tube, temperature increases and vaporizes the water flow. (Based on Martinez I, Almanza R, 2009, Análisis teórico-experimental de flujo bifásico anular en tuberías horizontales para un sistema solar-geotérmico. In *Ingeniería de la Energía Solar III*, Almanza R (ed), Instituto de Ingeniería, UNAM. http://aplicaciones.iingen.unam.mx/ConsultasSPII/Buscarpublicacion.aspx.)

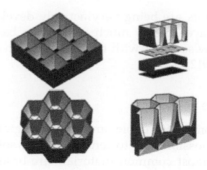

FIGURE 9.6
Examples of structures using angularly selective metamaterials.

for structures to achieve high angular selectivity (Figure 9.6), and, second, determining the optimal scale of said structures.

Design-wise, it is required to analyze and optimize structures with optoenergetic behavior as shown in Figure 9.6, in which the angular selectivity is to be achieved by the appropriate combination of geometry and material.

Ongoing developments will be implemented in the medium to long term because the design and optimization of these structures requires a combination of optical, thermal, and mechanical modeling of the processes occurring in these structures, both steady state and transient, when subjected to concentrated solar radiation. These processes are conductive, convective, radiative, and mechanical; take place in complex geometries; and combine surfaces and materials with very different properties. Moreover, the processes are coupled, an example of which is the variations in material properties with temperature. In addition, at small scales, simplifying assumptions of geometrical optics become invalid (Fu and Zhang 2006).

Storing Thermal and Electric Energy

Solar thermal power is dispatchable because it uses materials with great thermal inertia thus preventing sudden power generation lapses. Molten salt storage is a preferred solution. For example, a 280 MW plant located in Arizona will include 6 hours of molten salt storage capability.

As to electricity storage, although lithium-ion batteries have now captured the market of portable electronics and might make inroads into the hybrid electric vehicles and wind or solar energy markets, they need improved energy density. Lithium-air (Li-air) batteries, although still embryonic, have very high theoretical energy densities and can reach 13,000 Wh kg^{-1}, very close to gasoline (13,200 Wh kg^{-1}), one of the most energy-dense common liquids (Zhang et al. 2012) But as global Li reserves are quite limited, supercapacitors are being very actively developed to last longer, and take up and deliver energy much faster than batteries and fuel (up to 10 kW kg^{-1} in a few seconds, with full charge often obtained in around 1 second) (Lu et al. 2012).

Oxides and Polymers in PV Cells

PV cells have become affordable and technologically mature. There remain, however, possibilities to enhance the solar-electric conversion efficiency. The most common materials are based on silica (mono-, polycrystalline, and amorphous). An emerging alternative is dye sensitized solar cells (DSSCs), which promise high conversion efficiency, ease of production, low production costs, and the possibility of using a wide variety of semiconductors for their manufacture. They are made of thin semiconductor nanocrystal layers combined with dye layers that transfer charges. Titanium dioxide and zinc oxide are among the most promising oxides for the manufacture of photoanodes due to their good optoeletrical properties. Other alternatives for the fabrication of DSSCs are conductive polymers with good electrical properties, flexibility and mechanical resistance, and relative ease of manufacture. These include polypyrrole, poly (3, 4-ethylenedioxythiophene), and polyaniline (PANI) (Ameen et al. 2013).

Industrial Applications of Solar Thermal Energy

Process heat accounts for about 40% of the primary energy supply in the OECD. The major share of the energy needed by industries is below 250°C. This low temperature level can easily be reached using solar thermal collectors already on the market; in general, about 30% to 40% of the process heat demand could be covered with low to medium temperature solar collector systems (Faninger 2010).

Solar thermal technology is an especially good choice for industrial applications requiring process heat or steam power. Around the world there are already numerous facilities using heat or steam for industrial applications such as air-conditioning, heating, hot water, repowering conventional thermal power plants, enhanced oil recovery, cooking, and a multitude of other industrial processes. Users can lower their electricity bill while generating clean, pollution-free, and greenhouse gas-free energy on site, reducing actual consumption.

There is a growing need for industries to reduce their emissions and achieve greater energy independence in the face of increasing volatility in fossil fuel prices due to looming scarcity and a significant increase in energy demand from emergent economies. The latter, in addition, are increasing their CO_2 emissions, further diminishing the share of emissions that can come from any single economy.

The most important factor that sets concentrated solar power apart from other forms of renewable energy generation is its dispatchability, one of the most valuable features for electricity systems. Having the ability to adjust energy generation to the demand curve has the benefit of being able to sell electricity to the grid at peak generation hours, with a resulting increase in price, and the ability to compensate for the effects of intermittent sources. Both continuous operation and energy dependency are critical to some industries, and can be enhanced by combining CSP with conventional power plants. This hybridization lowers CO_2 emissions from hydrocarbon fuels or coal-fueled generating plants. For example, the 150 MW Hassi-R'Mel plant in Algeria has been operating since 2011 using this solution (Abengoa Solar 2013).

From an environmental point of view, the sustainability of a hybrid solar-combined cycle for a power generation system will essentially depend on the following:

- Reduction in natural gas consumption—This will also reduce green-house gas emissions into the atmosphere.

- Preservation and integration of biodiversity—The use of parabolic trough concentrators requires a large area, so it is important to locate the hub area while affecting as little regional flora and fauna as possible.

Another positive aspect is the local creation of highly qualified jobs, because during the construction and subsequent operation and maintenance over the course of the useful life of solar power plants, between four and five temporary jobs will be created for each megawatt during project execution and up to two permanent positions per megawatt during the operating period (Abengoa Solar 2013).

Sustainability of Solar Systems

The energy transition toward a global system based on renewable energy sources requires the synergistic implementation of all renewable sources in each region. This will compensate for the main limitation of solar and wind energy, which is its intermittency, and ensure capacity without large energy storage systems. To be sustainable, solar systems (and more generally renewable energy systems), need to be managed according to technical, environmental, and socioeconomic principles (Table 9.2). A system's sustainability

TABLE 9.2

Evidence of Sustainability in Solar Energy Systems

Systemic Attributes[a]	Operational Attributes	Evidence
Productivity—The system provides the required level of services over a long period	Efficient and synergistic use of natural and economic resources	Commercial CSP plants have been in use for over 30 years. PV technology now has 60 GW in operation and commercial plants in place since the 1980s (Abengoa Solar 2013).
Equity—The system delivers productivity in a fair manner	Fair distribution of costs and benefits to the different stakeholders, ensuring economic access and cultural acceptance of the system	Solar technologies in homes can lower the use of nonrenewable energy sources for space and water heating by a factor of three. In work environments, daylighting applications alone can reduce electricity use by 30% to 50% as well as improve worker safety and performance (Murphy 2012).
Reliability—The system can be maintained at about the usual equilibrium	Improves sustainability in the socioenvironmental setting of the system	Solar collectors for <120°C industry processes could reach an installed capacity of 3200 GWth (7.2 EJ) per year by 2050. Solar process heat will account for 20% of low temperature industrial heat by that time (IEA 2012).
Stability—Ownership of the system in a dynamic state of equilibrium	Responsible risk-sharing of externalities (pollution, accidents), use, restoration, and protection of resources. Degrowth of consumption embedded in the system	Most commercial and industrial processes are <250°C, a temperature low enough to reduce most combustion, explosion, and highly polluting emissions. This range is well suited for solar thermal technologies. Applications include food, textile, pharmaceutical and biochemical processes, desalination, and factory heating and cooling (Weiss and Mauthner 2012).
Resilience—Return to equilibrium or maintain productive potential after major disturbances	Access and availability of productive renewable resources; delivers at stable prices	CSP and PV are mature and commercially available, very cost-effective in areas with good solar insolation. Resilience is augmenting as prices dwindle with mass market deployment, and retrofitting gives way to solar systems integrated into building designs, optimizing energy use (Weiss and Mauthner 2012).
Adaptability and flexibility—New equilibria can be found in the face of long-term changes in the environment and active search for new levels of productivity	Adjusts to new socioeconomic and biophysical conditions through innovation, as well as learning and participative processes	If IEA's deployment scope in solar heating and cooling is achieved, solar systems can avoid 800 megatonnes (Mt) of CO_2 emissions per year by 2050 (IEA 2012). Major innovation roadmaps are in place in China and the United States (ESTTP 2008; IEA 2010).

TABLE 9.2 (*Continued*)

Evidence of Sustainability in Solar Energy Systems

Systemic Attributes[a]	Operational Attributes	Evidence
Self-reliance or self-management— The system endogenously defines goals and priorities, under regulation through interactions with organizational processes and socioenvironmental mechanisms	Responds to externally induced changes, maintaining its identity and values	Research and development priorities are low-cost materials and components, improved building integration and optimization, and affordable compact storage technologies (ESTTP 2008). PV is solid-state and epitomizes autonomy.

[a] From Masera O, Astier M, 1996, Metodología para la evaluación de sistemas de manejo incorporando indicadores de sustentabilidad, Mexico: Grupo Interdisciplinario de Tecnología Rural Aplicada.

depends on endogenous properties and structural linkages with other external systems.

As to exogenous properties, energy security is foremost. This is why diversification in the energy mix is a key element in energy policy. Solar energy lowers the dependency on oil, gas, and uranium in countries without these resources. Solar energy increases the proportion of renewable energy in the energy mix and, in the case of CSP, enhances the system reliability as the result of its dispatchability. Additionally, solar energy helps offset the volatility of fossil fuel prices.

Discussion

The transition toward a global system based on renewable energy sources, in which wind power will be dominant in the short term and solar power in the medium and long terms, has already begun. This trend is expected to continue in the coming years. But it will require a three-pronged course. First, the excessive consumption of energy will have to be reined in to reduce excessively large solar energy systems. Without reduced consumption, any additional energy source, renewable or otherwise, will likely result in a rebound effect augmenting energy consumption. Second, to further reduce energy waste and the size of the systems, specific tools are required to evaluate the efficiency of solar technology, in emergy terms (see Chapter 8), and within more integrative sustainability frameworks as set forth in this chapter. Third, to enhance the resilience of the global energy system and the autonomy of users, that is, its social sustainability, electric power generation must benefit all communities irrespective of their location, thus providing people with energy to increase productive capacities and self-management.

In the foregoing context, CSP faces the challenge of scaling down current systems for wider deployment and minimal maintenance. This seems all the more feasible as the CSP scientific and engineering community is multinational: hundreds of man-years accumulated in the cooperation between Spain, Germany, and the United States at the Plataforma Solar de Almeria have led to the incipient worldwide deployment of commercial CSP.

Skirting Roadblocks

The trend leading to renewable energies dominating fossil fuels clashes with timid or conservative discourses from multilateral organizations (e.g., the International Energy Agency and the International Atomic Energy Agency) and energy ministers from various countries. These discourses originate in think tanks and obey oil and nuclear energy vested interests; they are reminiscent of entrenched opposition to climate change evidence. At the core of the discourse is the claim that nonrenewable energy sources can prevail in the coming decades.

This is reflected in discourses from nondemocratic countries on sustainability, renewable energy, and environmental care, while their actual policies focus on the oil sector or nuclear programs without explicit renewable energy or energy transition programs (SENER 2013). In those countries, the solution seems to be a bottom-up energy transition that only decentralized energy systems can affect. Solar systems are prime examples of available technologies that can take households and communities into their own sustainability transition. As for democratic countries, they are still hesitant in implementing available technology and deployment roadmaps (European Solar Thermal Technology Platform 2008; IEA 2010, 2012). However, Germany and Japan, two technology powerhouses, are now taking the lead and making headway in their energy transitions, spurred by market and public opinion, respectively.

The foregoing bottom-up-driven transitions will have to wait for the acknowledgment that nonrenewable energy resources are finite. At the moment, however, the IEA has only recognized the need to curb climate change and has attributed it to fossil fuels (IEA 2010). Fossil fuels remain the dominant energy sources in 2035 in all IEA scenarios, though their share of the overall primary fuel mix varies markedly, from 62% in the 450 Scenario to 79% in the Current Policies Scenario. In the Current Policies Scenario, energy intensity continues to gradually decline over the projection period but at a much slower rate than in the other scenarios. Sustainability-wise, the current IEA stance is way above current estimates of maximum tolerable CO_2 atmospheric concentrations. Apart from ecological and socio-economic vaticinated consequences, this stance is conducive to a protracted hence costlier transition, carried out in an energy-scarcer environment. This may eventually lead to the technological dominance of some wealthier pioneering nations and a technological dependence in the energy sectors for laggard countries.

References

Abengoa Solar (2013) 2012 Annual Report Summary. http://www.abengoasolar.com/web/en/acerca_de_nosotros/informe_anual/index.html (accessed June 30, 2013).

Almanza R, Jiménez G, Lentz A, Valdés A, Soria A (2002) DSG under two-phase and stratified flow in a steel receiver of a parabolic trough collector. *Journal of Solar Energy Engineering* 124:140–144.

Ameen S, Shaheer M, Song M, Shik H (2013) Metal oxide nanomaterials, conducting polymers and their nanocomposites for solar energy. In *Solar cells, research and application perspectives*, Morales A (ed), http://www.intechopen.com/books/solar-cells-research-and-application-perspectives (accessed April 27, 2013).

Badescu V (2005) Spectrally and angularly selective photothermal and photovoltaic converters under one-sun illumination. *Journal of Physics D: Applied Physics* 38:2166.

Becker M, Meinecke W, Geyer M, Trieb F, Blanco M, Romero M, Ferriére A (2002) Solar thermal power plants. In *The future for renewable energy 2. Prospects and directions*, EUREC Agency (ed), 115–137. London: James & James.

Çengel Y, Boles M (2012) *Thermodynamics: An engineering approach*. New York: McGraw-Hill.

European Commission (2005) SOLGATE: Solar hybrid gas turbine electric power system, Final Publishable Report. European Communities. Luxembourg. http://ec.europa.eu/research/energy/pdf/solgate_en.pdf (accessed August 23, 2012).

European Solar Thermal Technology Platform (ESTTP) (2008) Solar heating and cooling for a sustainable energy future in Europe: Vision, potential, deployment roadmap, and strategic research agenda. http://esttp.org/cms/upload/SRA/ESTTP_SRA_RevisedVersion.pdf (accessed October 13, 2012).

Faninger G (2010) The potential of solar thermal technologies in a sustainable energy future. Results from 32 years of internacional R&D co-operation. Solar Heating and Cooling Programme. Brussels: International Energy Agency.

Flamant G, Kurtcuoglu V, Murray J, Steinfeld A (2006) Purification of metallurgical grade silicon by a solar process. *Solar Energy Materials and Solar Cells* 90:2099–2106.

Fu CJ, Zhang ZM (2006) Nanoscale radiation heat transfer for silicon at different doping levels. *International Journal of Heat and Mass Transfer* 49:1703–1718.

Gevorkian P (2007) *Solar power in building design*. New York: McGraw-Hill.

Hang Q, Jun Z, Xiao Y, Junkui C (2008) Prospect of concentrating solar power in China. The sustainable future. *Renewable and Sustainable Energy Reviews* 12:2505–2514.

Intergovernmental Panel on Climate Change (IPCC) (2005) Climate Research Unit (CRU) high resolution climate data, version 2.1. http://www.ipcc-data.org/observ/clim/cru_ts2_1.html (accessed December 14, 2012).

International Energy Agency (IEA) (2010) World Energy Outlook 2010. http://www.worldenergyoutlook.org/ (accessed May 19, 2012).

International Energy Agency (IEA) (2012) CO_2 emissions from fuel combustion, documentation for beyond 2020. http://data.iea.org/ieastore/product.asp?dept_id=101&pf_id=305&mscssid=XA455HQXMGJA9HTRQ9NFBGQDUFSMF3N4 (accessed March 15, 2012).

International Renewable Energy Agency (IRENA) (2012) Concentrating solar power. Renewable energy technologies: Cost Analysis Series, Vol. 1: Power Sector, Issue 2/5, Bonn: IRENA Innovation and Technology Centre. www.irena.org/Publications (accessed November 24, 2012).

Kalogirou S (2009) *Solar energy engineering: Processes and systems*. Boston: Academic Press.

Lu Y (2010) Science and technology in China: A roadmap to 2050. Springer and Chinese Academy of Sciences. http://www.bps.cas.cn/ztzl/cx2050/nrfb/201008/t20100804_2917262.html (accessed June 12, 2012).

Lu P, Xue D, Yang H, Liu Y (2012) Supercapacitor and nanoscale research towards electrochemical energy storage. *International Journal of Smart and Nano Materials* 4:2–26.

Martinez I, Almanza R (2009) Análisis teórico-experimental de flujo bifásico anular en tuberías horizontales para un sistema solar-geotérmico. In Ingeniería de la Energía Solar III. Almanza R (ed), Instituto de Ingeniería, UNAM. http://aplicaciones.iingen.unam.mx/ConsultasSPII/Buscarpublicacion.aspx (accessed January 14, 2013).

Masera O, Astier M (1996) Metodología para la evaluación de sistemas de manejo incorporando indicadores de sustentabilidad. Mexico: Grupo Interdisciplinario de Tecnología Rural Aplicada.

Morse F (2008) Concentrating solar power (CSP) as an option to replace coal. In Energy and Climate Mini-Workshop, Washington DC. May 14, 2008. http://ossfoundation.us/projects/energy/2008-overview/h_FredMorse_Solar_Thermal_approvedversion.pdf (accessed June 12, 2013).

Murphy P (2012) IEA Solar Heating and Cooling Programme, 2011 Annual Report. http://www.iea-shc.org (accessed September 16, 2012).

Osuna R, Fernández V, Romero M, Blanco M (2000) PS10: A 10 MW solar tower plant for Southern Spain. 10th International Symposium on Solar Thermal Concentrating Technologies. Sydney, Australia. March 8–10.

Patlitzianas K, Kagiannas A, Askounis D, Psarras J (2005) The policy perspective for RES development in the new member states of the EU. *Renewable Energy* 30:477–492.

Pitz-Pall R, Dersch J, Milow B (eds) (2005) Deliverable No. 7. Road Map Document. European Concentrated Solar Power Road-Mapping. ECOSTAR. SES6-CT-2003-502578. European Commission. February 16.

Purohit I, Purohit P (2010) Techno-economic evaluation of concentrating solar power generation in India. *Energy Policy* 38:3015–3029.

Quaschning V, Blanco M (2001) Solar power: Photovoltaic or solar thermal power? VGB Congress Power Plants 2001. Brussels. October 10–12.

Richter C, Teske S, Short R (2009) Concentrating solar power, Global outlook 2009. Amsterdam, Tabernas, Brussels: Greenpeace International, SolarPACES, ESTELA.

Rincón-Mejía EA (2010) Tecnologías solares de cero emisiones de carbono. In Memorias del Simposio de contaminación atmosférica y tecnologías de cero emisiones de carbon. Mexico: El Colegio Nacional. April 13–15.

Ruiz V (2005) Sistemas energéticos del futuro. Conferencia magistral en la XXIX Semana nacional de energía solar. Tuxtla Gutierrez: ANES. October 6, 2005.

SENER (2013) Estrategia nacional de energía [National energy strategy] 2013–2027. Mexico: Secretaría de Energía.

Serway R, Jewett J (2006) *Principles of physics.* Toronto: Thomson Learning.

Skocypec RD, Boehm RF, Chavez JM (1989) Heat transfer modeling of the IEA/SSPS volumetric receiver. *Transactions of the ASME* 111:138–143.

Walser RM (2001) Electromagnetic metamaterials. *Proceedings of the SPIE* 4467:1–15.

Weiss W, Mauthner F (2012) Solar heat worldwide, markets and contribution to the energy supply 2010. Graz: IEA. http://www.iea-shc.org/annual-reports (accessed July 22, 2012).

Zhang L-L, Wang Z-L, Xu D, Zhang X-B, Wang L-M (2012) The development and challenges of rechargeable non-aqueous lithium – air batteries. *International Journal of Smart and Nano Materials,* 4:27–46. DOI:10.1080/19475411.2012.659227.

Total precipitation (mm)
High: 2353
Low: 0

FIGURE 1.1
The main climatic limits for life on Earth's continental masses in the early Anthropocene. Water is delivered through precipitation (rainfall, snow, hail, fog, and dew) with very large differences: dry and cold deserts are defined by unpredictable rainfall (rain can go unrecorded there for years). Polar areas and continental areas (far from the sea) receive 80 to 150 mm per year. Meanwhile, temperate coastal areas receive 300 to 1500 mm and tropical forests might receive 3000 mm. Not surprisingly, in the early Anthropocene (the era dominated by mankind), human life is mostly concentrated in areas well endowed with solar energy and water (the Tropics) where crops can be harvested at low costs. (Data from Intergovernmental Panel on Climate Change [IPCC], 2005, Climate Research Unit [CRU] high resolution climate data, version 2.1, http://www.ipcc-data.org/observ/clim/cru_ts2_1.html.)

Months in a year with scarcity > 100%

0-1
2-4
5-9
10-12

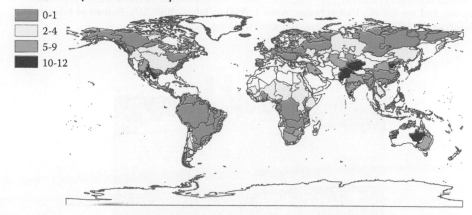

FIGURE 1.2
Blue water scarcity in the most important drainage systems of the world. Blue water is surface water and groundwater. Scarcity in the majority of the basins responds to human appropriation. Biological and chemical pollution come from domestic, agricultural, and industrial uses. Lack of wastewater treatment makes water a single-use commodity driving extraction and blue water scarcity upward. (Data from Hoekstra AY et al., 2012, *PLoS ONE* 7: e32688.)

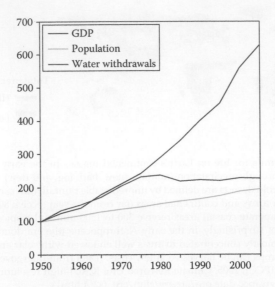

FIGURE 1.4

Peak water and decoupling of economic output, population, and water withdrawals after 1975 in the United States (1950 = 100). Water withdrawals stabilized their increase thereafter independently from socioeconomic factors. The explanation may be that maximum water extraction capacity has been reached either in technical or ecological terms, as suggested by Gleick and Palaniappan (2010). This runs counter to the more optimistic decoupling interpretation whereby GDP could grow without exhausting water resources. This decoupling also means that reducing water extraction may not jeopardize the evolution of well-being. GDP in U.S. dollars, billions; water withdrawals in billion gallons per day. (Data from Kenny JF et al., 2009, Estimated use of water in the United States in 2005, U.S. Geological Survey Circular 1344; U.S. Census Bureau, Population Division, 2011, Intercensal estimates of the resident population by sex and age for the United States: April 1, 2000 to July 1, 2010; U.S. Bureau of Economic Analysis, 2013, U.S. economic accounts, http://www.bea.gov/.)

FIGURE 2.1

The aerobic stage of an activated sludge wastewater treatment process. (© Photo by Margaret Wexler.)

FIGURE 2.2
Eutrophication of a lake as a result of sewage contamination. (© Photo by Jeremy Burgess.)

(a)

(b)

(c)

(d)

FIGURE 3.14
An artificial wetland in Mexico. (a) Sieving operation (foreground) and tanks allowing for flow regulation. (b) The reed plantation is irrigated from below ground; the high evapotranspiration rates of the reed help keep up with the wastewater flow of 2000 people. (c) Vents of the slotted underground PVC pipes allow for unclogging them with pressurized air. (d) Effluent water can be further treated and used in other ponds. (© Photo by Marina Islas-Espinoza.)

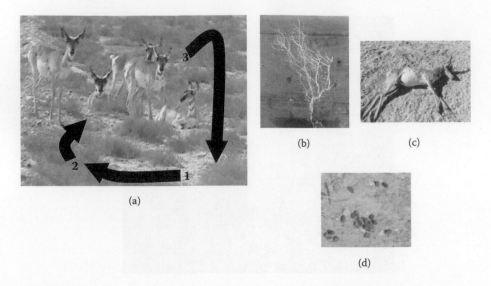

(a)

(b)

(c)

(d)

FIGURE 4.1
(a) Nutrient cycling in the pronghorn (*Antilocapra americana*) habitat. 1: Water and nutrients (e.g., phosphate) in the soil are taken up by plant roots. 2: Plants are eaten by pronghorns. 3: Organic and inorganic compounds are returned to the soil. Plant and animal necromass (b, c) as well as feces (d) are decomposed and mineralized in the soil. Plants take up these minerals and the cycle is closed. (© Photos by Marina Islas-Espinoza.)

FIGURE 4.2
Soil biological crust growing after a fire. The major components are cyanobacteria, green algae, fungi, mosses, liverworts, and lichens. (© Photo by Marina Islas-Espinoza.)

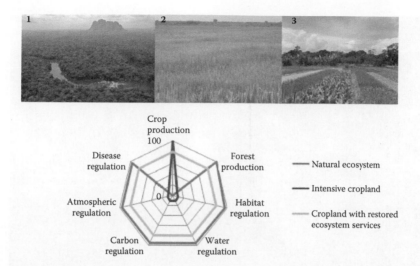

FIGURE 5.3
A conceptual framework for comparing land use and tradeoffs of ecosystem services in three hypothetical landscapes: (1) a natural ecosystem, (2) an intensively managed cropland, and (3) a cropland with restored ecosystem services. The provisioning of multiple ecosystem services under different land-use regimes is presented. The condition of each ecosystem service is demonstrated along each axis. The natural ecosystem provides many ecosystem services but not food production. The intensively managed cropland produces food (maybe in the short term) at the cost of diminishing other ecosystem services. A cropland that is managed to maintain other ecosystem services (cropland with restored ecosystem services) may provide a broader portfolio of ecosystem services. (Figure modified from Foley et al., 2005, *Science* 309:570. Photos 1, 2, and 3 courtesy of Trond Larsen©/Conservation International, Bernardo Strassburg/International Institute for Sustainability, Maureen Silos/the Caribbean Institute, respectively.)

FIGURE 5.6
Natural, unconverted habitats are home to a variety of species. Losing these habitats often leads to species loss, following a basic ecological relationship known as the species–area curve. The cheetah is an example of a threatened species according to the Red List compiled by the International Union for Conservation of Nature (IUCN). (Photo courtesy of Bernardo Strassburg, International Institute for Sustainability.)

FIGURE 6.3
"Safe food" market in Suriname. (Photo courtesy of Maureen Silos, the Caribbean Institute.)

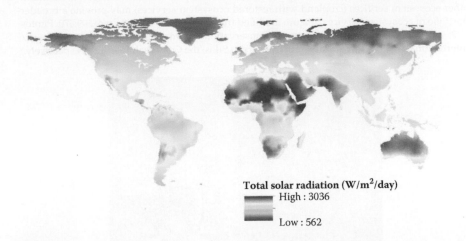

Total solar radiation (W/m²/day)
High : 3036

Low : 562

FIGURE 9.1
Average daily insolation estimated on the land surface (1961–1990). Median insolation is 1800 W/m²/day, in yellow. The insolation is strongly influenced by cloud cover in forested locations as well as by low solar elevation above the horizon at high latitudes. The intertropical zone has the highest insolation, especially in desert areas. Insolation is even higher on the larger ocean surface than on land. (Based on Intergovernmental Panel on Climate Change [IPCC], 2005, Climate Research Unit [CRU] high resolution climate data, version 2.1. http://www.ipcc-data.org/observ/clim/cru_ts2_1.html.)

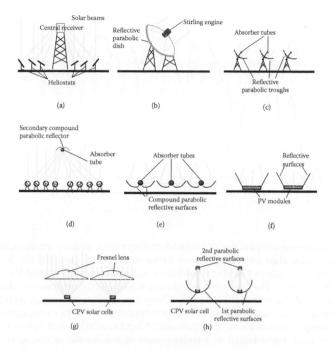

FIGURE 9.3
Solar power systems. (a) Central tower and heliostats. (b) Parabolic dish. (c) Parabolic trough.
(d) Fresnel mirrors system. (e) Compound parabolic concentrator (CPC). (f) PV concentrators
with plane mirrors. (g) Fresnel lenses system. (h) PV concentrators with plane mirrors and
parabolic mirrors. (Modified from Purohit I, Purohit P, 2010, *Energy Policy* 38:3015–3029.)

FIGURE 10.1
The possibility of methane-propelled spaceships. (a) The Morpheus lander take-off test.
(Reproduced from the National Aeronautics and Space Administration [NASA], 2012a,
Morpheus testing picks back up at Johnson Space Center. http://www.nasa.gov/centers/john-
son/exploration/morpheus/morpheus_tests_pickup.html.) (b) Methane release concentrations
in the northern summer on Mars. (Photo by Trent Schindler. Reproduced from NASA 2012b,
Mars: NASA explores the red planet. http://www.nasa.gov/mission_pages/mars/news/mars-
methane_media.html.) Other planets with abundant methane are Jupiter, Saturn (Atreya 2007),
Uranus, and Neptune, plus Titan, the moon of Saturn.

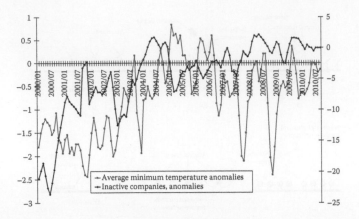

FIGURE 11.1
The relationship between temperature fluctuation and economic activity (construction industry). Temperature anomalies are the difference between observed data and the 2000 to 2010 average temperature in Toluca, Mexico, a midsized industrial city connected through train freight to the U.S. economy. The construction industry is a large economic sector whose behavior correlates well with the rest of the economy. Overall in the last decade, an increase in the monthly number of companies that went out of business coincided with a temporary decrease in the minimal temperature, without any pronounced lag time. In the first half of the decade, a temperature increase trend ended as a consequence of a domestic recession around 2006 followed by the economic slowdown in the United States and most of the world. (Data from INEGI, 2010, Banco de Información Económica, http://www.inegi.org.mx/Sistemas/BIE/Default.aspx; and SMN, 2010, Estaciones automáticas: Toluca. http://smn.cna.gob.mx/emas/.)

FIGURE 12.3
A vortex creates an ascendant flow to suspend the particles imitating wind. Then the suspended particles are trapped in the drops imitating rain, and absorbed in a filter that imitates deposition and soil filtration. (© Photo by Marina Islas-Espinoza.)

10

Bioenergy Solutions

Marina Islas-Espinoza and Bernd Weber

CONTENTS

Energy

Generally, energy is defined as the capacity to do work, and it is basic to understand life. Life is a state of activity controlled genetically and driven by energy to maintain and reproduce its cellular organization (Mendoza and Mendoza 2011). What is surprising is that all known living organisms share the same energy and cell synthesis processes.

The autotrophs produce their own chemical energy sources from solar energy as well as water and CO_2, whereas the heterotrophs consume organic matter derived from other organisms to obtain energy. Since most autotrophs obtain their energy from the sun, the solar energy is the main source of life.

This chapter deals first with the basic energy processes in living organisms (bioenergetics) and second with the most sustainable technological applications of bioenergy (also called biofuels), especially the production of biogas, which has shown to be one of the most abundant, viable energy sources now that fossil fuels have become scarce, expensive, and environmentally unsustainable.

Entropy and Life

Living beings are open systems that interchange matter and energy with the environment (Mendoza and Mendoza 2011). The flow of energy through an ecosystem is governed by two laws. The first law states that energy may not be created or destroyed, the second that no energy transformation is 100% efficient (Spellman and Drinan 2001). Energy flows from autotrophs to heterotrophs; between each of these trophic levels there is a loss of energy as heat (called entropy). Eventually, all energy flowing through the trophic levels is dissipated as heat losing its capacity to do work (Margalef 1998) be it chemical, osmotic, thermal, or mechanical. This state in which a system is unable to use energy is called thermodynamic equilibrium and is attained after one organism dies. However, living beings are highly organized systems, able to pass their genetic information to their progeny. This is one means whereby life challenges the second law and escapes thermodynamic equilibrium. The other means is through recycling dead organisms, a role assigned to degraders. Degraders transform dead cells into simple molecules, used as energy sources and other nutrients by autotrophs.

Every day we lose energy and our technologies do not replace it. Even recycling uses energy that comes automatically with an increase in entropy. Therefore, it is critical that we find ways to reduce energy loss (Schmitz 2007). One such way is to recycle waste using biological systems. In biogas systems as opposed to fossil fuels, living beings continuously reorganize the matter and energy mineralizing organic compounds making them available in the form of nutrients to new life and as a byproduct they make a fuel available to humans. Also, if solid waste is not transformed into fertilizer and biogas, humans have to resort to fossil fuels to obtain these products and this has to be done at a high energy cost, which includes lengthy prospection, long-haul transport, and, with some frequency, oil spill remediation (Adler et al. 2007). Producing biogas and fertilizer from solid waste leads to less entropy than an equivalent production using fossil fuels. Finally, nonbiological systems use energy and do not reproduce themselves, and so energy is lost replacing them. These systems are not able to recycle themselves as are biological systems.

Bioenergetics

Overall, the main functions in living beings in which energy transformations are carried out are movement, nutrient transport, and synthesis of new molecules. However, it must be emphasized that in every energy conversion one part is converted into heat.

The capture and use of energy in living systems includes two processes: photosynthesis and respiration. Photosynthetic organisms (autotrophs) capture the sunlight, water (H_2O), and carbon dioxide (CO_2) to form glucose and O_2 that is released into the atmosphere. Heterotrophic organisms get the energy from photosynthetic and other organisms (as food) releasing CO_2 to the atmosphere. The oxidation of glucose by organisms is called respiration and it can be aerobic (with oxygen) or anaerobic (without oxygen). Respiration converts the chemical bonds' energy of glucose to the usable energy of adenosine triphosphate (ATP) required for metabolic processes: (1) catabolism (degradation of chemical compounds to withdraw nutrients and release energy) and (2) anabolism (synthesis of new molecules).

Bioenergy

Biomass refers to material of biological origin (excluding material embedded in geological formations) and therefore is renewable. Bioenergy is the use of biomass to produce energy (O'Connell et al. 2009). When the energy stored in biomass is used, greenhouse gases such as CO_2 or methane (CH_4) are emitted, but the amount is the same as that produced by natural decomposition processes if the human rate of their use does not overshoot the natural rate. In contrast to the direct use of solar energy or wind power that is not always available, biomass can be produced and stored at all times, and be converted to electricity, heat, or fuel (German Solar Energy Society [DGS] and Ecofys 2005).

Biofuels today only supply 12.7% of the world's total primary energy consumption; 7% correspond to woodfuels. However, the global bioenergy potential could equal the current world energy consumption by 2050 (Food and Agriculture Organization of the United Nations [FAO] 2002; International Energy Agency [IEA] 2012).

Sustainable bioenergy follows the Earth's biogeochemical cycles to obtain energy and minimize the accumulation of waste in the environment; this is done by recycling domestic and industrial wastes. Bioenergy life-cycle analyses comprise production, harvesting, storage, transportation, distribution, greenhouse gas emissions, social and ecological impacts, pollutant emissions and their remediation, as well as comparative economic costs. Considering these elements, bioethanol from corn and sugarcane, or biodiesel from palm oil or oilseeds have been criticized due to their competition for land and water with food production and their lack of waste recycling. This is why

TABLE 10.1

Currently Available Biofuels

Source	Advantages	Disadvantages
Biomethane	It can be obtained from organic waste under anaerobic conditions. It can be used directly after the anaerobic process, separating it from other gases. CO_2 emissions are minimal compared with other biological processes.	Exerts 23 times the greenhouse effect of CO_2. High emissions when digesters are not sealed. It is abundant in the Earth atmosphere; however, there is no technology to store it yet.
Biodiesel	Algae, bacteria, or cyanobacteria may be used instead of oil grains.	Water demand. Conversion of natural areas to energy crops. Soil acidification, fuel and fertilizer use, biodiversity loss, toxicity from biocides.
Bioethanol	Cellulose from seaweed, switchgrass, jatropha, cyanobacteria, and green algae may be used instead of food crops.	Water use demand. Conversion of natural areas to energy crops. Soil acidification, fuel and fertilizer use, biodiversity loss, toxicity from biocides.
Woodfuels	Direct use of woodfuel for heat. Charcoal, pellets, or briquettes may be produced.	Production of greenhouse gases by combustion (CO_2 and H_2O). Generation of aerosol particulates, black carbon, or soot that increase irradiation sorption. Continuous reforestation is needed to close the cycle.

the generation of biogas from organic waste remains the most sustainable option until now (Table 10.1). Therefore, biogas is described with more detail in the following sections.

Biomethane: As Old as Life on Earth

Methane exists in the solar system and beyond, where it is abiotically produced. It is the simplest hydrocarbon, the main component of biogas and fossil natural gas, and probably the most abundant organic compound on Earth (Petrescu-Mag et al. 2011), where it is biogenically produced by the degradation of organic matter without oxygen. Methane generation on Earth involves different groups of organisms, mainly bacteria and methanogenic archaea. The latter seem to have metabolisms adapted to the early conditions of Earth. Methanogenic organisms are found in places where oxygen (O_2) is absent, and hydrogen (H_2) and carbon dioxide (CO_2) are present. Methanogens use mainly H_2 as an energy source and CO_2 as a carbon source for growth. Methanogens are found in marine and freshwater sediments, hot springs, swamps, wastewater treatment plants, digestive tracts of termites, ruminants, and humans. This shows that despite their intolerance to oxygen, the methanogens are still widely distributed on Earth.

Biogas Applications

Compared to other hydrocarbon molecules that have more carbon atoms, burning methane produces less CO_2 for each unit of heat released. The ratio of methane's heat of combustion (891 kJ/mol) to its molecular mass (16.0 g/mol, of which 12.0 g/mol is carbon) is 55.7 kJ/g, which is more heat per mass unit than more complex hydrocarbons.

It is estimated that a landfill with 5.6 million tons of solid waste can produce enough biogas to power a 5 MW plant for 10 years. The digesters used in the landfills must be airtight to avoid atmospheric emissions, which can be higher than 65% of the landfill-generated biogas in the United States (Themelis and Ulloa 2007). This means that all organic solid waste should be anaerobically digested rather than disposed of to a landfill, on pollution and energetic grounds.

In many cities, methane is already being piped into homes for domestic heating, cooling, and cooking purposes. There are also countries with sufficient volumes of methane to be used industrially as fuel (DGS and Ecofys 2005).

Land and Maritime Transportation

Methane in the form of compressed natural gas is used as a vehicle fuel and is more environmentally friendly than other fossil fuels such as gasoline/petrol and diesel. Some cities in European countries have started to distribute biomethane in gasoline stations as a solution for green transportation. In Sweden and Denmark, organic waste is imported from other countries to generate biogas, which is used for urban transport (Røstad 2012). In several countries liquid natural gas (LNG) has been introduced in railways, passenger cars, and vans. Since 1985 several ferries around the world have been powered by more than 70% natural gas and diesel. Japanese companies are investing in Mexico to generate biogas from organic waste to produce cars (EcoSecurities International 2009).

Data Centers

Several projects are underway to power data centers with biogas, in an attempt to curb Internet downtime episodes. Microsoft's Wyoming facility will rely on a system of pipes extracting methane from wastewater, cleaning it, and moving it into fuel cells. The fuel cell will chemically turn the biogas into electricity (LaMonica 2012). Apple's North Carolina data center will tap landfills for biogas, which will then be converted into electricity also using fuel cells (Kerr 2012).

Space Travel

NASA is working on the use of methane as fuel for rockets (Morring 2009). The assembly of a liquid oxygen/liquid methane rocket engine has been completed (Figure 10.1a). One advantage of methane is that it is abundant in many parts of the solar system (Figure 10.1b) and this may be collected on the surface of another planetary body providing fuel for a return trip (Lorentz and Mitton 2002).

Methane Production

Methanogenesis

Archaea generate CH_4 to obtain metabolic energy in the form of ATP and molecules for biosynthesis (Thauer and Shima 2006). It is thought that the biological conversion of solid wastes under anaerobic conditions follows four steps (Figure 10.2). The first step involves enzyme-mediated hydrolysis of high molecular mass compounds to make them transferable into cells for use

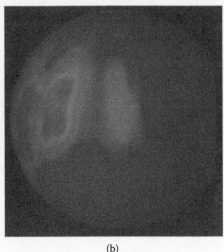

(a) (b)

FIGURE 10.1 (See color insert.)
The possibility of methane-propelled spaceships. (a) The Morpheus lander take-off test. (Reproduced from the National Aeronautics and Space Administration [NASA], 2012a, Morpheus testing picks back up at Johnson Space Center. http://www.nasa.gov/centers/johnson/exploration/morpheus/morpheus_tests_pickup.html.) (b) Methane release concentrations in the northern summer on Mars. (Photo by Trent Schindler. Reproduced from NASA 2012b, Mars: NASA explores the red planet. http://www.nasa.gov/mission_pages/mars/news/marsmethane_media.html.) Other planets with abundant methane are Jupiter, Saturn (Atreya 2007), Uranus, and Neptune, plus Titan, the moon of Saturn.

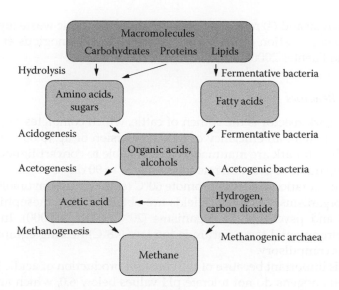

FIGURE 10.2
Methanogenesis and its different stages.

TABLE 10.2
Biogas Composition

Compound	Percentage
CH$_4$	50%–80%
CO$_2$	20%–50%
H$_2$	<1%
NH$_3$	<1%
H$_2$S	<1%
N$_2$	<1%

Source: German Solar Energy Society (DGS) and Ecofys, 2005, *Planning and installing bioenergy systems: A guide for installers, architects and engineers*, London: James & James.

as an energy source. The second step involves bacterial conversion to lower molecular mass sugars, amino and fatty acids. In the third step, compounds with a methyl group (–CH$_3$) are produced. Finally, methanogens produce biogas whose main components are CH$_4$ and CO$_2$ (Table 10.2). Almost two-thirds of the methane produced in manmade digesters are produced this way.

The benefits of anaerobic digestion include: (1) biogas to generate electricity, heat, or engine fuel; (2) organic fertilizer with a mineral content similar to fresh excreta and equally useful for soils, crops, and plankton; (3) destruction of pathogenic microorganisms, parasite eggs, and weed seeds contained in fresh excreta; and (4) reduction of pollution and organic waste (up to 50%) during the degradation of the organic material (Tchobanoglous et al. 1994; Masera and Fuentes 2006).

Anaerobic Reactors

The benchmark reactor is the rumen of cattle, which completes digestion in only 24 hours with an efficiency of food conversion of up to 80%. Far away from that benchmark are manmade digesters able to convert lignocellulose, with a comparable conversion rate following 30 to 90 days.

The main operation task is to promote 60°C temperatures favorable to thermophilic organisms. Decreased yields are obtained from mesophilic (36°C optimum) and psychrophilic organisms (20°C) (Ortega 2000). In regions where environmental temperature is too low (<8°C) heating equipment for digesters is compulsory.

The pH is important because of the constant production of acidic intermediates; methanogens do not tolerate pH values below 6.0, which are linked to excess substrate (feed) input. Observed organic dry matter loading rates for different substrates are shown in Table 10.3. Sometimes hydrolysis and acidification are carried out in a first reactor while in a second reactor acetogenic and methanogenic activities prevail. As in the first reactor the pH can decrease below 5, its output needs to be mixed thoroughly with the content of the second digester to neutralize the pH. The operation of this kind of digesters usually is more stable.

When complete mineralization (i.e., digestion) of organic matter occurs, the quantity and composition of the resulting biogas tends to follow the Symons and Buswell relation (1933): 0.79 m^3 biogas kg^{-1} substrate and 50% methane concentration for the conversion of carbohydrates; 1.25 m^3 kg^{-1} and 68% methane concentration for the conversion of lipids; and approximately 0.7 m^3 kg^{-1} and 71% methane concentration for the conversion of proteins. Recalcitrant organic matter lowers the biogas yield compared to the theoretical value from Symons and Buswell. Biogas and CH_4 experimental yields from various substrates are a useful reference to assess the performance of an anaerobic reactor (Table 10.3).

Compared to other biofuels where energy-intensive separation is needed, like extraction to produce biodiesel or distillation to produce anhydrous ethanol, the biogas bubbles from the liquid phase of the anaerobic digestion (also called wet biogas), and are in some cases ready to use. Alternatively, adaptations to equipment for natural gas now allow the use of biogas with methane concentrations around 60%. This makes on-site utilization for cooking, hot water, and electricity production possible. Also, energy-efficient conversion of biogas from wastewater treatment plants using fuel cells has been demonstrated in various plants working successfully over many years (Kishinevsky and Zelingher 2003).

TABLE 10.3

Biogas Conversion Rates from Various Substrates

Substrate	DM (% Wet Matter)	oDM (% Dry Matter)	Biogas (m^3/kg oDM)	Methane (% m^3 CH_4/m^3_{Biogas})
Cattle manure on wet basis	7.5–13	6.4–10	0.17–0.63 (0.38)	53–62 (55)
Pig manure on wet basis	2.3–11	1.3–7.1	0.–0.88 (0.42)	47–68 (60)
Broiler manure	25	22	0.15–0.53 (0.50)	42–68 (55)
Wheat straw			0.25–0.4 (0.4)	52
Corn straw			0.5	
Grass silage	27–57	25–46	0.21–0.7 (0.60)	52–56
Corn silage	25–37	24–36	0.3–1.13 (0.65)	47–69
Sudan grass	33–46	14–36	0.33–0.38 (0.33)	54–62 (63)
Grease traps residues	48	85	0.7–1.3	60–77
Digestive tract viscera	11–19	9–16	0.2–0.4	60
Rumen contents			0.45–0.55 (0.48)	44–55 (50)
Glycerin	100		(0.85)	50
Ethanol production wastes (wheat)	4.6–76	4.3–71	0.39–0.72 (0.70)	53–55 (55)
Molasses	5		(0.68)	54
Canteen residues	14–19	12–16	0.15–0.68 (0.68)	43–77 (60)

Source: KTBL, 2012, Biogasrechner. http://daten.ktbl.de/biogas/navigation.do?selectedAction = Startseite#start, zuletzt aktualisiert am 03.07.2012.

Notes: DM, dry matter; oDM, organic dry matter. Values in parentheses refer to state-of-the-art digesters and are lower than maximum potential yields.

CO_2 Separation

The first step of biogas purification focuses on removing water and hydrogen sulfide (H_2S). In small-scale units H_2S is captured in activated carbon filters, which have to be replaced periodically. In bigger units biological desulfurization is applied with great success. The secondary products of desulfurization are sulfur and sulfate (depending on oxidation conditions), which can be used as fertilizers in agriculture. This kind of purification benefits the biogas facility with a longer equipment lifetime and a lower failure rate.

Further biogas purification produces the more calorific biomethane, equivalent to fossil natural gas. Biomethane has an increased diversity of applications and allows for reduced greenhouse gas emissions as well as a relative independence from fossil fuels (Weber et al. 2012); it is more efficient than bioethanol (Insam 2010); and can be introduced into the natural gas grid allowing for cogeneration systems whereby small-scale producers contribute to the grid.

A requirement for introducing biomethane into the natural gas grid consists of removing CO_2 and H_2S. In biogas plants, a biological pretreatment for the removal of H_2S may be favorable. Further process steps focus on CO_2 removal by physicochemical treatments like separation in membranes, cryogenic rectification columns, adsorption of CO_2 and H_2S on activated carbon, physical absorption in water or organic solvents like polyethylene-glycol-dimethyl-ether (which, however, is a suspected compound with central nervous system toxic effects), and chemical absorption with some alcoholamines. The characteristics of the processes are different depending on method selectivity, pressure, purity, temperature, energy consumption, and methane leaks. The strategy for purifying depends also on the end use. For example, when liquefied gas is produced, cooling helps in separation as well. A high pressure level is required for using membranes, activated carbon or water absorption, introducing biomethane into grid, or filling the tanks of cars. Technological knowledge of all these processes exists due to the processing of natural gas for many decades. The challenge is to make it economically feasible at a small scale. The technologies based on chemical absorption with alcoholamines may be favorable for producing biomethane because desorption is carried out by heat, lowering the demand of electric energy. Moreover, this process produces the lowest methane slip with 0.1%, whereas in other processes the emission of methane is near 2% (Table 10.4).

Safety

Biogas is flammable and can be burned in gas torches or internal combustion engines. The range of explosion is narrower than for natural gas, which makes it easier to prevent. Danger is mainly attributed to the H_2S content, since levels above 100 ppm are immediately dangerous to life and health (IDLH). The main focus on cleaning biogas consists of removing H_2S before burning because of the formation of corrosive exhausts.

Methane is a greenhouse gas that remains in the atmosphere for approximately 9 to 15 years and is 23 times more effective in trapping heat than CO_2 over a period of 100 years (Intergovernmental Panel on Climate Change [IPCC] 2007). There is a debate as to whether methane is increasing due to human activities (industrial processes, livestock, paddy rice fields, biomass burning, landfills, extraction of oil and natural gas, agriculture, coal mining, and wastewater treatment plants) (Atmospheric Infrared Sounder [AIRS] 2011). Ironically, while petroleum is being depleted as a source of energy, methane, which is another fuel, is being released to the atmosphere in hazardous concentrations, largely above the last 650 thousand years (Wolff and Spahni 2007).

TABLE 10.4

Comparison of Various Technologies for Producing Biomethane

Criteria	Adsorption on Activated Carbon	Water Absorption	Absorption on Tetraethylene-Glycol-Dimethyl-Ether	Chemical Absorption with Methanol-Amine	Chemical Absorption with Diethanol-Amine
Prepurification	Yes	No	No	Yes	Yes
Pressure (bar)	4–7	4–7	4–7	Atmospheric	Atmospheric
Methane slip	<3%	<1%	2%–4%	<0.1%	<0.1%
Methane concentration product	>96%	>97%	>96%	>99%	>99%
Electric energy demand (kWh/Nm³)	0.25	<0.25	0.24–0.33	<0.15	<0.15
Thermal energy demand (kWh/Nm³)	Not required	Not required	0.5 at 55°C–80°C	0.7 at 160°C	0.7 at 160°C
Dynamic operation % of nominal power	10%–15%	50%–100%	50%–100%	50%–100%	50%–100%

Source: Fraunhofer, 2009, Verbundprojekt: Biogaseinspeisung http://www.biogaseinspeisung.de/download/Kurzbroschuere_Biogaseinspeisung.pdf.

Oncoming Technologies

Artificial Photosynthesis

Artificial photosynthesis refers to any scheme that captures and stores solar energy in the chemical bonds of a fuel (Concepcion et al. 2012). Photocatalytic water splitting is a research area that aims to convert water into protons (and eventually hydrogen fuel) and oxygen (Sun et al. 2001). Light-driven carbon dioxide reduction is another process that replicates natural carbon fixation. These fields of research encompass the design and assembly of devices for the direct production of solar fuels, photoelectrochemistry and its application in fuel cells, and the engineering of enzymes and photoautotrophic microorganisms for microbial biofuel and biohydrogen production from sunlight (Royal Society of Chemistry 2012). Another area of research is the selection and manipulation of photosynthetic microorganisms, namely, green microalgae and cyanobacteria, for the production of solar fuels. Many strains are able to naturally produce hydrogen. Biofuels from algae such as butanol and ethanol are produced both at the laboratory and commercial scales. This approach is also being explored to develop organisms capable of producing biofuels (J. Craig Venter Institute 2012).

Biohydrogen

Hydrogen is attractive due to its potentially higher energy efficiency of conversion to usable power and low generation of pollutants. Hydrogen is conventionally produced from natural gas by steam reforming, coal gasification, and electrolysis of water; however, these methods use nonrenewable energy sources and are not sustainable. Biohydrogen production comprises renewable processes, which are classified into four categories: (1) biophotolysis of water (direct and indirect) using algae and cyanobacteria; (2) photodecomposition of organic compounds by photosynthetic bacteria (photofermentation and dark fermentation); (3) biocatalyzed electrolysis; and (4) fermentative hydrogen from organic wastes or energy crops (Manish and Banerjee 2009). However, still low yields and production rates have been major barriers to the practical application of biohydrogen technologies (Hallenbeck and Ghosh 2009).

Animal Power

Animals have allowed mechanical work since animal domestication started 10,000 years ago. Cattle, horses, mules, llamas, camels, dogs, and donkeys provide transport, pull implements, lift water, and enable other processing activities (Karekezi and Kithyoma 2002). The integration of crop and animal production is an efficient symbiosis where animals increase production, income generation, and the sustainability of cropping systems in small-scale agriculture, with complementarity resources from one sector to the other (Devendra and Thomas 2002). Using renewable resources from the local environment, animals are self-recruiting, produce manure, leather, meat, and hide; and through a well-planned grazing and stocking rate they can contribute to the maintenance of a multifunctional landscape (Rydberg and Jansén 2002). The level of investment on farm implements is lower with minimum tillage. Soil compaction from farm equipment traffic is less with animal traction compared to even small tractors. Recent developments in computer-based simulations help to optimize the implement designs and operational conditions of animal use (Gebregziabher et al. 2006). Moreover, some implements have been developed to use biomass and animal force to produce energy, as for example, an animal traction piston press for producing biomass briquettes as household fuel (Mazzù 2007). Animal power is more versatile than machine power on steep and difficult terrains, typical of poor isolated rural areas.

Microbially Produced Electricity

Recent discoveries of microorganisms capable of converting chemical energy into electric current using fuel cells suggest an important biotechnological avenue. Some examples of such bacteria belong to the δ-proteobacteria, mainly the Geobacteraceae family (Esteve-Núñez et al. 2008). Microbial fuel cells (MFCs) are microorganisms that biologically oxidize organic matter and transfer electrons to an anode. Enriched anodic biofilms have generated power densities as high as 6.9 W per m^2 (Logan 2009).

Termite Gut

Current research on the termite gut aims to identify the microorganisms that can be useful to produce biofuels in the future (Brune 2007). Microbes in the termite gut have the ability to convert cellulose into sugars that can be turned into ethanol, hydrogen, or methane. Recently, the microbial community of the termite gut was sequenced and a number of novel cellulases have been identified that break down cellulose into sugar. The next goal is to understand and reconstruct a diverse range of metabolic processes that could be scaled up for industrial fuel production (Singer 2007).

References

Adler PR, Del Grosso SJ, Parton WJ (2007) Life-cycle assessment of net greenhouse-gas flux for bioenergy cropping systems. *Ecological Applications* 17:675–691.

Atmospheric Infrared Sounder (AIRS) (2011) Methane. http://airs.jpl.nasa.gov/maps/maps_in_motion/methane/ (accessed July 21, 2012).

Atreya SK (2007) The mystery of methane in Mars and Titan. *Scientific American* 296:43–51.

Brune A (2007) Woodworker's digest. *Science* 450:487–488.

Concepcion JJ, House RL, Papanikolas JM, Meyer TJ (2012) Chemical approaches to artificial photosynthesis. PNAS 109:15560–15564.

Devendra C, Thomas D (2002) Crop–animal interactions in mixed farming systems in Asia. *Agricultural Systems* 71:27–40.

EcoSecurities International (2009) Aguascalientes EcoMethane landfill gas to energy project. CDM Monitoring Report. http://cdm.unfccc.int/filestorage/2/V/O/2VOGY41D0W5STLMACNR8P369XZHEIB/Monitoring%20Report.pdf?t = Z298bXFlb3E1fDCIPn_7BizEwIaQrveIbxFV (accessed April 14, 2013).

Esteve-Núñez A, Sosnik J, Visconti P, Lovley DR (2008). Fluorescent properties of c-type cytochromes reveal their potential role as an extracytoplasmic electron sink in *Geobacter reducens*. *Environmental Microbiology* 10:497–505.

Food and Agriculture Organization of the United Nations (FAO) (2002) Wood Energy Information System (WEIS). http://www.fao.org/forestry/FOP/FOPH/ENERGY/databa-e.stm/ (accessed May 12, 2012).

Fraunhofer (2009) Verbundprojekt: Biogaseinspeisung http://www.biogaseinspeisung.de/download/Kurzbroschuere_Biogaseinspeisung.pdf (accessed March 12, 2013).

Gebregziabher S, Mouazen AM, Van Brussel H, Ramon H, Nyssen J, Verplancke H, Behailu M, Deckers J, De Baerdemaeker J (2006) Animal drawn tillage, the Ethiopian Ard plough, Maresha: A review. *Soil & Tillage Research* 89:129–143.

German Solar Energy Society (DGS) and Ecofys (2005). *Planning and installing bioenergy systems: A guide for installers, architects and engineers*. London: James & James.

Hallenbeck PC, Ghosh D (2009) Advances in fermentative biohydrogen production: The way forward? *Trends in Biotechnology* 27:287–297.

Insam H, Franke-Wittle I, Goberna M (2010) Microbes in aerobic and anaerobic waste treatment. In *Microbes at work: From wastes to resources*. Insam H, Franke-Wittle I, Goberna M (eds), 1–34. Heidelberg: Springer-Verlag.

Intergovernmental Panel on Climate Change (IPCC) (2007) Fourth assessment report, Working Group 1, Chapter 2. https://www.ipcc-wg1.unibe.ch/publications/wg1-ar4/ar4-wg1-chapter2.pdf (accessed January 15, 2012).

International Energy Agency (IEA) (2012) World Energy Outlook. Executive summary. Paris: OECD.

J. Craig Venter Institute (2012) Synthetic biology and bioenergy: Overview. http://www.jcvi.org/cms/research/groups/synthetic-biology-bioenergy/ (accessed June 21, 2012).

Karekezi S, Kithyoma W (2002) Renewable energy strategies for rural Africa: Is a PV-led renewable energy strategy the right approach for providing modern energy to the rural poor of sub-Saharan Africa? *Energy Policy* 30:1071–1086.

Kerr D (2012) Apple looks to double its N.C. biogas fuel cell farm. http://news.cnet.com/8301-13579_3-57557187-37/apple-looks-to-double-its-n.c-biogas-fuel-cell-farm/(accessed April 15, 2012).

Kishinevsky Y, Zelingher S (2003) Coming clean with fuel cells. *IEEE Power & Energy Magazine*. http://www.science.smith.edu/~jcardell/Courses/EGR325/Readings/FuelCells.pdf (accessed August 13, 2012).

KTBL(2012)Biogasrechner.http://daten.ktbl.de/biogas/navigation.do?selectedAction = Startseite#start, zuletzt aktualisiert am 03.07.2012 (accessed May 24, 2012).

LaMonica M (2012) Microsoft data center fueled by waste gases. http://www.technologyreview.com/view/507626/microsoft-data-center-fueled-by-waste-gases/ (accessed May 24, 2012).

Logan BE (2009) Exoelectrogenic bacteria that power microbial fuel cells. *Nature Reviews/Microbiology* 7:375.

Lorentz R, Mitton J (2002) *Lifting Titan's veil: Exploring the giant moon of Saturn*. Cambridge: Cambridge University Press.

Manish S, Banerjee R (2009) Comparison of biohydrogen production processes. *International Journal of Hydrogen Energy* 33:279–286.

Margalef R (1998) *Ecología*. Barcelona: Ediciones Omega.

Masera O, Fuentes G (2006) Introducción. In La Bioenergía en México, un catalizador del desarrollo sustentable. Omar Masera Cerutti (ed). México: Comisión Nacional Forestal, Mundi–Prensa.

Mazzù A (2007) Study, design and prototyping of an animal traction cam based press for biomass densification. *Mechanism and Machine Theory* 42:652–667.

Mendoza LA, Mendoza E (2011) *Biología I*. México: Trillas.

Morring Jr F (2009) Lunar engines. *Aviation Week & Space Technology* 171:16.

National Aeronautics and Space Administration (NASA) (2012a) Morpheus testing picks back up at Johnson Space Center. http://www.nasa.gov/centers/johnson/exploration/morpheus/morpheus_tests_pickup.html (accessed October 15, 2012).

National Aeronautics and Space Administration (NASA) (2012b) Mars: NASA explores the red planet. http://www.nasa.gov/mission_pages/mars/news/marsmethane_media.html (accessed November 14, 2012).

O'Connell D, Braid A, Raison J, Handberg K, Cowie A, Rodriguez L, George B (2009) Sustainable production of bioenergy: A review of global bioenergy sustainability frameworks and assessment systems. RIRDC Publication No. 09/167. Canberra: Rural Industries Research and Development Corporation and CSIRO.

Ortega RM (2000) *Energías renovables*. Madrid: Paraninfo.

Petrescu-Mag IV, Oroian IG, Păsărin B, Petrescu-Mag RM (2011) Methane in outer space: The limit between organic and inorganic. *International Journal of the Bioflux Society. ELBA Bioflux* 3:89–92. http://www.elba.bioflux.com.ro/docs/2011.3.89-92.pdf (accessed July 18, 2013).

Royal Society of Chemistry (2012) Solar fuels and artificial photosynthesis: Science and innovation to change our future energy options. http://www.rsc.org/ScienceAndTechnology/Policy/Documents/solar-fuels.asp (accessed June 23, 2013).

Røstad AS (2012) Biogas: The power of manure. SFFE lunch seminar. http://sffe.no/colloquia/arkiv/120911_NordgaardBiogas.pdf (accessed May 11, 2013).

Rydberg T, Jansén J (2002) Comparison of horse and tractor traction using emergy analysis. *Ecological Engineering* 19:13–28.

Schmitz JEJ (2007) *The second law of life: Energy, technology and the future of earth as we know it*. Norwich: William Andrew Publishing.

Singer E (2007) Why termite guts could bring better biofuels. *MIT Technology Review*, Jan 17. http://www.technologyreview.com/news/407190/why-termite-guts-could-bring-better-biofuels/ (accessed January 22, 2013).

Spellman FR, Drinan JE (2001) *Stream ecology and self-purification: An introduction*. Boca Raton, FL: CRC Press.

Sun L, Hammarström L, Åkermark B, Styring S (2001) Towards artificial photosynthesis: Ruthenium–manganese chemistry for energy production. Review Article. *Chemical Society Reviews* 30:36–49.

Symons GE, Buswell AM (1933) The methane fermentation of carbohydrates. *Journal of the American Chemical Society* 55:2028–2036.

Tchobanoglous G, Theisen H, Vigil S (1994) *Gestión integral de residuos sólidos*. Volumen II, 758–769. Madrid: McGraw-Hill.

Thauer RK, Shima S (2006) Biogeochemistry: Methane and microbes. *Nature* 440:878–879.

Themelis NJ, Ulloa PA (2007) Methane generation in landfills. *Renewable Energy* 32:1243–1257.

Weber B, Rojas M, Torres M, Pampillón L (2012) Producción de biogás en México. Estado actual y perspectivas. México: Red Mexicana de Bioenergía and Imagia Comunicación.

Wolff E, Spahni R (2007) Methane and nitrous oxide in the ice core record. *Philosophical Transactions of the Royal Society A* 365:1775–1792.

11

Air Pollution and Anthropogenic Climate Change

Alejandro de las Heras

CONTENTS

Global and Regional Air Pollution: Damage

Global atmospheric change compounds the damage of several pollution manifestations. Global warming is poised to enhance the conditions for malaria, dengue, plague, and viral encephalitis spread. Contact with pathogens is likely to augment due to immunosenescence in aging populations. As elderly populations have been exposed to now rarer pathogens, they could become a reservoir of reemerging pathogens, airborne and otherwise.

The heat waves of 1995 in the United States and 2002 in Europe in conjunction with urban heat islands caused temperatures several degrees warmer than rural areas and increased the tropospheric ozone. Cardiac arrhythmias and cardiopulmonary dysfunctions especially among the elderly were rife. The addition by pollution of ozone is linked to cognitive deficits in middle-aged individuals in the United States, equivalent to aging 4 years. Traffic-related air pollution is the main cause of heart attacks in 36 studies. Air pollution in Los Angeles determines carotid thickening. Vehicular smog in Los Angeles and Hong Kong augments hospital admissions due to cardio-pulmonary conditions.

Mineral and pollen particles readily enter lung alveoli leading to chronic inflammation and pulmonary, vascular and cardiac dysfunctions, such as in Spain where 2.5 μm particles from the Sahara seasonally cause mortality peaks in the elderly population (Finch 2012).

Similarly, damage across animal taxa are observable and reviewed elsewhere (Lechter 2009), manifesting ecosystemic damage worldwide. This makes global warming effects some of the most pervasive.

The stratrospheric ozone layer (15–50 km above the sea level) has protected life on Earth from solar ultraviolet (UV) radiation. UV in the 100 to 280 nm wavelength range rapidly damage living matter and 280 to 315 nm UV provoke damage after prolonged exposure. The low ozone concentrations in the layer (2 to 8 parts per million by volume) filter out 200 to 315 nm UV. The ozone layer depletion is most marked above Antarctica. Depletion also takes place over Europe and North America, and in 1999 depletion over western Europe was as pronounced as it usually is over Antarctica.

Acid rain damages forest ecosystems through increased cation nutrients leaching from the foliage, altered foliage physiology and altered plant growth. Tropospheric ozone, toxic metals, and natural stressors combine with excess nutrients like nitrogen from rain bringing cation deficiencies (VanLoon and Duffy 2005). Leaf injury occurs with very acid fogs in the Adirondack Mountains in the state of New York. Acidic deposition in forest soils releases aluminum ions from silicates, which can damage fine roots; it makes toxic heavy metal ions available; important soil colloids are destroyed. Fungal and bacterial symbionts are damaged. Long-term but slow acid deposition can be neutralized by different soils. However, sudden release leads to acid shock, meaning a large amount of water infiltrates without time for soil reactions with ions to occur and trap the latter. The ultimate sinks for these polluting compounds are soil, water bodies, and groundwater (VanLoon and Duffy 2005). Acid deposition flows, accumulating acidity. In the eastern United States 1200 lakes and 4700 rivers can hardly retain any life, 1200 lakes in Ontario are devoid of life, similar to 6500 lakes in Norway and Sweden (Aguado and Burt 2010). Acid rain also accelerates metal corrosion and stone degradation in buildings. The damage most often occurs hundreds of kilometers away from the industrialized pollution sources.

In South Asia and North Africa, powerful effects on monsoons yielding less rainfall are due to changes in several monsoon drivers: surface–atmosphere heating, sea surface temperature gradients, convective instability of the troposphere, evaporation, and the Hadley circulation. In addition, atmospheric brown clouds (ABCs) mix with rain clouds diminishing the precipitation efficiency of the latter (Ramanathan and Feng 2009; Kennel et al. 2012). Observable results have been weaker monsoons and ensuing diminished crops.

Associated with smog, tropospheric ozone entails notably reduced lung capacity and asthma-related fatality. Conifers and crops, such as grapes, are also impacted by ozone and other photooxidants (Aguado and Burt 2010). Eight percent of all lung cancers in nonsmokers in the United States could be due to the respirable fraction of total suspended particles from vehicles.

Chronic inhalation of biomass combustion products (carbon particles, endotoxins) has caused lung and vascular damage since the emergence of urban environments 8000 to 9500 years before present time, as testified by particles present in lung tissues from the Neolithic. Indoor cooking using fossil fuels and biomass (the hut lung syndrome) is the single most hazardous pollution risk globally, especially when people sleep where cooking is carried out. Browning and charring during cooking may rid us of pathogens but they also release inflammagens that are new to our species (Finch 2012). Indoors, ultraviolet A (UV-A) from fluorescent tubes efficiently penetrate the skin leading to its aging. Sunblock is ineffective against UV-A (VanLoon and Duffy 2005).

Same Pollutants, Different Manifestations

The same air pollutants elicit very different global manifestations, are directly toxic to organisms, or both. In particular, NOx (NO and NO$_2$) are present in virtually every manifestation (Table 11.1). Greenhouse gases (GHGs), volatile organic compounds (VOCs), O$_3$, and aerosols have global physicochemical and toxic effects (Table 11.1). Natural sources generally dwarf when compared with anthropogenic sources. In general, the latter can be attributed to the same set of human activities: agriculture, industry,

TABLE 11.1

Look-Up Table of the Main Anthropogenic Pollutants Involved in Global and Regional Environmental Change

	Global		Regional		
Pollutant	Climate Change	Ozone Layer Depletion	ABCs	Acid Rain	Smog
NO	Form tropospheric O_3	Catalytically decompose O_3		With O_3 forms NO_2	With sunlight forms NO_2; participates in O_3 formation
NO_2				Forms nitric acid	Produces PAN; participates in O_3 formation
N_2O	GHG	Enhances NOx, and O_3 depletion			
SO_2			Drives the ABC trend	Forms sulfuric acid and SO_3	Hydrocarbon combustion produces classical smog
SO_3				Forms sulfuric acid	Intermediary in sulfuric acid formation
VOCs	Form tropospheric O_3			Their combustion emits NOx	With sun, NOx, and O_2 they produce smog; some are precursors of PAN
O_3	Tropo- and stratospheric GHG			Reagent in the formation of sulfuric acid	Participates in PAN formation
CO_2	GHG			Acidifies rain	A final stable carbon compound
CH_4	GHG	Catalytically decompose O_3			
HOx				React with O	Participate in the oxidation of hydrocarbons
CFCs and successors	GHG				

Continued

TABLE 11.1 (*Continued*)

Look-Up Table of the Main Anthropogenic Pollutants Involved in Global and Regional Environmental Change

	Global		Regional		
Pollutant	Climate Change	Ozone Layer Depletion	ABCs	Acid Rain	Smog
Aerosols	Sulfate cooling and Arctic haze		Black carbon and SONs	Possible Arctic precipitation acidification	Soot

Notes: ABCs, atmospheric brown clouds; NO, nitrogen oxide; NO_2, nitrogen dioxide; N_2O, nitrous oxide; SO_2, sulfur dioxide; SO_3, sulfur trioxide; VOCs, volatile organic compounds. They result from the evaporation of hydrocarbons. O_3, ozone; PAN, peroxyacyl nitrates; a reservoir of NO species. CO_2, carbon dioxide; CH_4, methane, one of the most common VOCs. HOx, hydrogen-containing species. CFCs, chlorofluorocarbons. They are being replaced by HCFCs (hydrochlorofluorocarbons, also being phased out), HFCs (hydrofluorocarbons), and bromine-containing compounds that no longer have ozone layer depletion effects but are GHG. Aerosols: liquid or solid particulate matter suspended in the atmosphere. SONs: sulfates, organic pollutants, and nitrates.

transportation, heating, having in common combustion and vegetation change as key drivers. Therefore, global and regional air pollution ought to be amenable to a relatively small set of solutions, basically aiming at reducing fossil fuel consumption. Fertilizers and coolants should also be reduced. Emission reductions, however, have to be accompanied with measures tending to lower the extant concentrations of accumulated atmospheric pollutants.

Climate Change

Solar Energy and the Greenhouse Effect

The Earth–sun geometry has changed as the planet's orbit and inclination is modified by the other planets' gravitational pull. Today's configuration is most likely responsible for the current ice age (Berger 2001). The climate contains five subsystems: atmosphere, oceans, cryosphere, land, and biosphere. These systems have effects called radiative forcings that modify the sun–Earth solar radiation balance and therefore the climate (Intergovernmental Panel on Climate Change [IPCC] 2007). They are linked by their energy, mass, and momentum (motion) exchanges.

The atmosphere reflects higher energy solar radiation back into space and absorbs lower energy reflected from the planet surface (greenhouse effect). Without greenhouse gases (GHGs) the planet's surface would stand at −15°C instead of its current +18°C. GHGs comprise water vapor and trace gases (listed in Table 11.1), whose atmospheric concentration is increasing.

Greenhouse gas molecules vibrate and rotate as they absorb infrared energy. Atmospheric gases are selective absorbers and emitters: a gas atom is restricted to absorbing certain energies given by the wavelength of incoming photons. Upon being absorbed, a photon shifts a gas electron to an excited state. Upon return to the ground state, energy is released to the atmosphere in the form of a photon. Different GHGs have different absorbances and so taken together, they make a more efficient trap for infrared radiation. Although GHG molecules do not last in excited states, there is still a conjecture according to which GHGs store infrared energy during the day and release it at night.

CO_2 is a less powerful energy absorber but much more concentrated in the atmosphere than the other GHGs. By 2000, CO_2 equated to 50% of the global GH effect. The CO_2 lifetime is 10 years in the atmosphere before being dissolved into the sea, the largest sink of CO_2, which can concentrate 60 times more CO_2 than the atmosphere but much of the ocean-absorbed CO_2 is returned to the atmosphere.

Aerosols

Aerosols are solid and liquid particles suspended in the atmosphere. Aerosols seem to have dimmed and so cooled to a modest extent the atmosphere around 1960 to 1990; thereafter, air pollution control may have slightly contributed to additional atmospheric global warming.

Evidence of Past Climate Change

Instrumental records, chemical ratios, stable isotopes ratios, ice cores, ocean sediment cores, laminated lake sediments, palynology in lake bottoms, cave deposits, relict soils, tree rings, corals, and packrat nests document past climate variation. These records have shown that GHGs like CO_2 and CH_4 have covaried with the atmospheric temperature over the last 650,000 years (Berger 2001; Maslin 2009). In the current period or Holocene, a Climatic Optimum in Europe occurred 9 to 6 ka BP (thousand years before present) coinciding with large changes in the monsoon regions, mainly in North Africa where the Saharan desert substantially subsided. Thereafter an abrupt moisture decrease about 5.5 ka BP led the vegetated Sahara to revert to desert.

These changes may have had strong impacts on human societies such as those documented in more recent times: mass migration in central and northern Europe about 1600 years ago associated with cool and wet climate; colonization of Iceland and Greenland 1100 years ago followed by the loss of Viking settlements; and crop failure, famine, and population decrease in Europe during the Little Ice Age (1350–1850) (Lemke 2006).

BOX 11.1   THE CLIMATE ROLES OF OCEANS, CRYOSPHERE, LAND, AND BIOSPHERE

The oceans are the largest heat reservoir: released in winter it reduces the Earth's seasonal temperature cycle. A mere 0.3% of the oceans' water evaporates yearly but the oceans gain thermal energy called latent heat. The latter is gained by the atmosphere during water condensation. Oceanic heat transport makes west coasts of continents milder: western Europe is 15°C warmer than eastern North America thanks to the Gulf Stream, which annually transports energy equivalent to 1 million nuclear power plants. The Kuroshio is a similar stream in Asia. The oceans are also the largest physicochemical CO_2 sink, thanks to $CaCO_3$ (lime)-absorbing CO_2 from the atmosphere.

The cryosphere is the second largest component of the climate system, in terms of mass and heat capacity. Ice covers currently 11% of the land area and on an average year 6.5% of the oceans. Because of the low incoming solar radiation in polar regions and the high ice albedo (light reflection), oceans there cool down making seawater sink. Seawater turning into ice releases salt into deeper water, which becomes denser and sinks farther, taking gases to the bottom and entering the thermohaline circulation around the globe oceans. Snow covers an even larger area (50% of the Northern Hemisphere continents) and has higher albedo than ice. Snow covered areas are large energy sinks.

As for land, the continental drift considerably modified the distribution of the continents. It is the boundary for one-third of the atmosphere. Atmospheric and ocean circulations are affected by continents, mountains, and ocean passages. The current period is cool due to the isolation of Antarctica from Tasmania, the opening of the Drake and Northeastern passages, the Himalayas uplift, and the closure of the Panama Ishtmus (Maslin 2009). Volcanoes eject particulate matter and sulfate-bearing gases.

With regard to the biosphere, the Earth's living organisms seem involved in the carbonate–silicate cycle, a thermostat keeping the Earth from freezing. Photosynthetic organisms interact chemically with the atmospheric tracts of the N_2, O_2, CO_2, water vapor, and CH_4 cycles as well as physically (with a low albedo, and strongly participating in evapotranspiration) (Lemke 2006).

Evidence of Current Climate Change

Today's orbital will remain constant for the next 50 ka so that other effects than solar forcing will be conspicuous in climate modifications. Fast or abrupt changes will likely be attributable to human activity (Berger 2001).

Evidence of present climate change is most visible in the melting of ice caps, glaciers, and tropical volcanoes. Instrumental data show evidence of global warming and climate change in all climate subsystems (oceans, cryosphere, atmosphere, land, and biosphere). The last three decades of the 20th century were the warmest in the last millennium, the 1990s were the warmest, 1998 was the warmest year on record since 1861, and then the years 1995 to 2006 were the warmest in the record of global surface temperature since 1850. This hints at accelerated warming. Widespread changes in extreme temperatures have been observed over the last 50 years. Heat waves, hot days, and nights have become more frequent. Oceanic average temperatures since 1961 have increased to depths of at least 3000 m, causing seawater expansion and consequently sea level rise. The global snow cover and Arctic ice cover have decreased by 10% since 1960 and almost all glaciers have subsided. The sea-ice draft (the thickness of submerged ice) has decreased by 40% in volume. Greenland has shrunk by over 1000 billion metric tonnes since 2003. The permafrost active layer (melting layer) has become deeper in Alaska: it has gained 3°C to at least 1 m underground. Present-day CO_2 concentration is already 60% above the average value of the last glacial–interglacial cycle. GHGs have already altered the climate and can influence the glacial–interglacial cycles in the next 50 thousand years (Berger 2001; IPCC 2007; Maslin 2009).

As for extreme weather events, massive storms and floods have affected China, Italy, England, Bangladesh, Venezuela, and Mozambique. In 1900 to 2005 precipitation has increased in eastern North and South America, northern Europe, and northern and central Asia. The amount of rain during extreme events has increased in the United States, Europe, the former USSR, and China. The frequency of heavy precipitation events has increased over most land areas, consistent with warming and increases in atmospheric water vapor that warm air can hold. Conversely, drying (decreased precipitations and increased temperatures) is occurring in the Sahel, the Mediterranean, southern Africa, and parts of southern Asia. More intense and longer droughts over wider areas straddling the tropics have occurred since the 1970s (IPCC 2007; Maslin 2009).

Depletion of the Ozone Layer

O_2 chemistry explains the existence of the ozone layer but also predicts that creation and destruction of O_3 would be in balance (Berner and Berner 1996). This is not currently the case due to 3 types of free radicals acting as catalysts in O_3 destruction: mostly NOx, ClOx, and HOx at about 30 km and 50 km for the latter.

The largest O_3 layer destruction is due to two NOx reactions:

$$NO + O_3 \rightarrow NO_2 + O_2 \tag{11.1}$$

and

$$NO_2 + O \rightarrow NO + O_2 \tag{11.2}$$

(whereby NO reacts again with O_3). In addition, ground-level N_2O diffuses into the stratosphere where

$$N_2O + O(^1D) \rightarrow NO + O_2 \tag{11.3}$$

^1D is an electronically activated state of oxygen resulting from photoactivation.

ClOx are among the most reactive catalysts of O_3 destruction; they include chlorine and chlorine-containing free radicals (.Cl and .ClO; only their bromine analog are more reactive). The O_3 layer depletion due to chlorine-containing chlorofluorocarbons (CFCs) was predicted by 1974. CFCs have low reactivity allowing them to reach the stratosphere where they release Cl species in a sequence possibly involving all Cl atoms in a CFC molecule, reacting some 100,000 times releasing ClO before entering the HCl sink (HCl forming with CH_4 is the main Cl-radical deactivation route). With CFC life-times around 100 years, despite progress in reducing stratospheric Cl since 1997, pre-1980 levels would be reached again between 2060 and 2075 (Aguado and Burt 2010).

HOx are sources of hydrogen that combine with O_2 in the dry and cold stratosphere. For instance, H is photochemically extracted from CH_4 to produce water and thence hydroxyl radicals, which catalytically destroy O_3.

Atmospheric Brown Clouds

Less than 15 years ago, a global change issue was identified: atmospheric brown clouds (ABCs) are aerosols made up of black carbon (elemental carbon in soot), and SONs (sulfates, organic pollutants. and nitrates) that are emitted by megacities. Due to atmospheric transport, ABCs circulate eastward from Asia to America, America to Europe, and Europe to Asia. ABCs have a solar dimming effect at global and regional scales. The dimming comprises a light scattering effect and an absorption effect, which result in atmospheric cooling and warming, respectively.

ABCs are simultaneously a major climate change, ozone depletion, and air pollution issue. After CO_2, black carbon is the second most important radiative forcing affecting climate change. As to methane, ozone, and aerosols, they interact chemically and mutually modify their atmospheric mixing ratios thereby impacting the stratospheric and tropospheric chemistry (Ramanathan and Feng 2009; Kennel et al. 2012).

Acid Rain

Acid rain has a pH below 5.6, the normal acidity due to the atmospheric CO_2 content. Acid rain pH can be as low as 2 in the troposphere. To form acid rain, gaseous SO_2, NO_2, and HCl undergo reactions leading to sulfuric, nitric, and hydrochloric acids, respectively. Furthermore, the SO_2, NO, NO_2, and HCl gases produce hydrogen ions H⁺, directly expressed in pH measures (VanLoon and Duffy 2005).

SO_2 reacts with OH radicals to form, after intermediate stages, sulfuric acid H_2SO_4 aerosol. In another process, SO_2 in cloud droplets or as aerosol reacts with H_2O_2 (or with O_3) to form H_2SO_4. H_2SO_4 is a strong acid and in water it completely dissociates into hydrogen and sulfate ions. Sulfur trioxide SO_3 derives from an SO_2 reaction with OH and then O_2; it then readily reacts with water droplets in acid fog (inhalable) or acid rain (VanLoon and Duffy 2005; Aguado and Burt 2010).

NO forms in combustion above 2000°C (where N_2 and O_2 react). NO rapidly degrades but reacts with O_3 (or with HO_2) to form NO_2:

$$NO + O_3 \rightarrow NO_2 + O_2 \tag{11.4}$$

and

$$NO + HO_2 \rightarrow NO_2 + OH \tag{11.5}$$

NO_2 and OH react to form the strong nitric acid HNO_3:

$$NO_2 + OH \rightarrow HNO_3 \tag{11.6}$$

which completely dissociates in water releasing H+ and acidifying rain (Berner and Berner 1996; Schwedt 2001; VanLoon and Duffy 2005).

Smog

Two types of smog can be distinguished: classical coal smog (London-type) and photochemical smog (Los Angeles type). They differ in the time of the day determining sunlight and temperature; in cloud cover and moisture; in chemical characteristic (reductant and oxidant, respectively); and in content, soot and SO_2 (classic) and O_3 and PAN among many others (photochemical).

Photochemical reactions include the generation of acids from SOx, NOx, and photooxidants like tropospheric O_3 and PAN. Tropospheric O_3 in sunlit conditions comes from the following reactions:

$$NO_2 + h\nu \rightarrow NO + O_2 \qquad (11.7)$$

(*h*ν in this context stands for solar ultraviolet energy with 290 to 430 nm wavelength) and

$$O_2 + O + M \rightarrow O_3 + M \qquad (11.8)$$

(*M* stands for a nonreacting molecule that collides with the reagents and helps take away some energy.) PAN forms in photochemical smog from partially oxidized hydrocarbons, O_2, O_3, and NO_X.

During the day nitric oxide NO is oxidized by O_2, O_3, or ROO (R being an alkyl group, i.e., part of a hydrocarbon), for example,

$$NO + O_3 \rightarrow NO_2 + O_2 \qquad (11.9)$$

NO_2 then contributes to O_3 formation initiating photochemical smog. At night nitric acid is produced through the nitrate radical and can accumulate, in absence of photolysis.

Attribution

Attributing a pollutant to its natural and anthropogenic sources has helped pinpoint the responsibility of emitters, and circumvent economic and political resistance to mitigation actions. Attribution at times is made difficult by the number of diffuse events, such as forest fires around the world. Contrariwise, "smoking guns" directly link human activity to pollution trends and events. Such was the case for the 1973 oil embargo and the ensuing international recession, transiently interrupting the U.S. NOx upward emission trend subsequent to 1950 (Berner and Berner 1996). Similarly, the 1973 and 1981 oil shocks coincided in 1972 to 1986, with a 40% SO_4 decrease in Swedish precipitation (VanLoon and Duffy 2005). Routine economic and meteorological monitoring also seems to link business cycles and temperature variation (Figure 11.1).

Greenhouse Gas Emissions

Carbon dioxide (CO_2) is naturally emitted by the respiration of living beings. It is used in photosynthesis by algae and plants to produce both biomass

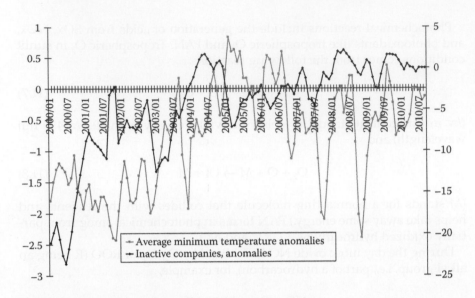

FIGURE 11.1 (See color insert.)
The relationship between temperature fluctuation and economic activity (construction industry). Temperature anomalies are the difference between observed data and the 2000 to 2010 average temperature in Toluca, Mexico, a midsized industrial city connected through train freight to the U.S. economy. The construction industry is a large economic sector whose behavior correlates well with the rest of the economy. Overall in the last decade, an increase in the monthly number of companies that went out of business coincided with a temporary decrease in the minimal temperature, without any pronounced lag time. In the first half of the decade, a temperature increase trend ended as a consequence of a domestic recession around 2006 followed by the economic slowdown in the United States and most of the world. (Data from INEGI, 2010, Banco de Información Económica, http://www.inegi.org.mx/Sistemas/BIE/Default.aspx; and SMN, 2010, Estaciones automáticas: Toluca. http://smn.cna.gob.mx/emas/.)

and energy. C is also naturally present in the atmosphere as methane CH_4, the most volatile hydrocarbon.

The human addition to fossil C to the atmosphere is patent in the current trend of decreasing $^{14}C/^{12}C$ ratio in tree rings (fossil C emitted to the atmosphere by hydrocarbon combustion is ^{14}C-free). As to atmospheric $\delta^{13}C$ its −6.4% decrease in the preindustrial era accelerated to −8% in the 1990s (fossil fuel C has an isotopic value of −25%). Fossil fuel emissions have accelerated from 5.2 PgC/year (5.2 million billion grams) in the early 1980s to 6.1 in the early 1990s (Fung 2001).

More than 90% of the fossil fuel CO_2 is emitted in the Northern Hemisphere as shown by a mean annual concentration 3 ppmv excess compared to the Southern Hemisphere (IPCC 2007; Maslin 2009), a hemispheric pattern similar to the CH_4 global distribution. This concurs with the concentration of human population, agriculture, and industries in the Northern Hemisphere. Eighty percent of the global CO_2 emissions are attributable to energy production, industrial processes, and transport; and 20% to land change, especially tree felling and forest fires, which reduce CO_2 storage and emit CO_2. Developed countries have been consuming more energy and have deforested to a greater extent. China is the largest emitter since 2007 but still amounts to only one-fourth of the U.S. per capita emissions (Maslin 2009), despite the massive Chinese industrialization and the deindustrialization of the United States. This means that the Chinese CO_2 emissions are predominantly industrial and household consumption substantially lower than in the United States. The global CH_4 increase is attributable to anaerobic respiration in rice paddies, waste dumps, and livestock intestines, among others (Maslin 2009).

Consensus is emerging that changes in temperature in the atmosphere, land surface, and several hundred meters deep in the ocean, as well as sea level rise can be attributed to humans. In addition, tropospheric warming and stratospheric cooling can be linked to anthropogenic greenhouse gas increases and stratospheric ozone depletion (IPCC 2007).

Hydrofluorocarbons such as CFCs and HCFCs (hydrochlorofluorocarbons) have accumulated or still are accumulating in the atmosphere. They are very potent anthropogenic GHGs with long lifetimes or fast emission rates.

Sources of Stratospheric Ozone-Depleting Substances

Methyl chloride (CH_3Cl) is the most important natural O_3 catalyst. It is biogenically produced in the oceans and to a smaller degree in burning vegetation and volcanism. Volcanism also produces hydrochloric acid and chloride ions from sea-salt spray, which may mitigate and worsen O_3 depletion, respectively. CH_4 is increasingly anthropogenic in the troposphere. In the 1970s to 1980s, persistent depletion of the O_3 layer occurred over large urban areas in the Northern Hemisphere, which was further exacerbated by the 1991 Pinatubo eruption. Stratospheric sulfate aerosols from the volcano allowed reactive Cl species to lead to enhanced O_3 depletion (Berner and Berner 1996; Maslin 2009).

The production of long-lived catalyst gases is either natural but anthropogenically increased (N_2O, CH_4, bromine compounds), or simply anthropogenic (CFCs and their replacement compounds). The largest O_3 layer destruction is due to NOx from fuel combustion in vehicles (although the vastly deleterious supersonic transport has now disappeared), industries, and nuclear weapons.

CFCs are a very large anthropogenic source of Cl. CFCs are nontoxic and nonflammable and had become very desirable in industrial applications as coolants, propellants, and solvents, prior to their ban. Chlorine radicals and nitric acid are largely due to humans and have reduced O_3 layer concentrations by 5% to 7% globally. Halons are the bromine analogs of CFCs, widely used as fire extinguishers or fire retardants in textiles. Br-containing compounds deplete O_3 even more readily than chlorine ones. CH_3Br comes from vegetation burning and agricultural pesticides, as well as marine plankton (Berner and Berner 1996). Methyl bromide is injected under a plastic sheet onto the soil to fumigate all life forms in strawberry fields. Despite provisions for phase-out by 2005, opposition is strong to the ban of methyl bromide among manufacturers and agriculturalists.

Atmospheric Brown Cloud Attribution

ABCs can be attributed to a four- to fivefold increase in soot and SO_2 emissions since the 1960s. ABCs are evidence that anthropogenic aerosols are no longer solely emitted in extratropical countries. Since the 1970s, tropical countries have emitted two types of black carbon: a brownish one emitted by biomass combustion and a black one, 100% more efficient as a global warming agent, emitted by inefficient engines and backward technologies in industrial processes (Ramanathan and Feng 2009; Kennel et al. 2012).

SOx and NOx Attribution

In natural rain acidification, microbiological processes favored by high temperatures in the tropics produce many partially oxidized or reduced sulfur compounds, and release SO_2 and ultimately sulfuric acid. Denitrification occurs naturally but is largely enhanced by agricultural fertilizers. Dimethyl sulfide is biogenically produced in the oceans and contributes to SO_2 atmospheric emissions (Berner and Berner 1996). Half the precursors of nitric and sulfuric acid are natural (geochemical and biogenic).

The anthropogenic sources of nitric and sulfuric acid are mostly related to fossil fuel combustion and smelting of sulfide-based ores (including copper, nickel, lead, and zinc). Industrial nitration and the processing of nitric and sulfuric acids also contribute. Two-thirds of the NOx sources are anthropogenic (fossil fuel and biomass combustion) (Berner and Berner 1996). Automobiles contribute largely to NOx but modestly to SOx emissions. Combustion from supersonic aircraft engines, exclusively military nowadays, directly egresses unknown large quantities of NOx in the stratosphere.

The combustion of petroleum derivatives above 1100°C forms 90% of NO in internal combustion engines and large industrial plants. NO produced at lower temperature comes from forest fires or domestic cooking and heating. NO_2 is a major toxic and corrosive smog component. It degrades rapidly so that its concentrations vary with urban traffic (VanLoon and Duffy 2005; Aguado and Burt 2010).

Excess sulfate in rain comes from SO_2 gas emitted by coal and oil combustion, which contain pyrite FeS_2 and organic sulfur compounds. In a large measure, pyrite in coal explains SO_2 emissions (Berner and Berner 1996). All fossil fuels contain sulfur, but coal contains less than 1% to up to 10%, whereas diesel contains 1000 to 5000 ppm and gasoline 10 to 500 ppm. Substantial sulfur amounts may be contained in natural gas but are removed in refineries (VanLoon and Duffy 2005). Coal in the recent past has mostly been used for electricity generation and smelting. Coal will likely expand to other uses as petroleum reserves deplete.

High smokestacks and atmospheric circulation are involved in the long-haul transport of pollutants from the industrial to the deposition areas. In Norway and Sweden, pollution mostly crosses the Baltic from northern Central Europe (the Ruhr Valley in Germany, the Czech Republic, Poland, Holland) and the British Isles, and adds to emissions from southern Scandinavia. In the eastern United States and Canada, 50 large emitters are responsible for 50% of the acid rain. This is very different from the western United States and Canada, where acid deposition seems primarily due to NOx attributable to vehicles. In Beijing, the severe rain acidity is to an extent decreased by $CaCO_3$, dust, NH_3, and sulfate pollution in the air (Berner and Berner 1996; VanLoon and Duffy 2005; Aguado and Burt 2010).

Sources of Photochemical Oxidants

Produced in smog processes, photochemical oxidants like O_3 and PAN have many hydrocarbon precursors. Globally, most hydrocarbons come from natural emissions enhanced by humans (such as CH_4 in husbandry and rice cultivation), but in industrialized nations, emissions from industries and cars dominate due to incomplete fuel combustion and gasoline evaporation leading to emissions of other ozone-forming chemicals (nitrogen oxides and carbon monoxide). In addition, O_3 is used as disinfectant (water, shoes, etc.), but it strongly oxidizes paints, plastics, and textiles. O_3 has increased by 36% in the troposphere since 1750 (IPCC 2007; Maslin 2009).

Aerosol Sources

Natural sources of atmospheric suspended particulate matter include volcanism and mineral weathering. Artificial sources are combustion engines and furnaces, fires, and the construction industry, in particular cement manufacture. Metallic particles are released by cement works (in particular the highly toxic and suspected carcinogen metal thallium), glass production, fossil fuel, and solid waste combustion. Land change releases dust and suspends cement particles.

After incineration, 1000 kg of solid waste yields 300 to 400 kg of solid residues, 5000 to 7000 m^3 of flue gases, as well as fly ash containing heavy metals, dioxins, and furans. Especially toxic are polychlorinated dibenzodioxins and dibenzofurans, polycyclic aromatic hydrocarbons (PAHs), and polychlorinated biphenyls (PCBs). Total suspended particles can be as high as 60 to 200 µg/m^3 of air in highly populated areas. Pollutants can be in the waste itself, be generated in the flue gases during incomplete combustion (PAHs), or in the presence of halogen organic compounds in the ashes in the 270°C to 470°C cool-down phase (Schwedt 2001).

Human influence is manifest in over 50 Mt (50 million metric tonnes) sulfate aerosols deposited in Greenland per year around the year 2000 (Maslin 2009). Anthropogenic aerosols collectively might have produced a cooling effect, which, along with volcanic aerosols, could have offset some global warming (IPCC 2007).

Solutions

No panacea exists but a series of feasible actions, combining prevention (to curb contaminant byproducts of combustion locally to reduce regional and global impacts), monitoring and modeling (to ascertain risks before damage is incurred), and mitigation through global atmospheric governance.

Prevention

Decarbonizing or the Transition to Sustainable Energy Production and Use

Fossil fuel combustion appears to be the common source of most GHGs; the origin of NOx-related O_3 layer depletion; black carbon in ABCs; SOx- and NOx-related acid rain; soot in classic smog; and NOx, PAN, and CO in photochemical smog. Combustion has a limited number of sources, including, but not limited to, engines, furnaces, waste incineration, and forest fires.

Decarbonizing has two acceptations: on the one hand phasing out fossil fuel production and use, and on the other hand taking carbon out of fossil fuels. Reducing CO_2 emissions and global warming has recently become more urgent in light of the demonstration that the global oceans, which are now the largest thermal and CO_2 stores reducing accelerated climate change, might release heat and CO_2, and maintain a high atmospheric temperature for a thousand years (Kennel et al. 2012). Decarbonization may not mean the end of all useful combustion and so renewable CO_2 sorption and separation materials (D'Alessandro et al. 2010) may allow for postcombustion CO_2 capture and the use of CO_2 as green chemistry solvent and input in the production of new plastics (see Chapter 17). Several decarbonization steps with co-benefits toward mitigating several atmospheric phenomena are necessary:

- Decreasing energy use through efficiency measures. However, this often lowers energy costs and so offsets decontamination gains (rebound and backfire effects). This requires economic measures to compensate for technological efficiency gains (Sorrell 2007; U.K. Department of Energy and Climate Change [DECC] 2011).

- Producing noncombustion energy via renewable energy sources.
- Preventing the production of noncarbon polluting gases (changing the combustion temperature and configuration of combustion chambers, trapping the exhaust particles, or removing pollutant precursors).
- Removing gases after their production (using catalyzers, injecting reagents into the combustion chambers, or retrofitting pollutant removal devices like slurry scrubbers). This, however, may produce large amounts of pollutants for whose proposed geological sinks are hotly debated.
- Producing synfuels, for example, combining carbon oxides (CO and CO_2) with hydrogen H_2.
- Recycling combustion byproducts (use of CO_2 in green chemistry).

It is possible to replace all fossil fuel with renewable energies, mostly wind, water, and sun (WWS), which would take up 0.5% of the land area. The infrastructure needed to supply future additional energy demand could be in place by 2030, since 70% of the needed hydropower is already operational, and all energy could be WWS-generated by 2050. Future energy costs can be similar to current costs. Political and social barriers, rather than technological and economic, have to be overcome: demand has to be reduced, and the development of the new infrastructure has to be planned and managed. Economic policies include eliminating subsidies for fossil fuels, and taxing fossil fuel use and production to finance climate change mitigation. Feed-in tariffs are subsidies to finance early differences between renewable generation costs and wholesale energy prices; although effective to transition to renewable energy sources, they still are subsidies that raise demand. A complementary approach is to substitute high-energy activities with low-energy ones (Delucchi and Jacobson 2011; Jacobson and Delucchi 2011).

It is, however, desirable to reduce the contribution of hydropower and land requirement for renewable energy, while generating more mobile energy sources. This can be done by making the economy circular, as in the recycling of solid waste for biomethane generation. The precautions needed to prevent leaks of CH_4 in the case of biomethane are easier to implement than those advocated for to curb CH_4 emissions in the current natural gas industry (Center for Biological Diversity, Clean Air Task Force, Western Environmental Law Center 2012).

Monitoring

Identifying the pollutants is the first crucial action toward controlling them, followed by tracing them to their sources. Thereafter, each emission source is monitored and the immission or deposition points sampled (Figure 11.2). The lack of systematic monitoring at all emission points and

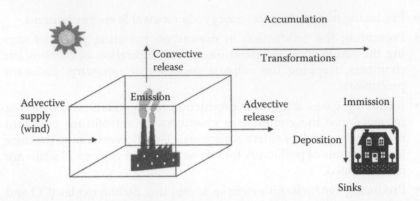

FIGURE 11.2
The Crutzen box model helps understand regional and global pollutant circulation. Through the interplay of convection, advection, pollutant lifetimes, and reactions, the contaminant plumes are transported regionally or globally. Those that do not react to enter sinks are transiently deposited or accumulated in the atmosphere.

sampling at sufficient immission points gives a prominent role to catastrophic events in prompting legal, political, and technological actions. This has been the case for the accidents at Bhopal (Dunnivant and Anders 2006) and Chernobyl. However, sudden catastrophic regional or global atmospheric events affecting humans have only been attributed so far to classical smog. Skin carcinogenicity hazard in temperate countries prompted adequate reactions to control the depletion of stratospheric ozone. For the rest of the phenomena in this chapter, all the disquieting evidence of sharp biophysical environmental modifications, and increased health hazards, has been insufficient to prompt sufficient regulatory reactions.

In this context, even a comprehensive monitoring and sampling approach might prove fruitless. This seriously undermines the effectiveness of prevention measures.

Modeling

In a 1971 report, the Federal Republic of Germany estimated at 60,000 the number of environmental chemicals produced worldwide. In 1980, the European Union compiled a list of 100,000 chemicals to which 5000 may have been added annually. One estimate of the current number of man-made chemicals exceeds 250,000, to which metabolites (the products of the biological degradation of chemicals, at times more dangerous that the original pollutants) and interactions within mixtures have to be added (Schwedt 2001; Hayward and Fuerhacker 2012).

It seems that chemical innovation is caught in a loop of development, mass production, damage, subsequent research to identify pollutants, attribution to specific pollution emitters, and micro and macro coping mechanisms are negotiated and implemented and eventually coping measures. This has been the sequence leading to classical smog control. However, the attribution of the ozone layer depletion to chlorofluorocarbons coincided with the observation of the Antarctic ozone hole, and led to the hitherto only international agreement successful in banning a vast set of pollutants. More recently, climate change reverted to the usual sequence of investigation, manifesting our reticence to proactive change. The sustainability logic would call for a precautionary approach whereby effects are modeled and probed *in vitro* and *in silico*, similar to pharmaceutical trials, prior to real-life scaling experiments (Figure 11.3).

The large number of new potential atmospheric pollutants calls for high-throughput *in vitro* ecotoxicological and chemical assays that consider several key endpoint (effects) and model the sequence of atmospheric pollution from resource exploitation to key long- and short-term endpoints, using the concepts of fugacity (the propensity of a chemical to distribute globally), accumulation potential, and degradability (based on life duration and mixing ratio).

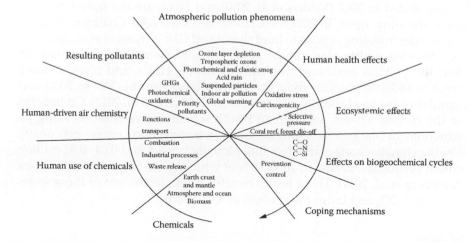

FIGURE 11.3

The logical sequence of atmospheric pollution investigation. This sequence moves away from ex-post facto monitoring to ex-ante atmospheric modeling, experimenting, and trialing (similar to ex-ante toxicity modeling already initiated in green chemistry). Humans use chemicals from the Earth. The resulting pollutants combine into pollution manifestations, which affect humans first as these are closer to the emission points, and subsequently alter survival and competition in the ecosystems. Eventually, several biogeochemical cycles are altered, calling for a combined view on biogeochemical cycles instead of the usual separate approach to them (Schwedt 2001). As this cycle suggests, the further the effects are left to develop, the deeper and more pervasive they become and the harder to tackle: prevention and control become insufficient when pollutants have accumulated beyond the capacity of natural sinks to remediate pollution.

Mitigating

Mitigation boils down to restoring atmospheric systems to their preindustrial states via pollution-neutral and pollution-negative measures. When this fails to succeed, adaptations are warranted.

Focusing on Co-Benefits

We may already be incurring dangerous anthropogenic interference with climate, since ABCs and other pollutants with cooling effects are masking the full extent of global warming. This warrants fast-action regulation that can be enacted within 2 to 3 years, implemented inside of a decade, and produce substantial results within decades. Salient fast-action measures are the reduction of tropospheric ozone precursor gases, short-lived GHGs, and phasing out long-lived, high global warming potential hydrofluorocarbons (HFCs) initially intended to curb ozone layer depletion. Regulatory measures could extend the Montreal Protocol and biosequestration through reforestation and biochar (Molina et al. 2009). Extending the Montreal Protocol on Substances That Deplete the Ozone Layer to HFCs, with null ozone depletion potential but very large global warming potential, is a way forward as the Kyoto Protocol on Climate Change ended in 2012 (Velders et al. 2010) and HFCs are the fastest-growing global warming agent. In 2012 the Climate and Clean Air Coalition included 25 countries initiating actions to limit short-lived GHGs (Kennel et al. 2012).

Co-benefits within the climate system would stem from reductions in short-lived GHGs (methane, tropospheric ozone, HFCs, and black carbon): forecasts indicate possible reductions in sea level rise (by 24% to 50%) and cumulative sea level rise (by 22% to 42%) by 2100 (Hu et al. 2013). Co-benefits for the climate system and human health would derive from focusing on reductions in black carbon and methane (a tropospheric ozone precursor), slowing climate change, and avoiding globally 0.6–4.4 and 0.04–0.52 million annual premature deaths in 2030, 80% of which could be prevented in Asia (Anenberg et al. 2012). These health benefits are independent of those stemming from CO_2 and indoor-air pollution reductions.

Adapting

Climate change mitigation is proceeding too slowly to limit temperature rise in the first half of the 21st century. This gives credence to the calls for sea level rise adaptations such as coastal defenses, albeit limited in their effectiveness (Kennel et al. 2012), or nutritional and pharmaceutical adaptations (Finch 2012). Adaptations are last-resort measures that do not address the causes of atmospheric pollution, on the contrary. A typical example is that to adapt to climate change we have used long-lived CFC and HFC coolants in electric fridges and air-conditioning, while unwittingly incurring ozone depletion and further climate change.

References

Aguado E, Burt JE (2010) *Understanding weather and climate*. Upper Saddle River, NJ: Prentice Hall.

Anenberg SC, Schwartz J, Shindell D, Amann M, Faluvegi G, Klimont Z, Janssens-Maenhout G, et al. (2012) Global air quality and health co-benefits of mitigating near-term climate change through methane and black carbon emission controls. *Environmental Health Perspectives* 120:831–839.

Berger A (2001) The role of CO_2, sea level, and vegetation during Milankovitch-forced glacial-interglacial cycles. In *Geosphere-biosphere interactions and climate*, Bengtsson LO, Hammer CU (eds), 119–143. Cambridge: Cambridge University Press.

Berner EK, Berner RA (1996) *Global environment. Water, air and geochemical cycles*. Upper Saddle River, NJ: Prentice Hall.

D'Alessandro DM, Smit B, Long JR (2010) Carbon dioxide capture: Prospects for new materials. *Angewandte Chemie International Edition* 49:6058–6082.

Delucchi MA, Jacobson MZ (2011) Providing all global energy with wind, water, and solar power, Part II: Reliability, system and transmission costs, and policies. *Energy Policy* 39:1170–1190.

Center for Biological Diversity, Clean Air Task Force, Western Environmental Law Center (2012). Before the Secretary of the Interior. Petition to update the Bureau of Land Management's regulations, notices, and orders to reduce emissions of natural gas from oil and gas operations.

Dunnivant FM, Anders E (2006) *A basic introduction to pollutant fate and transport: An integrated approach to chemistry, risk assessment, and environmental legislation.* Hoboken, NJ: John Wiley & Sons.

Finch CE (2012) Evolution of the human lifespan, past, present, and future: Phases in the evolution of human life expectancy in relation to the inflammatory load. *Proceedings of the American Philosophical Society* 156:9–44.

Fung I (2001) Atmospheric CO_2 variations. Response to natural and anthropogenic Earth system forcings. In *Geosphere-biosphere interactions and climate*, Bengtsson LO, Hammer CU (eds), 38–51. Cambridge: Cambridge University Press.

Hayward K, Fuerhacker M (2012) The challenge of emerging contaminants: Setting the course for a future of *in vitro* testing. *Water21* (February):38–39.

Hu A, Xu Y, Tebaldi C, Washington WM (2013) Mitigation of short-lived climate pollutants slows sea-level rise. *Nature Climate Change* 3:1–5.

INEGI (2010) Banco de Información Económica. http://www.inegi.org.mx/Sistemas/BIE/Default.aspx (accessed August 12, 2010).

Intergovernmental Panel on Climate Change (IPCC) (2007) Climate Change 2007: Synthesis report. Working group contributions of the Fourth Assessment Report. Valencia, November 12–17.

Jacobson MZ, Delucchi MA (2011) Providing all global energy with wind, water, and solar power, Part I: Technologies, energy resources, quantities and areas of infrastructure, and materials. *Energy Policy* 39:1154–1169.

Kennel CF, Ramanathan V, Victor DG (2012) Coping with climate change in the next half-century. *Proceedings of the American Philosophical Society* 156:398–415.

Lechter TM (ed) (2009) *Climate change: Observed impacts on planet Earth*. Amsterdam: Elsevier.

Lemke P (2006) Dimensions and mechanisms of global climate change. In *Multilevel governance of global environmental change: Perspectives from science, sociology and the law*, Winter G (ed), 37–66. Cambridge: Cambridge University Press.

Maslin M (2009) *Global warming: A very short introduction*. Oxford: Oxford University Press.

Molina M, Zaelke D, Sarma KM, Andersen SO, Ramanathan V (2009) Reducing abrupt climate change risk using the Montreal Protocol and other regulatory actions to complement cuts in CO_2 emissions. *PNAS* 106:20616–20621.

Ramanathan V, Feng Y (2009) Air pollution, greenhouse gases and climate change: Global and regional perspectives. *Atmospheric Environment* 43:37–50.

Schwedt G (2001) *The essential guide to environmental chemistry*. Chichester: John Wiley & Sons.

SMN (2010) Estaciones automáticas: Toluca. http://smn.cna.gob.mx/emas/ (accessed August 15, 2012).

Sorrell S (2007) The rebound effect: An assessment of the evidence for economy-wide energy savings from improved energy efficiency. London: UK Energy Research Centre.

U.K. Department of Energy and Climate Change (DECC) (2011) Decarbonisation. Planning our electric future: A White Paper for secure, affordable and low carbon electricity. London: TSO.

VanLoon GW, Duffy SJ (2005) *Environmental chemistry: A global perspective*. Oxford: Oxford University Press.

Velders GJM, Ravishankara AR, Miller MK, Molina MJ, Alcamo J, Daniel JS, Fahey DW, et al. (2010) Climate benefits by limiting HFCs. *Science* 335:922–923.

12

Airborne Toxic Pollutants

Marina Islas-Espinoza

CONTENTS

Introduction

Most of the deaths linked to environmental exposure are attributed to air pollution (World Health Organization [WHO]/IER 2009). Atmospheric emissions are theoretically able to disperse maximally when in gaseous state. This is a property that has been put to use in chemical warfare using sarin (Murakami 2003) or pentafluorophenylarsenic oxide among many other toxic war gases (Rettenmeier 2004). However, the atmosphere is not a perfect gas, and precipitation, wind, and topography lead to a heterogeneous distribution of both pollutants in the air and deposition.

More than 2 million premature deaths in 2008 were due to indoor smoke from solid fuels and 1.34 million deaths were attributed to urban outdoor air pollution (Figure 12.1). The main causes of death by outdoor air pollution are cardiopulmonary failure, lung cancers, and respiratory infections (WHO 2011c). Several of the air pollutants may occur naturally, but most are the result of human activities. Fine particle pollution often originates from

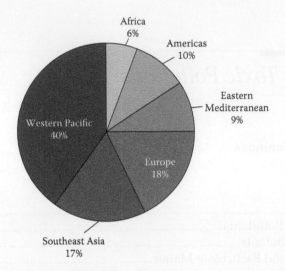

FIGURE 12.1
Percentage of deaths due to outdoor air pollution in 2008 by region. (Data calculated from World Health Organization [WHO], 2011, Public health and environment: Outdoor air pollution, number of deaths, 2008. http://gamapserver.who.int/gho/interactive_charts/phe/oap_mbd/atlas.html.)

combustion sources such as power plants and motor vehicles. The great majority of urban populations have an average annual exposure to PM10 particles (diameter <10 μm) in excess of the WHO Air Quality Guidelines recommended maximum level of 20 μg/m³ (WHO 2011d).

Exposure to Pollutants

Air pollutants can enter the human body through the skin and mucosal tissues (including the mouth) from different environmental sources: water, soil, gases, dust, and radiation, as well as the food chain (Florea et al. 2004; Stoltenburg-Didinger 2004). Similarly, their bodily storage, metabolism, and excretion may follow different paths (Hirner et al. 2004) and cause interactions (Fender and Wolf 2004; Hartwig et al. 2004; Stoltenburg-Didinger 2004). Hitherto, environmental monitoring and follow-ups of human subjects have seldom factored in this diversity of paths and interactions of mixtures. This likely has resulted in underestimates of exposure; the most regrettable consequence is probably that damages from this exposure may be attributed to nonenvironmental risk factors such as individual frailty and behavior. Of particular concern is exposure in the working environment and added exposures from outdoor and indoor air pollution.

Toxicity, or the amount needed for damage to occur at the organism, cell, or molecular levels, is measured as concentration or dose of a toxicant (micrograms per deciliters [μg/dl] of blood, nanogram [ng] or microgram per kilogram [μg/kg] of bodily mass). Inhibition of biological processes and lethality are the most widely used indicators of toxicity and are often summarized as LC_{50} (lethal concentration affecting 50% of the population under study), on account of variation in resistance among individual cells or organisms. Responses to acute and chronic exposures may differ considerably. Lead acute effects are localized in the cerebellum, whereas chronic effects are a subtle axodendritic disorganization in the hippocampus leading to hyperactivity, aggression, and seizures (Stoltenburg-Didinger 2004). There are other factors that make the dose-response vary such as age (mainly <5 years or >80 years), physical condition or genetic problems, and interactions of different compounds.

Health effects are changes in an individual or population, identifiable either by clinical diagnosis or statistical epidemiological methods. Risk analysis is a process that incorporates three components: risk assessment, risk management, and risk communication. The assessment of human health risk requires identification, compilation, and integration of information on the health hazards of a chemical; human exposure to the chemical; and relationships among exposure, dose, and adverse effects (WHO 2010b). The description of these steps is presented in Table 12.1; however, multiple exposures to different pollutants may alter the exposure (International Programme on Chemical Safety [IPCS] 2004). For example, in rural areas usually it is possible to find a mix of compounds such as cooking fuel,

TABLE 12.1

Risk Assessment Process

Step	Description	Content
Hazard identification	Identifies the type and nature of adverse health effects	Human studies
		Animal-based toxicology studies
		In vitro toxicology studies
		Structure–activity studies
Hazard characterization	Qualitative or quantitative description of inherent properties of an agent having the potential to cause adverse health effects	Selection of critical data set
		Modes/mechanisms of action
		Kinetic variability
		Dynamic variability
		Dose–response for critical effect
Exposure assessment	Evaluation of concentration or amount of a particular agent that reaches a target population	Magnitude
		Frequency
		Duration
		Route
		Extent
Risk characterization	Advice for decision making	Probability of occurrence
		Severity
		Given population
		Attendant uncertainties

Source: World Health Organization (WHO) (2010b), WHO human health risk assessment toolkit: Chemical hazards, IPCS harmonization project document No. 8, Geneva: WHO.

herbicides, pesticides, fertilizers, and combustion of vegetation and garbage that may generate a combination of symptoms in the population.

Main Air Pollutants

Aerosols and Particulate Matter

Recently, the difference between aerosols in global change studies and particulate matter in urban and outdoor air pollution research has vanished. There is now evidence of the global atmospheric circulation as a factor in dispersing and concentrating toxic pollutants. This is clearly demonstrated by aviation, which accounts for 8000 premature deaths per year (1% of all air pollution deaths and 10% of all deaths linked to particulate matter intercontinental transport). Due to atmospheric circulation, the United States incurs 7 times fewer casualties than expected from their aircraft emissions, whereas India's mortality is sevenfold the amount attributable to its aircraft emissions. India and China emit 2% and 8% of global aviation emissions, respectively, but combined represent 35% of aviation-attributable deaths. Long-term exposure to black carbon and particulate organic carbon less than 2.5 μm in diameter (also called PM2.5 or fine particulate matter) are deemed responsible (Barrett et al. 2010). The PM2.5 particles are more dangerous since, when inhaled, they may reach the peripheral regions of the bronchioles and interfere with gas exchange inside the lungs (WHO 2011a).

PM10 can penetrate into the lungs and the bloodstream, can cause heart disease, lung cancer, asthma, and acute lower respiratory tract infections (WHO 2011d). The major components of PM are sulfate, nitrates, ammonia, sodium chloride, carbon, mineral dust, and water suspended in the air (WHO 2011a).

Urban air pollution is mostly derived from fossil fuel combustion sources releasing a complex mixture of thousands of compounds to the atmosphere, which includes carcinogens such as benzo(a)pyrene, benzene, 1,3-butadiene, and lead. Considering 3200 cities with more than 100,000 inhabitants around the globe, urban PM air pollution causes 3% of cardiopulmonary disease in adults; 5% of cancers of trachea, bronchus, and lung; and 1% of acute respiratory infections in children 4 years old and younger. This is equivalent to 0.80 million premature deaths (1.4% of the global total) and occurs largely in the western Pacific Rim, Southeast Asia, and eastern Europe. Energy use is held responsible for particulate matter emissions and an associated 14,000 preventable deaths in Europe (Bollen et al. 2009). Similarly, mortality in Latin American cities such as Mexico City and Sao Paulo increase in response to higher PM concentrations (Cohen et al. 2004).

Age-wise, almost half the premature deaths (1 million) by indoor air pollution are due to pneumonia in children under 5 years of age. Exposure is difficult to pinpoint as the sensitivity of children and their number differ across regions in response to nutrition and health conditions and fertility trends; these differences in the populations of children are compounded in

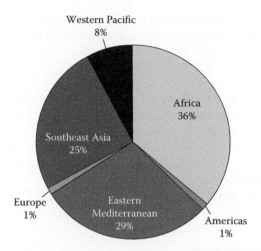

FIGURE 12.2
Percentage of deaths of children <5 years old due to outdoor air pollution in 2008 by region. (Data calculated from World Health Organization [WHO], 2011, Public health and environment: Outdoor air pollution, number of deaths, 2008. http://gamapserver.who.int/gho/interactive_charts/phe/oap_mbd/atlas.html.)

the regional pattern of deaths in children compared to the general regional patterns of deaths due to outdoor air pollution (Figure 12.1 and Figure 12.2). There are international, national, and local institutions that have established different guideline values to limit the air pollutant emissions in various parts of the world. The WHO Air Quality Guidelines represent the most widely accepted and up-to-date assessment of health effects of air pollution and recommends targets to reduce health risks. However, exposure to air pollutants is largely beyond the control of individuals and requires public support at the local, national, and international levels (WHO 2011a).

Heavy industries illustrate the link between global atmospheric change emissions and toxic air pollutants. The cement industry atmospheric emissions include but are not limited to benzene, tin, chromium, lead, chlorine, mercury, cadmium, and nonmethane organic volatile compounds. A salient feature of emission inventories is the heterogeneity with which different cement plants declare pollutants (Chen et al. 2010). This complicates the estimates of global pollution and personal exposure.

Notable among PM10 particles are the carcinogenic polycyclic aromatic hydrocarbons (PAHs), which have been the focus of several molecular epidemiological investigations. Several studies of populations exposed to PAHs showed increased levels of several markers of genotoxicity, including DNA adducts, chromosome aberrations, sister chromatid exchanges, and *ras* oncogene overexpression. These results indicate that at least in some regions PAHs are a major source of the genotoxic and embryotoxic activities of organic mixtures associated with air pollution. In addition to the formation of DNA damage by exogenous exposures to carcinogens, such as PAHs, it is known that DNA is modified, often oxidatively, by radicals (Farmer et al. 2003).

Ozone (O₃)

Tropospheric O_3 is one of the major constituents of photochemical smog. It is formed by the reaction with sunlight of pollutants such as nitrogen oxides (NOx) and volatile organic compounds (VOCs) emitted by vehicles, solvents, and industry. The highest levels of ozone pollution occur during periods of sunny weather (WHO 2011a). European studies have reported that the daily mortality rises by 0.3% and heart diseases by 0.4% per 10 $\mu g/m^3$ increase in ozone exposure. The main effects of ozone and other pollutants are described in Table 12.2.

Volatile Organic Compounds

Volatile organic compounds (VOCs) can generally be understood as solvents containing carbon that evaporate easily at 20°C, such as used in dry cleaning and painting; in the plastic coating industries; in adhesives, inks, and printing; in the footwear industry, and in vegetal and animal fat processing. They contribute to the formation of photochemical oxidants such as O_3 and include carcinogens, mutagens, or agents toxic to human reproduction (Council of the European Union 1999). VOCs are one of the reportable categories of criteria pollutants in the United States, along with PM10, PM2.5, and Pb (U.S. Environmental Protection Agency [EPA] 2008).

Nitrogen Dioxide (NO₂)

NO_2 is a toxic gas, highly reactive, oxidant, and corrosive. The primary sources are combustion processes in vehicles, gas stoves, vented appliances with defective installations, as well as welding and tobacco smoking. NO_2 acts mainly as an irritant affecting the eyes, nose, throat, and respiratory tract. Extremely high-dose exposure to NO_2 may result in pulmonary edema and diffuse lung injury. Continued exposure to high NO_2 levels can contribute to the development of acute or chronic bronchitis. Low NO_2 exposure may cause increased bronchial reactivity in some asthmatics, decreased lung function in patients with chronic obstructive pulmonary disease, and increased risk of respiratory infections, especially in young children (EPA 2012a).

Sulfur Dioxide (SO₂)

SO_2 is a highly reactive colorless gas. It can be oxidized to sulfur trioxide, which in the presence of water vapor is readily transformed to sulfuric acid aerosol. SO_2 is a precursor of sulfates, which are one of the main components of respirable particles in the atmosphere. It also damages trees and crops.

TABLE 12.2

Source and Effects of Selected Air Pollutants

Pollutant and Sources	Harmful Effects in Humans
Particulate matter (PM): PM10 and PM2.5—Indoor air pollution due to biomass fuel and coal for heating and cooking.	Premature death by pneumonia in children <5 years of age, and deaths by chronic obstructive pulmonary disease (e.g., chronic bronchitis) among adults. Risk of acute lower respiratory infections and associated mortality among young children. Major risk factor for chronic obstructive pulmonary disease and lung cancer among adults. Asthma, cataracts, stroke, tuberculosis, adverse pregnancy outcomes, in particular low birth weight, ischemic heart disease, interstitial lung disease, and nasopharyngeal and laryngeal cancers (WHO 2006, 2010a, 2011a, 2011b).
Particulate matter (PM): PM10 and PM2.5. Black carbon—Rural and urban outdoor air pollution, generated mainly from motor transport, small-scale manufacturers and other industries, burning of biomass and coal for cooking and heating, as well as coal-fired power plants. Residential wood and coal burning for space heating, especially in rural areas in colder months. PM can also be generated by mechanical grinding processes during industrial production, and by natural sources such as natural wind-blown dust.	Chronic exposure to particles contributes to the risk of developing cardiovascular and respiratory diseases, as well as lung disease. PM2.5 are more dangerous since, when inhaled, they may reach the peripheral regions of the bronchioles, and interfere with gas exchange inside the lungs. Animal data suggest that exposure to chemicals including tributyltin, bisphenol A, organochlorine and organophosphate pesticides, lead, diethylstilbestrol, perfluorooctanoic acid, monosodium glutamate, and nicotine can lead to altered cholesterol metabolism and weight gain later in life. In rodents also, pesticides and air pollutants have been shown to contribute to diabetes-related effects following adult exposures. Similarly, adult male rats exposed to fine particulate matter (PM2.5) in conjunction with a high-fat diet developed insulin resistance. Allergies, asthma, and airway disorders, endometriosis, and autoimmune thyroid disease in humans may have roots in endocrine disrupting chemical exposure (WHO 2011b; UNEP/WHO 2013).
Ozone (O₃)—Vehicles, solvents, and industry.	Breathing problems, triggered asthma, reduced lung function, lung and heart diseases, and mortality. Increased incidence of type 1 diabetes (WHO 2011a; UNEP/WHO 2013).
Nitrogen dioxide (NC₂)—Combustion processes (heating, power generation, and engines in vehicles).	Inflammation of the airways. Symptoms of bronchitis in asthmatic children increase in association with long-term exposure. Reduced lung function (WHO 2011a).

Continued

TABLE 12.2 (*Continued*)

Source and Effects of Selected Air Pollutants

Pollutant and Sources	Harmful Effects in Humans
Sulfur dioxide (SO_2)—Burning of fossil fuels (coal and oil) and the smelting of mineral ores that contain sulfur. Burning of sulfur-containing fossil fuels for domestic heating, power generation, and motor vehicles.	Affects the respiratory system and the function of the lungs, and causes irritation of the eyes. Irritation of the respiratory tract causes coughing, mucus, secretion, aggravation of asthma and chronic bronchitis and makes people more prone to infections of the respiratory tract. Cardiac disease and mortality. Increases the incidence of type 1 diabetes. People with asthma or chronic lung or heart disease are the most sensitive to SO_2 (WHO 2011a; Ontario Ministry of the environment 2010; UNEP/WHO 2013).
VOCs—DEHP-phthalate, formaldehyde, indoor PVC-phthalate release.	Increases the risk for bronchial obstruction, asthma, and wheezing in children (UNEP/WHO 2013).
Polybrominated diphenyl ethers (PBDEs)—Flame retardants in textiles, electronics, electric articles, furniture, and building materials.	Limited evidence for cognitive disorders, earlier age at menarche and cryptorchidism. Strong experimental evidence for suppression of thyroid hormone (UNEP/WHO 2013).
PAHs—Cigarette smoke, motor vehicle exhausts, industrial activities, and fossil fuel and wood burning.	Carcinogenic (Farmer et al. 2003; UNEP/WHO 2013).
Bacteria, fungi, algae, and some protozoa (their β-glucans, toxins, spores, pollen, cell fragments, methane, and VOCs emissions)—Microbial indoor air pollution when sufficient moisture is available.	Increases prevalence of respiratory symptoms, wheezing, cough, infections (bronchitis, fungal), allergies (alveolitis) and asthma, upper respiratory tract symptoms as well as perturbation of the immunological system (WHO 2009a).
Organometals and organometalloids—Metal and waste industries.	Mutagenic, carcinogenic, cytotoxic, neurotoxic (Hirner et al. 2004).
Radioactive materials—Radiation from nuclear power plants.	High doses (>100 mSV): Affect fetal development or outcome of pregnancy, spontaneous abortion, miscarriage, perinatal mortality, congenital defects, cognitive impairment and death. Also, significant associations between radiation exposure, and both cerebrovascular and cardiac disease mortality. Low doses (<100 mSV): Leukemia, breast cancer, thyroid cancer, and all solid cancers (WHO 2013).
Radon—Indoor radon exposure in buildings.	Lung cancer (WHO 2009b).

SO_2, along with nitrogen oxides, are the main precursors of acid rain. As such it contributes to the acidification of lakes and streams, accelerated corrosion of buildings, and reduced visibility. SO_2 also causes the formation of microscopic acid aerosols, which have serious health implications and contribute to climate change (Ontario Ministry of the Environment 2010). The largest sources of SO_2 emissions are from fossil fuel combustion at power plants (73%) and other industrial facilities (20%). Smaller sources of SO_2 emissions include industrial processes such as extracting metal from ore, and the burning of high-sulfur-containing fuels by locomotives, large ships, and nonroad equipment. SO_2 is linked with a number of adverse effects on the respiratory system (EPA 2012b).

Organometal(loid)s

Organometals and organometalloids are volatile, mobile, and toxic metals with a molecule that has a carbon moiety, most often a methyl group, rendering them lipophilic and so allowing them to bind more easily to living cells and become bioaccumulated.

Most organometal(loid)s yield metabolites that are both hydrophilic and lipophilic, facilitating their transport in bodily fluids and cell wall penetration. Interference with enzymes, proteins, hemoglobin, cytochromes, DNA, and the immune system ensue, but neurotoxic diseases dominate, including harm to muscle and sensory functions (paralysis and seizures), and harm to mood and cognitive functions (learning, memory, speech, and behavior) (Gruner et al. 2004).

Some organometals have been used as industrial catalysts, gasoline antiknock agents, biocides and antitumoral agents (Yonchev et al. 2004), preservatives, and antifouling paints (Gruner et al. 2004). Natural and anthropogenic areas (wetlands and ponds, waste dumps, and sewage systems) also host anaerobic microorganisms able to methylate heavy metals. Organometal(loid)s enter the atmosphere as gases or aerosols, and are exchanged with the hydrosphere and soil (Florea et al. 2004). Metal recycling plants are also hazardous working places (Fender and Wolf 2004).

Metals interfere with cell functions. Methylation in the human liver and kidney or by anaerobic gut microorganisms yields more cytotoxic compounds (affecting cells) than those entering it (Hirner et al. 2004); metals accumulate along the food chain (Stoltenburg-Didinger 2004).

High levels of arsenic exist in smelting, glass production, and other working environments. Arsenic inhalation increases the risk of lung cancer and may inhibit DNA repair and generate reactive oxygen species leading to elevated oxidative DNA damage at very low doses. Some of the induced lesions are mutagenic and lead to carcinogenesis (Hartwig et al. 2004).

Increased chromosomal aberrations have been found in workers in most exposure studies. Chromosomal aberrations and cancer risk have been linked. Organometal(loid) exposure includes tin, antimony, arsenic, and lead in landfill gases, two order of magnitudes higher than in the background environment (Fender and Wolf 2004). Waste site workers had frequent multiaberrant cells, without influence of the duration of employment. In copper smelters, elevated lead and dioxin/furans (from plastics) were found, as well as aberrations, but significant only in cells with dicentric chromosomes. More sister chromatide exchanges were found than in waste site workers.

The brain concentrates metal ions such as lead (Pb) and mercury (Hg) in astroglia. Pb enters the human body mainly through inhalation and ingestion. It kills cells and interferes with normal neuronal development. Lead often mimics and inhibits the action of calcium. Lead produces irreversible learning and memory deficits in children at 15 µg/dl blood but not in adults so the limit tolerable serum level was fixed at 10 µg/dl blood by WHO and the U.S. Centers for Disease Control. Astroglia seem to be the main accumulation site in the brain, as well as for other neurotoxic metals (Stoltenburg-Didinger 2004). Astroglia participate in neurotransmitter metabolism in the blood–brain barrier, stress, and injury responses.

Lead can be found in all parts of our environment—the air, the soil, the water, and even inside our homes. Much of our exposure comes from human activities including the use of fossil fuels including past use of leaded gasoline, some types of industrial facilities, and past use of lead-based paint in homes. Lead and lead compounds have been used in a wide variety of products found in and around our homes, including paint, ceramics, pipes and plumbing materials, gasoline, batteries, ammunition, and cosmetics (EPA 2012c).

Lead emissions from volcanoes or fires are dwindled by anthropogenic emissions: 2000 to 18,000 t per year as compared to 300,000 to 450,000 t. Lead has been banned from gasoline and paints in many countries but even after production declined in developed countries, dust, water, and blood still contain lead beyond WHO recommended limits in 5% to 15% of the samples. Children absorb 50% of the lead intake and adults 10%, meaning possible intakes of up to 170 µg per day per person. Until the late 1960s, 80 µg were allowed by law (60 µg in children). WHO established admissible limits in 1987 when it was already known since 1979 that these levels were conducive to impaired development, memory, and behavior. Furthermore, it was known since 1974 that lead accumulates in the hippocampus, an essential brain structure for memory and learning. Reduction of hemoglobin synthesis at the set admissible levels was known since 1982. It was already known since 1984 that 25 µg/dL shortens pregnancy and reduces birth weight. In fact, lead neurotoxicity in children was clinically known since 1897. Many other cellular processes are disturbed by lead at low doses. As to lead from gasoline, it is metabolized in the human body to even more toxic products that depress several cellular processes (Hirner et al. 2004). According to a New York State Department of Health 1993 report, more patients were treated for lead than alcohol toxicity. Cities in developing countries are most at risk (Büsselberg 2004).

Aluminum has been linked to dementia and Alzheimer's disease. Zinc plays several important roles in mammal cells with reports of toxicity at even low doses. Glutamate is the most important excitatory neurotransmitter and its receptors are involved in learning, memory, development, and synaptic plasticity, which are blocked or disturbed by lead and aluminum. Mercury is used in tooth fillings and as antiseptic in vaccines, despite known cognition, tremor, coordination, and reflex issues of mercury intake, the mechanisms of which have been widely studied (Hirner et al. 2004). According to the 2008 U.S. EPA National Emissions Inventory, coal-fired electric power plants cause 48% of the total anthropogenic mercury emissions, followed by electric steel furnaces (8%), industrial boilers (7%), waste combustion in cement manufacturing (7%), and gold mining (3%).

Subclinical nerve injuries were found in Swedish copper smelter workers due to long-term exposure. Chronic exposure to combustion of coal containing elevated arsenic led to loss of hearing, taste, blurred vision, and limb numbness. Neurotoxic pesticide effects, ranging from chronic subclinical to acute clinical, have been reported around the world (Gruner et al. 2004).

Radiation

Radiation is a general term referring to any sort of energy that can travel through space either as a wave or a particle. In considering related health risks, radiation may be classed as: (1) nonionizing radiation (lower energy, such as UV, visible light, infrared, microwaves, radio and radar waves, wireless Internet connections, mobile phone signals) and (2) ionizing radiation (higher energy, such as α-particles, β particles, γ-rays, x-rays, cosmic protons, neutrons in accidental emissions of nuclear power plants) (NHS 2013).

Radiation damage to tissue or organs has been shown to depend on the type of radiation, the sensitivity of different tissues and organs, the dose, and the dose rate. Adverse health effects of radiation result from two mechanisms: (1) cell killing, which may cause functional impairment of the exposed tissue or organ only if a sufficient number of cells are affected; and (2) nonlethal changes in molecules of a single cell, most commonly in the DNA molecule, which may result in an increased risk of disease long after exposure. The first type of effects is determined mostly at the time of radiation. The second type of effect occurs through a random process that is not entirely determined at the time of radiation. These effects include cancer and inheritable effects. At low doses, radiation risks are primarily related to the second type, rather than to the deterministic effects characteristic of higher-dose exposure (WHO 2013).

Most radiation exposure is from natural sources (85% of the annual human radiation dose), mainly radioactivity in rocks and soil; radon, a radioactive gas given out by many volcanic rocks and uranium ore; and cosmic radiation. Radiation arising from human activities typically accounts for up to 15% of the public's exposure every year. This radiation is no different from natural radiation except that it can be controlled. X-rays and other medical procedures account for most exposure from this quarter. Less than 1% of exposure is due to the fallout from past testing of nuclear weapons or the generation of electricity in nuclear plants, as well as coal and geothermal power plants (World Nuclear Association 2012). This does not account for local exposure following nuclear plant accidents.

Radon is a radioactive gas that emanates from rocks and soils, and tends to concentrate in enclosed spaces like underground mines or houses. Soil gas infiltration is the most important source of residential radon; other sources, including building materials and water extracted from wells are of lesser importance. The proportion of all lung cancers linked to radon is estimated between 3% and 14%, depending on the average radon concentration in the country and on the method of calculation. Radon is the second most important cause of lung cancer after smoking in many countries. Radon is much more likely to cause lung cancer in people who smoke or who have smoked in the past than in lifelong nonsmokers. However, it is the primary cause of lung cancer among people who have never smoked. There is no known threshold concentration below which radon exposure presents no risk. Addressing radon measures in construction of new buildings (prevention) and in existing buildings (mitigation or remediation) is needed. The primary radon prevention and mitigation strategies focus on sealing radon entry routes and on reversing the air pressure differences between the indoor occupied space and the outdoor soil through different soil depressurization techniques (WHO 2009b).

Endocrine Disrupting Chemicals

There is global transport of endocrine disrupting chemicals (EDCs) through air currents, leading to worldwide exposure of humans and wildlife. EDCs have the capacity to interfere with tissue and organ development and function, and therefore they may alter the susceptibility to different types of diseases throughout life. Nondescended testes in young boys are linked with exposure to diethylstilbestrol (DES) and polybrominated diphenyl ethers (PBDEs), and with occupational pesticide exposure during pregnancy. Exposures to polychlorinated dioxins and certain polychlorated biphenyls (PCBs) are risk factors in breast cancer. Prostate cancer risks are related to occupational exposures to pesticides (of an unidentified nature), to some PCBs and to arsenic. Developmental neurotoxicity with negative impacts on brain development is linked with PCBs. Attention deficit/hyperactivity disorder (ADHD) is overrepresented in populations with elevated exposure to organophosphate pesticides. An excess risk of thyroid cancer was observed among pesticide applicators and their wives, although the nature of the pesticides involved was not defined (UNEP-WHO 2013).

Halogenated (chlorin-containing) organic substances include dioxins and furans and the most toxic compound yet known to man:

2,3,7,8-Tetrachlorodibenzodioxin (TCDD). Dioxins, furans, heavy metals, some PCBs, and pentachlorinated phenol (PCP) are carcinogens found in incineration atmospheric emissions (Council of the European Union 2000).

Dampness and Mold

Indoor environments contain a mixture of live and dead microorganisms, fragments thereof, toxins, allergens, volatile microbial organic compounds, and other chemicals, which in some concentrations are known or suspected to be elevated in damp indoor environments and may affect the health of people living or working there. In particular, it has been suggested that dust mites and fungi play a major role, since these produce allergens, toxins, and irritants known to be associated with allergies, asthma, and respiratory diseases. Dampness may also promote bacterial growth and the survival of viruses, but this has received little attention in the literature. The prevalence of home dampness indicated by occupants and inspectors is in the order of 10% to 50% in the most affluent countries. For less affluent countries, the prevalence sometimes exceeds 50%. Climate change and the rise in sea level and increased frequency and duration of floods are likely to increase the proportion of buildings with damp problems. In addition, inadequate design will prevent adequate ventilation in winter in many houses, leading to increased condensation and indoor dampness. Preventing persistent dampness and microbial growth on interior surfaces (through ventilation) is the most important means of avoiding harmful effects on health (WHO 2009a).

There are many evidences of health effects in humans due to air pollutants, which are summarized in Table 12.2. This list is not exhaustive and the guideline values of the pollutants must be consulted in the WHO reports. The variation range of exposure in the scientific literature is high. The limits or guideline values change according to new evidence (WHO 2011a).

Air-Cleaning Technologies

The most immediate method of improving air quality would be the use of biogas, solar, and wind energy, and hybrid vehicle technologies which apart from reducing toxic pollutants may help to reduce greenhouse emissions (WHO 2010a). Particle filtration and sorption of gaseous pollutants are among the most effective air-cleaning technologies (Figure 12.3), although used particle filters can be a source of pollution and there is insufficient information regarding long-term performance and proper maintenance (Zhang et al. 2011). Cool roofs and pavements, and urban trees can have a substantial effect on urban air temperature and, hence, can reduce cooling-energy use and smog (Akbari et al. 2001).

FIGURE 12.3 (See color insert.)
A vortex creates an ascendant flow to suspend the particles imitating wind. Then the suspended particles are trapped in the drops imitating rain, and absorbed in a filter that imitates deposition and soil filtration. (© Photo by Marina Islas-Espinoza.)

References

Akbari H, Pomerantz M, Taha H (2001) Cool surfaces and shade trees to reduce energy use and improve air quality in urban areas. *Solar Energy* 70:295–310.

Barrett SRH, Britter RE, Waitz IA (2010) Global mortality attributable to aircraft cruise emissions. *Environmental Science & Technology* 44:7736–7742.

Bollen J, Hers S, van der Zwaan B (2009) An integrated assessment of climate change, air pollution, and energy security policy. Nota di lavoro 105. Fondazione Eni Enrico Mattei.

Büsselberg D (2004) Actions of metals on membrane channels, calcium homeostasis and synaptic plasticity. In *Organic metal and metalloid species in the environment*, Hirner AV, Emons H (eds), 260–281. Berlin: Springer.

Chen C, Habert G, Bouzidi Y, Jullien A (2010) Environmental impact of cement production: Detail of the different processes and cement plant variability evaluation. *Journal of Cleaner Production* 18:478–485.

Cohen AJ, Anderson HR, Ostro B, Pandey KD, Krzyzanowski M, Kuenzli N, Gutschmidt K, et al. (2004) Mortality impacts of urban air pollution. In *Comparative quantification of health risks: Global and regional burden of disease due to selected major risk factors*, Vol. 2, Ezzati M, Lopez AD, Rodgers A, Murray CUJL (eds). Geneva: World Health Organization.

Council of the European Union (1999) Council Directive 1999/13/EC of 11 March 1999 on the limitation of emissions of volatile organic compounds due to the use of organic solvents in certain activities and installations. *Official Journal of the European Communities* 22.

Council of the European Union (2000) Directive 2000/76/EC of the European Parliament and of the Council of 4 December 2000 on the incineration of waste. *Official Journal of the European Communities* 22.

Farmer PB, Singh R, Kaur B, Sram RJ, Binkova B, Kalina I, Popov TA, et al. (2003) Molecular epidemiology studies of carcinogenic environmental pollutants. Effects of polycyclic aromatic hydrocarbons (PAHs) in environmental pollution on exogenous and oxidative DNA damage. *Mutation Research/Reviews in Mutation* 544:397–402.

Fender H, Wolf G (2004) Cytogenetic investigations in employees from waste industries. In *Organic metal and metalloid species in the environment*, Hirner AV, Emons H (eds), 235–246. Berlin: Springer.

Florea AM, Dopp E, Obe G, Rettenmeier AW (2004) Genotoxicity of organometallic species. In *Organic metal and metalloid species in the environment*, Hirner AV, Emons H (eds), 205–219. Berlin: Springer.

Gruner J, Kruger K, Binding N, Madeja M, Mushoff U (2004) Effects of organometal(loid) compounds on neuronal ion channels: Possible sites for neurotoxicity. In *Organic metal and metalloid species in the environment*, Hirner AV, Emons H (eds), 283–316. Berlin: Springer.

Hartwig A, Schwerdtle T, Walter I (2004) Current aspects on the genotoxicity of arsenite and its methylated metabolites: Oxidative stress and interactions with the cellular response to DNA damage. In *Organic metal and metalloid species in the environment*, Hirner AV, Emons H (eds), 221–233. Berlin: Springer.

Hirner AV, Hartmann LM, Hippler J, Kresimon J, Koesters J, Michalke K, Sulkowski M, Rettenmeier AW (2004) Organometal(loid) compounds associated with human metabolism. In *Organic metal and metalloid species in the environment*, Hirner AV, Emons, H (eds), 181–204. Berlin: Springer.

International Programme on Chemical Safety (IPCS) (2004) IPCS risk assessment terminology. Part 1: IPCS/OECD key generic terms used in chemical hazard/risk assessment. Part 2: IPCS glossary of key exposure assessment terminology. Geneva, World Health Organization, International Programme on Chemical Safety. Harmonization Project Document No. 1. http://www.who.int/ipcs/methods/harmonization/areas/ipcsterminologyparts1and2.pdf (accessed August 22, 2010).

Murakami H (2003) *Underground: The Tokyo gas attack and the Japanese psyche*. London: Vintage.

NHS (2013) Radiation. http://www.nhs.uk/conditions/Radiation/Pages/Introduction.aspx (accessed March 12, 2013).

Ontario Ministry of the Environment (2010) Sulfur dioxide. http://www. airqualityontario.com/science/pollutants/sulphur.php (accessed March 12, 2013).

Rettenmeier AW (2004) Panel discussion: Toxicological aspects. In *Organic metal and metalloid species in the environment*, Hirner AV, Emons, H (eds), 319–321. Berlin: Springer.

Stoltenburg-Didinger G (2004) Neurotoxicity of metals. In *Organic metal and metalloid species in the environment*, Hirner AV, Emons H (eds), 245–257. Berlin: Springer.

United Nations Environment Program (UNEP)/World Health Organization (WHO) (2013) State of the science of endocrine disrupting chemicals 2012. Bergman Å, Heindel JJ, Jobling S, Kidd KA, Zoeller RT (eds). http://unep.org/pdf/9789241505031_eng.pdf (accessed May 12, 2013).

U.S. Environmental Protection Agency (EPA) (2008) EPA's 2008 National Emissions Inventory version 2. Technical support document. Methods (p. 157). Research Triangle Park, NC: US EPA. http://www.epa.gov/ttn/chief/net/2008inventory.html#inventorydoc (accessed August 17, 2012).

U.S. Environmental Protection Agency (EPA) (2012a) An introduction to indoor air quality (IAQ) Nitrogen dioxide (NO2). http://www.epa.gov/iaq/no2.html (accessed August 17, 2012).

U.S. Environmental Protection Agency (EPA) (2012b) Sulfur dioxide. http://www.epa.gov/air/sulfurdioxide/ (accessed August 17, 2012).

U.S. Environmental Protection Agency (EPA) (2012c) Learn about lead. http://www.epa.gov/lead/learn-about-lead.html (accessed August 17, 2012).

World Health Organization (WHO) (2006) Fuel for life: Household energy and health. http://archive.org/details/fuelforlifehouse00generich (accessed August 24, 2012).

World Health Organization (WHO) (2009a) WHO guidelines for indoor air quality: Dampness and mould. Copenhagen: WHO Regional Office for Europe.

World Health Organization (WHO) (2009b) WHO handbook on indoor radon: A public health perspective. Zeeb H, Shannoun F (eds). Geneva: World Health Organization. http://www.nrsb.org/pdf/WHO%20Radon%20Handbook.pdf (accessed March 12, 2013).

World Health Organization (WHO) (2010a) Health in the green economy. Public Health & Environment Department (PHE). Health Security & Environment Cluster (HSE). www.who.int/hia/green_economy/en/index.html (accessed May 18, 2013).

World Health Organization (WHO) (2010b) WHO human health risk assessment toolkit: Chemical hazards. IPCS Harmonization Project Document No. 8. Geneva: WHO.

World Health Organization (WHO) (2011a) Air quality and health. Fact sheet No. 313. http://www.who.int/mediacentre/factsheets/fs313/en/ (accessed November 14, 2012).

World Health Organization (WHO) (2011b) Indoor air pollution and health. Fact sheet No. 292. http://www.who.int/mediacentre/factsheets/fs292/en/ (accessed March 19, 2013).

World Health Organization (WHO) (2011c) Public health and environment: Outdoor air pollution, number of deaths, 2008. http://gamapserver.who.int/gho/interactive_charts/phe/oap_mbd/atlas.html (accessed October 12, 2012).

World Health Organization (WHO) (2011d) Tackling the global clean air challenge. Media Centre. http://www.who.int/mediacentre/news/releases/2011/air_pollution_20110926/en/ (accessed March 12, 2013).

World Health Organization (WHO)/IER (2009) Global estimates of environmental burden of disease. http://www.who.int/quantifying_ehimpacts/global/envrf2004/en/index.html (accessed March 12, 2013).

World Health Organization (WHO) (2013) Health risk assessment from the nuclear accident after the 2011 Great East Japan earthquake and tsunami based on a preliminary dose estimation. Geneva: World Health Organization. http://apps.who.int/iris/bitstream/10665/78218/1/9789241505130_eng.pdf (accessed July 18, 2012).

World Nuclear Association (2012) Nuclear radiation and health effects. England and Wales, number 01215741. http://www.world-nuclear.org/info/Safety-and-Security/Radiation-and-Health/Nuclear-Radiation-and-Health-Effects/ (accessed July 18, 2013).

Yonchev R, Rehage H, Kuhn H (2004) Molecular modeling studies of specific interactions between organometallic compounds and DNA. In *Organic metal and metalloid species in the environment*, Hirner AV, Emons H (eds), 167–180. Berlin: Springer.

Zhang Y, Mo J, Li Y, Sundell J, Wargocki P, Zhang J, Little JC, et al. (2011) Can commonly-used fan-driven air cleaning technologies improve indoor air quality? A literature review. *Atmospheric Environment* 45:4329–4343.

13

Evolution of Human Features Driving Current Unsustainability

Alejandro de las Heras

CONTENTS

No element sets *Homo sapiens* apart from the other animals, but rather a combination of elements found in other species. Our biological, technological, and organizational inheritance links us to nature and also helps explain the long-term evolution of unsustainability features in present-day human cultures. This chapter reviews literature on the evolution of such features. An apparent result is that they seem to be relatively recent and so amenable to rapid corrective actions: while our brain–hand–eye connections seem to have coevolved with technology and human sociability since at least 2.5 million years ago, our evolution in the last 200,000 years has been essentially social. Moreover, elevated rates of human population growth and technologically led encroachment on the planet's resources have only occurred in the last 10,000 years.

Nonhuman Culture, Technology, and Economy

Human biology undeniably makes us animals. This is why culture (human organization, technology, and communication) believed to be a uniquely human feature, has been posited as a prime mover of unsustainability. This section shows that culture precedes humans. Social organization exists in microbes: Some genes control altruistic roles in collective strategies. Other genes, required for reproduction, require an altruistic role, negate reproductive advantages to cells in selfish state, and keep cheaters in check (Robinson et al. 2005). Technology can arguably be found in amoeba *Diffulgia corona*'s 0.15 mm "house" built out of sand grains, or in the African village weaver bird's nest, which typically uses six knot types closely resembling human knots (Boyd et al. n.d.).

Birds, fishes, and mammals have traditions—behaviors shared by several individuals—which persist over time and are acquired by new individuals partly through learning in the social unit. Social learning makes adaptation faster than genetic change. Fishes have some of the most conspicuous animal traditions regarding food, mating, and predators. They show social learning, behavioral innovation, and diffusion just as primates do.

Nonhuman primates have local cultures with multiple traditions, which include tool use, communication akin to language, lethal intergroup aggression, and anticipation of future events (Kappeler et al. 2010). Chimpanzees have over 30 local traditions and a large capacity for culture (Whiten et al. 2007; Laland et al. 2011; Whiten 2011). Their tools are used for food, water, comfort or protection (hat making and seat making), and hygiene (dabbing and wiping feces, blood, and semen). Adult great apes seem particularly keen to avoid contact with excreta (Finch 2012). Some chimpanzee cultures are technologically oriented to solving their environmental problems. Tools also have social communication uses in some chimpanzee groups to convey mating intentions or frustration. But there is some sort of conservatism in the technology they adopt: simpler solutions are preferred even when the more complex ones were invented within the group (Whiten 2010).

In reference to language, dolphins, sea lions, African grey parrots, and other species can be taught to understand and produce verbs, modifiers, and prepositions. Nonhuman primates and other animals can think in terms of simple sentences. Forest monkeys and chimpanzees can modify the meaning of calls by combining them. Old World monkeys and apes display some understanding of the motives of others, can attribute them knowledge, share attention, and recognize social differences (Cheney and Seyfarth 2010). Bonobos (the species closest to chimpanzees) seem able to understand human language to a surprising extent (Savage-Rumbaugh et al. 1993).

Although chimpanzees and human children have very similar cognitive skills to deal with the environment, however, the former seem to have less sophisticated social skills (Herrmann et al. 2008). However, similar to humans, social intelligence in chimpanzees includes tactical deception, understanding of others' perceptions, regard for others, social learning, trading, and cooperation (Hirata 2009). Social learning is present in lemurs, the most basal primates. In addition, lemurs understand the outcome of a simple arithmetic operation (Fichtel and Kappeler 2010). Economically rational feeding strategies have been recorded in a variety of organisms. Capuchin monkeys trained to pay with coins spent more resources on a cheaper food than an equally valuable but more expensive alternative. They also seemed to share human economic biases, which depart from the view that humans ought to maximize their benefits at all times (Lakshminarayanan and Santos 2010). As part of social intelligence, cooperation is important in all primates. In some species, grooming is associated with food distribution, access to infants to cuddle them, support, participation in border patrols, and mating opportunities (dominant male chimpanzees trade support from other males in exchange for mating opportunities with the females). Depending on rearing conditions and cultural contact, chimpanzee social-cognitive competencies can resemble a *H. sapiens* way of thinking (Bjorklund et al. 2010).

Derived from social intelligence, social tolerance may explain differences in transmission and retention of tool use across chimpanzee and orangutan populations. It is therefore likely that mutual tolerance facilitated the rapid evolution of technology among hominin species (the human lineage that separated from chimpanzees 5 million to 8 million years ago), along with a food-processing and food-sharing lifestyle (van Schaik et al. 1999). A combination of intelligence types has been found in 62 nonhuman primate species, whose behavioral innovation, social learning, tool use, extractive foraging, and tactical deception were closely related, hinting at coevolution of social, technical, and ecological abilities in primates. Human general intelligence seems to reflect "processes that evolved in common ancestors and are thus shared in our extant primate relatives" (Reader et al. 2011).

The Brain

The foregoing discussion shows human culture as a further evolution of primate coevolved skills. In this section, we delve deeper in human biology, to try and assess whether their evolved brain drives a particular relationship of humans with nature.

Humans do not have the largest animal brain nor the most convoluted cortex (convolution of the cortex, the outermost brain layer, which has evolved more recently, increases along with cortex size) (Roth and Dicke

2005). Frontal lobes, often credited for advanced brain processes, are no larger in humans than expected in a primate with human brain size.

How We Differ

So what has eventually made humans so adept at transforming their environment? The answer is probably neurogenesis, or the process whereby the brain develops to full adult functionality (a bigger brain takes longer), combined with neurological reorganization (brain structural changes compared to chimpanzees) involving coordinated evolution of neurons, brain areas activated by a given function, and brain chemistry.

Humans have the largest primate brain, apparently due to overall enlargement of the nonolfactory brain. Brain size increase was gradual, starting with *Australopithecus*, and kept gradual in the *Homo* genus. Neocortices (outer layers) and brain volumes are now thrice larger than expected for nonhuman primates scaled to the human body size. As hominid brains enlarged and reorganized, areas involved with vision became more connected with cortices involved in associations. An asymmetrical brain is the norm in human adults and is linked to language and handedness (enlarged right frontal and left occipital lobes) (Falk 2012).

During evolution, the numbers of neurons, neuron types, and connections have augmented massively. More complex functions were possible. Particularly in humans, neurons and dendritic arborizations (dendrites collect information from other neurons, glial cells, and hormones) have allowed "complex manipulations of signals representing internal and external worlds" (Korogod and Tyc-Dumont 2010). The horizontal spacing between neurons increased in human brain areas BA10 and BA44/45 (Broca's speech area) but not BA4 (primary motor), BA3 (primary somatosensory) nor BA17 (primary visual cortex) compared to chimpanzee (Falk 2012).

The Human Social Brain

Human social intelligence seems to have depended on empathy, and on a simulation/theory of the others' minds. Empathy is the inner experience of similarity and some consideration of the other's emotions. Empathy enables altruism. Humans inherently seek to engage other humans, to the point that most actions are entangled with cognition and affect about others. Empathy is often elicited voluntarily and so is amenable to social learning (Decety and Jackson 2004). In the brain, the bilateral anterior insula (AI), rostral anterior cingulated cortex (ACC), brain stem, and cerebellum are activated by pain or pain in a loved one. AI and ACC activation correlates with empathy, that is, only that part of the pain network associated with its affective qualities, but not its sensory qualities, mediate empathy. Decoupled representations—independent from external sensory stimuli—are seemingly needed to understand someone else's thoughts (Singer et al. 2004). This points to empathy as a possible starting point of symbolic and abstract thought, that is, independent from immediate bodily stimuli.

We may know other minds by simulation or a theory of mind. Simulation depends on mirror neurons as shown in infants mimicking facial expressions. The theory of mind might process lexical information preferentially (Adolphs 1999).

The brain default network (BDN) is involved in both simulation and theory of mind. The BDN is a neurological system like the motor or the visual ones. It connects several interacting subsystems. Other people's social and emotional situations activate ventral medial prefrontal cortex regions. The medial intertemporal lobe supplies memories and associations to carry out simulations. The medial prefrontal subsystem involves this information in flexible, self-relevant, mental simulations. The BDN activates when attention is not on the environment but on internal tasks (autobiographical memory retrieval, use of past experiences to envision future and alternative perspectives and scenarios, conceiving the perspectives of others, navigation of social interactions, free wandering, personal thoughts and experiences, generation and manipulation of mental images, deciding on a moral dilemma). Experiments on theory of mind activate the BDN. The BDN's main function could be to allow for self-relevant explorations or simulations to assess oncoming events. Task-unrelated activity could be a default self-aware operation since all but 4% of the people report daydreaming daily. Stimulus-independent thoughts occur at rest, during tasks, and even under heavy load of external information, and BDN-active states can emerge in seconds or less. Brain glucose metabolism and blood flow at rest imply substantial brain activity, similar to solving a math problem. This high BDN metabolism distinguishes it from other brain regions (Buckner et al. 2008).

Apart from these social simulations, in real-life emotionally relevant situations, three structures are activated in parallel to guide social behavior. First, the amygdala provides an automatic reaction by identifying aversive stimuli. Second, the ventromedial frontal cortex associates similar elements in past and current situations, and prompts reenactment of corresponding emotions. It links emotional experiences with decision making and participates in monitoring punishment and reward. And third, the right somatosensory-related cortices are activated insofar as a detailed representation of the body related to emotions and behavior becomes necessary (Adolphs 1999).

Current knowledge on brain functions and evolution does not support a particular view on human relationships with the environment. It does, however, point to the role of social life as a mover of brain evolution. In the following section, recent literature is reviewed to examine whether fitness or the ability to survive and reproduce may have driven relationships with the environment.

Breeding and Survival

Biological success for a species depends on reproduction and survival. Humans have transited from catastrophic mortality events in the archeological record of early hominins and boom and bust in ethnographic records (Steele and Shennan 2009) to support of reproduction by males and older individuals, prolonged juvenile dependence, and long life expectancy (Kaplan 2000).

A recent hypothesis on social support is that cooperative breeding by several females may have promoted greater social and cognitive abilities, social tolerance, and cooperation. Cooperative breeding, especially in the maternal lineage, provides more energy to allow brain enlargement, social learning capacity, and cooperative problem-solving skills (Mulder 2010; van Schaik and Burkart 2010).

A similar view is emerging that explains a decrease in aggressiveness and increase in size of human societies. Early hominins were likely territorial (like chimpanzees) and had group alliances, that is, the tribal level of organization needed pacification of adult males. For this, early hominins needed to recognize their kin on both maternal and paternal sides (chimpanzees only recognize kinship on the mother side), and the kin had to recognize the offspring and stable partner of its members. For this to happen, pair-bonding was essential and meetings at the common border of two groups must have taken place without aggressions (as do bonobos occasionally), through reduced food competition and increased benefits of cooperation and sharing (Chapais 2010).

Warfare: Cooperation Meets Competition

Competition and scarcity are the norm, as organisms tend to rapidly propagate when resources are abundant (Gat 2010). In all primates, violence is an instance of both competition and coalition (strength in numbers) intended to ensure resource access (Crofoot and Wrangham 2010). Chimpanzee intergroup violence has been suggested as foreshadowing lethal conflict in early human societies. In the chimpanzee social organization, when parties of several males encounter a wandering individual, the dominant party can afford intense violence while incurring very little risk of injury (Crofoot and Wrangham 2010). In humans, competition over resources and women are the underlying activators of deadly violence, whereas esteem, prestige, and leadership are only used as proximate justification (Gat 2010). In chimpanzees and humans, violence and casualties are maximal when the opposing party is small, fleeing, or has been captured, that is, when risks are minimal (Crofoot and Wrangham 2010).

This section does not support the hypothesis that fitness in early humans may have depended on a particular type of relationship to the environment.

BOX 13.1   PAST BIOLOGICAL AND MINERAL POLLUTION

Mortality is dominated by biological contamination in neonate chimpanzees and humans. Chimpanzees have lower exposure to feces, lower incidence of degenerative ailments (like cancer and vascular diseases), and no Alzheimer neurodegeneration. Carrion scavenging for at least 2 million years has exposed hominin populations to infections. The tapeworm (*Taenia*) has coevolved with hominins for 1.7 million years. Exposure to feces in hunter-gatherers and traditional farmers is extensive at all ages.

In the last 3 million years, humans were also increasingly exposed to pollen as they became established in grassland habitats and as East Africa aridified; dust also became more important. Volcanic aerosols from surges of volcanism 4, 2.5, and 1.5 million years ago probably caused mortality from particulate matter and sulfur dioxide inhalation. Subsequently, inflammagens such as glutens from cereals entered the human diet (Finch 2012).

It does instead appear that intraspecific competition over resources and reproduction may have been inherited from the primate lineage, while enhanced human cooperation contributed to survival and larger societies. The question addressed immediately is whether human technology may have driven the particular relationship of humans to nature.

Technology: Storage of Information

Unprecedented short-term variability in climate during the Plio-Pleistocene has been held responsible for evolution of human cognitive and especially technological abilities (Potts 1998). Hominids seem to have modified stone tools since 2.3 to 2.5 Ma BP (million years before present) (Kimbel et al. 1996). Hearths are presumed to have started 1.6 Ma BP in Africa for lighting, heating, and as a predator deterrent (Bellomo 1994). Clearer evidence pinpoints fire in Eurasia between 465 and 380 ka BP (thousand years ago) and 500 ka in China. But the systematic use in Africa, Asia, and Europe started 200 ka ago. Fire may have allowed for a shorter intestinal tract to use less energy, which was then diverted to the brain (Aiello and Wheeler 1995; Carbonell 2008). Concomitant with stone technology, larger brains (1.8 Ma BP) and widespread use of fire (200 ka BP), was the decreasing size of teeth (Brace 1999).

Human extrasomatic evolution may have followed three major cognitive transformations:

1. An early motor adaptation, which in *H. habilis* increased visual evaluation and produced a two-handed, tool-mediated referential realm. Body memories of say, tool making, were reenacted by imitation using eyes, hands, feet, posture, and voice.
2. An evolution of language in the *H. erectus-sapiens* line accompanied cerebral increase in cortex, hippocampus, and cerebellum, and the descent of the larynx. Phonology allowed a virtually unlimited number of easily retrievable sounds for symbolic use.
3. Visual symbols developed into symbolic literacy combining external memory and interaction skills. Prewriting and prearithmetic evidence in Europe dates to 12 ka BP (Donald 1993).

To-date humans share memories stored in body, language, and tools.

This evolution led to exclusively human cultural features: tool construction methods, which include components fitted together, and cumulative technological change (Whiten 2010). This allowed for human occupation of all land ecosystems (culminating with arctic foragers 30,000 years ago), and considerable spatial and temporal cultural variability: the earliest evidence of iron dates to 3.5 ka BP in southwest Asia, poison to 11 ka BP in Zambia, and the bow to 11 ka BP in central Europe. The bow's success may have led to increased meat consumption and population growth rates, shrinking game populations, and promoting the adoption of agriculture (Marlowe 2005). Today, human foragers extract and hunt much more food than chimpanzees: 90% and 5%, respectively, the remainder being simply collected (Silk and Boyd 2010).

Technology may have oriented the biological evolution of hominins. In addition, technology may have responded to or allowed for larger

---

BOX 13.2   TECHNOLOGICAL CONSEQUENCES OF POPULATION DENSITY AND CLIMATE CHANGE

Population size improves the chances of technological invention and diffusion. This is because inventions usually obey small additions, recombinations, and lucky errors more than groundbreaking ideas. With larger populations, more of these small changes occur and diffuse more widely (Henrich n.d.). Conversely, the Irish Mesolithic and Tasmania are instances wherein complex technology was lost and replaced by a much simpler one as population decreased in response to climate change in the late Pleistocene. In turn, technology loss may have hindered population growth (Steele and Shennan 2009).

populations. Population size, the aforementioned intraspecific competition, and power may in turn have combined to produce inequalities.

Onset of Inequalities

Power is universal in gregarious primates. It includes dominance (the fact of consistently winning an *agonistic* interaction) and leverage (the ability to control resources other than by force). Politics only applies to species with individual knowledge of relationships between others in the group. Individuals behave politically when trying to maintain or increase their power by manipulating social relationships (their own and those of others). Without language, or symbols standing for meaning, nonhuman primates cannot engage in publicity and propaganda nor invent ideologies (Watts 2010).

Early human foraging societies allegedly suppressed older primate dominance drives and status or resource accumulation, through enforcement of egalitarian norms. Subsequent societies became more complex and unequal over the last 10,000 years (Rogers et al. 2011).

From Dominance to Inequalities

Male chimpanzees are responsible for 71% to 90% of the hunting. This is also the case in humans with the exceptions of Australian and Aka foragers whose gender contributions are more balanced (Marlowe 2005). This division of labor and the control over carcasses may have been among the first gender differences and a source of social inequality, as access to large prey meat could be bartered. Moreover, whereas male chimpanzees contribute slightly more total calories than females, on average across human foragers males contribute 56% to 66% of the calorie production, depending on their living ecosystem. This may have altered the power balance in favor of males despite the fact that female production is also somatic (reproduction of the species) (Gangestad et al. 2006).

At some point in human history, generational inequality stemmed from parents taking control over the reproduction of their offspring. Children seeking spouses and their parents have different strategies: typically the former are interested in the biological side, the latter in social alliances and circulation of and control over resources. Institutions such as initiations may have arisen to control the timing of marriage and choice of mate of the children (Wiessner 2009).

In many societies, birth-order inequalities arise from inheritance rules and lead to imbalanced endowments to the other children. This has been counterbalanced in kin-based societies, which tend to provide a wider network of investment for children. However, even in societies where parents

BOX 13.3   COMPETITION AND INEQUALITY RESULTING FROM POPULATION AND CLIMATE

Demographic change, including population growth, population aggregation, and immigration, seems in many ways to have triggered the adoption of modern behaviors (Powell et al. 2009) including increased territorial competition, technological innovation, and intensification of resource use. Agriculture in particular was a response to population growth; it generated surplus production and its accumulation by the ruling elite. This coincided with ranked societies and competition for prestige. Competition was further activated by environmental changes such as a shift from interglacial to glacial conditions or by expansion into extreme climates (Henshilwood and Marean 2003; Kennett et al. 2009; Plourde 2010; Rogers et al. 2011).

are inclined to allocate equal shares of resources to their children, inequality may emerge as a result of the following: first, a difficulty for living organisms to divide equally resources of different types. Second, only about one-third of the mothers may treat their children equally. Third, cumulative resource allocation over the lifetime of the children depends on their birth order, generating earlier-born, middle-born, and later-born resource handicaps even in equalitarian societies. Fourth, inequality perceived by the children may alter their development in life as importantly as tangible inequality. And finally, resource allocation may respond to siblings vying for resources (Hertwig et al. 2002).

Elements of Power Leverage

Across primates there is a strong positive association between relative size of the neocortex and frequency of deceptive acts in nature (Trivers 2010). Chimpanzees use their social intelligence for selfish motives, except in mother–infant interactions (Hirata 2009). Humans use an array of social tools based on symbols and language to exert power in addition to physical strength and weapons. These tools include manipulation, political rhetoric, moral sanctions, delegating authority to punish (essential for large-scale hierarchical societies), moral justification of aggression, attribution of blame to victims, and symbolically based norms (Watts 2010). These tools have been allegedly used by aggrandizers, a type of individual prevalent across all cultures, whose natural tendencies toward selfishness, competition, and dominance are stronger than average. Because of patchy landscapes, resources are unevenly distributed over space, and aggrandizers are able to claim private property and implement inheritance mechanisms leading to the accumulation of differences in wealth over time (Plourde 2010; Rogers et al. 2011).

Human language has allowed for power negotiation, trade, exchange systems to manage resource distribution, alliance networks, and the sophisticated systems of egalitarian checks on wealth accrual (Henshilwood and Marean 2003). And similarly, language has allowed for the creation of ideologies through which individuals and some groups manipulate exchanges of material goods—both utilitarian and symbolic—as well as matrimonial alliances and coalitions. Economic flows have included not only goods but also labor. Individuals with leadership traits, most often adult males, have exploited their aggrandizer features and competitive skills (bellicose might or hunting skills) to coerce and subordinate. Subordinates are mustered to enforce submission of the majority. Compliance is attained for fear of eviction or aggression from other groups (Kennett et al. 2009). More generally, human communication is more about manipulation than information. Even apparently innocuous exchanges are attempts at enhancing one's relative position or inducing a particular behavior in the listener. The latter must then decipher the speaker's intentions and derive a possible use from them.

Material goods became used as prestige objects (ornaments or monuments) to symbolize social standing or competitive ability in political contests (Plourde 2009). The first known symbolic objects are shell bead ornaments from 82,000 years ago. They demonstrate that humans understood that a material object could conventionally represent something else or at least that an object could be a sign detached from a direct meaning. In addition, these beads show nonutilitarian behavior. They had been transported over more than 200 km and are similar to others found 5000 km away. This is suggestive of exchanges over long distances (Bouzouggar et al. 2007).

Technology has been a chief mover of social inequality. This is because technology is accumulative in humans (Henrich n.d.; Marlowe 2005; Whiten 2010) and has tended, along with labor productivity, to generate increasing surplus and the need for storage. The latter has allowed for appropriation by an elite (Henshilwood and Marean 2003).

Discussion: Features Driving Unsustainability

The anatomically modern *Homo sapiens* is 100 to 200 ka old and its cultural evolution has taken precedence over its biological evolution: salient features like the occupation of all terrestrial ecosystems in the last 30 ka BP, systematic warfare from 15 ka BP, agriculture in the last 10 ka, and the state 5 ka BP. These changes are too recent to have affected our genome or our basic neurological setup.

Human biology did not seem to inherently drive our domineering relationship to nature. Instead, current knowledge on brain functions and evolution point to the role of social life as a mover of human evolution. Prosociality, along with our technological skills, have driven our biological evolution and demographic expansion, through enhanced fitness.

Warfare has not yet altered our prosocial psychology (Gat 2010; van Schaik and Burkart 2010). Neither is the emergence of states imprinted in our makeup; it coincides however with the earliest evidence of widespread human dominance over the environment 5000 years ago (Ellis et al. 2013). States emerge as a consequence of technologically and socially induced inequality. In turn, intraspecific competition and warfare in particular, may reinforce inequalities. Overall, social inequality results in deprivation and excess environmental encroachment: instead of n units of land or fish, $n + k$ are required, where k is the accumulation by ruling elites. Resource accumulation by large human populations and their environmental toll have probably depended on agricultural technology.

References

Adolphs R (1999) Social cognition and the human brain. *Trends in Cognitive Sciences* 3:469–479.

Aiello LC, Wheeler P (1995) The expensive-tissue hypothesis: The brain and the digestive system in human and primate evolution. *Current Anthropology* 36:199–221.

Bellomo RV (1994) Methods of determining early hominid behavioral activities associated with the controlled use of fire at FxJj 20 Main, Koobi Fora, Kenya. *Journal of Human Evolution* 27:173–195.

Bjorklund DF, Causey K, Periss V (2010) The evolution and development of human social cognition. In *Mind the gap: Tracing the origins of human universals*, Kappeler PM, Silk JB (eds), 351–371. Heidelberg: Springer.

Bouzouggar A, Barton N, Vanhaeren M, D'Errico F, Collcutt S, Higham T, Hodge E, et al. (2007) 82,000-year-old shell beads from North Africa and implications for the origins of modern human behavior. *PNAS* 104:9964-9969.

Boyd R, Richerson P, Henrich J (n.d.) The cultural evolution of technology: Facts and theories. http://www2.psych.ubc.ca/~henrich/pdfs/Boyd Richerson Henrich The cultural evolution of technology 7.pdf (accessed June 24, 2012).

Brace CL (1999) Comments on the raw and the stolen: Cooking and the ecology of human origins. *Current Anthropology* 40:577–579.

Buckner RL, Andrews-Hanna JR, Schacter DL (2008) The brain's default network. Anatomy, function, and relevance to disease. *Annals of the New York Academy of Sciences* 1124:1–38.

Carbonell E (2008) *Hominidos: Las primeras ocupaciones de los continentes*. Madrid: Ariel.

Chapais B (2010) The deep structure of human society: Primate origins and evolution. In *Mind the gap: Tracing the origins of human universals*, Kappeler PM, Silk JB (eds), 19–51. Heidelberg: Springer.

Cheney D, Seyfarth RM (2010) Primate communication and human language: Continuities and discontinuities. In *Mind the gap: Tracing the origins of human universals*, Kappeler PM, Silk JB (eds), 283–298. Heidelberg: Springer.

Crofoot MC, Wrangham RW (2010) Intergroup aggression in primates and humans: The case for a unified theory. In *Mind the gap: Tracing the origins of human universals*, Kappeler PM, Silk JB (eds), 171–195. Heidelberg: Springer.

Decety J, Jackson PL (2004) The functional architecture of human empathy. *Behavioral and Cognitive Neuroscience Reviews* 3:71–100.

Donald M (1993) Précis of origins of the modern mind: Three stages in the evolution of culture and cognition. *Behavioral and Brain Sciences* 16:737–791.

Ellis EC, Kaplan JO, Fuller DQ, Vavrus S, Klein Goldewijk K, Verburg PH (2013) Used planet: A global history. *PNAS* 110:7978–7985.

Falk D (2012) Hominin paleoneurology: Where are we now? In *Progress in Brain Research*, Hofman MA, Falk D (eds), 195:255–227.

Fichtel C and Kappeler PM (2010) Human universals and primate symplesiomorphies: Establishing the lemur baseline. In *Mind the gap: Tracing the origins of human universals*. Kappeler PM, Silk JB (eds), 395–426. Heidelberg: Springer.

Finch CE (2012) Evolution of the human lifespan, past, present, and future: Phases in the evolution of human life expectancy in relation to the inflammatory load. *Proceedings of the American Philosophical Society* 156:9–44.

Gangestad SW, Haselton MG, Buss DM (2006) Evolutionary foundations of cultural variation: Evoked culture and mate preferences. *Psychological Inquiry* 17:75–95.

Gat A (2010) Why war? Motivations for fighting in the human state of nature. In *Mind the gap: Tracing the origins of human universals*, Kappeler PM and Silk JB (eds), 197–220. Heidelberg: Springer.

Henrich J (n.d.) Why societies vary in their rates of innovation. The evolution of innovation-enhancing institutions. http://www2.psych.ubc.ca/~henrich/Website/Papers/InventionInnovation05.pdf (accessed December 12, 2012).

Henshilwood CS, Marean CW (2003) The origin of modern human behavior: Critique of the models and their test implications. *Current Anthropology* 44:627–651.

Herrmann E, Call J, Hernández-Lloreda MV, Hare B, Tomasello M (2008) Humans have evolved specialized skills of social cognition: The cultural intelligence hypothesis. *Science* 317:1360–1366.

Hertwig R, Davis JN, Sulloway FJ (2002) Parental investment: How an equity motive can produce inequality. *Psychological Bulletin* 128:728–745.

Hirata S (2009) Chimpanzee social intelligence: Selfishness, altruism, and the mother-infant bond. *Primates* 50:3–11.

Kaplan H, Hill K, Lancaster J, Hurtado AM (2000) A theory of human life history evolution: Diet, intelligence, and longevity. *Evolutionary Anthropology* 9:156–185.

Kappeler PM, Silk JB, Burkart JM, van Schaik CP (2010) Primate behavior and human universals: Exploring the gap. In *Mind the gap: Tracing the origins of human universals*, Kappeler PM, Silk JB (eds), 3–15. Heidelberg: Springer.

Kennett DJ, Winterhalder B, Bartruff J, Erlandson JM (2009) An ecological model for the emergence of institutionalized social hierarchies on California's Northern Channel Islands. In *Pattern and process in cultural evolution*, Shennan S (ed), 297–314. Berkeley, CA: University of California Press.

Kimbel WH, Walter RC, Johanson DC, Reed KE, Aronson JL, Assefa Z, Marean CW, et al. (1996) Late Pliocene *Homo* and Oldowan tools from the Hadar formation (Kada Hadar Member), Ethiopia. *Journal of Human Evolution* 31:549–561.

Korogod SM, Tyc-Dumont S (2010) *Definition of the neuron: Electrical dynamics of the dendritic space*. Cambridge: Cambridge University Press.

Lakshminarayanan V, Santos LR (2010) Evolved irrationality? Equity and the origins of human economic behavior. In *Mind the gap: Tracing the origins of human universals*, Kappeler PM, Silk JB (eds), 245–259. Heidelberg: Springer.

Laland KN, Atton N, Webster MM (2011) From fish to fashion: Experimental and theoretical insights into the evolution of culture. *Philosophical Transactions of the Royal Society* B 366:958–968.

Marlowe FW (2005) Hunter-gatherers and human evolution. *Evolutionary Anthropology* 14:54–67.

Mulder MB (2010) The unusual women of Mpimbwe: Why sex differences in humans are not universal. In *Mind the gap: Tracing the origins of human universals*, Kappeler PM, Silk JB (eds), 85–106. Heidelberg: Springer.

Plourde AM (2009) Prestige goods and the formation of political hierarchy. A costly signaling model. In *Pattern and process in cultural evolution*, Shennan S (ed), 265–276. Berkeley, CA: University of California Press.

Plourde AM (2010) Human power and prestige systems. In *Mind the gap: Tracing the origins of human universals*, Kappeler PM, Silk JB (eds), 139–152. Heidelberg: Springer.

Potts R (1998) Variability selection in hominid evolution. *Evolutionary Anthropology* 7:81–96.

Powell A, Shennan S, Thomas MG (2009) Late Pleistocene demography and the appearance of modern human behavior. *Science* 324:1298–1301.

Reader SM, Hager Y, Laland KN (2011) The evolution of primate general and cultural intelligence. *Philosophical Transactions of the Royal Society B* 366:1017–1027.

Robinson GE, Grozinger CM, Whitfield CW (2005) Sociogenomics: Social life in molecular terms. *Nature* 6:257–271.

Rogers DS, Deshpande O, Feldman MW (2011) The spread of inequality. *PLoS ONE* 6:1–10.

Roth G, Dicke U (2005) Evolution of the brain and intelligence. *TRENDS in Cognitive Sciences* 9:250–257.

Savage-Rumbaugh S, Murphy J, Seveik R, Brakke D, Williams S, Rumbaugh D (1993) Language comprehension in ape and child. *Monographs of the Society for Research in Child Development* 58:1–222.

Silk JB, Boyd R (2010) From grooming to giving blood: The origins of human altruism. In *Mind the gap: Tracing the origins of human universals*, Kappeler PM, Silk JB (eds), 223–244. Heidelberg: Springer.

Singer T, Seymour B, O'Doherty J, Kaube H, Dolan RJ, Frith CD (2004) Empathy for pain involves the affective but not sensory components of pain. *Science* 303:1157–1162.

Steele J, Shennan, S (2009) Introduction: Demography and cultural macroevolution. *Human Biology* 81:105–118.

Trivers R (2010) Deceit and self-deception. In *Mind the gap: Tracing the origins of human universals*, Kappeler PM, Silk JB (eds), 373–393. Heidelberg: Springer.

Van Schaik C, Deaner R, Merrill M (1999) The conditions for tool use in primates: Implications for the evolution of material culture. *Journal of Human Evolution* 36:719–741.

Van Schaik CP, Burkart JM (2010) Mind the gap: Cooperative breeding and the evolution of our unique features. In *Mind the gap: Tracing the Origins of Human Universals*, Kappeler PM, Silk JB (eds), 477–496. Heidelberg: Springer.

Watts DP (2010) Dominance, power, and politics in non human and human primates. In *Mind the gap. Tracing the origins of human universals*, Kappeler PM, Silk JB (eds), 109–138. Heidelberg: Springer.

Whiten A (2010) Ape behavior and the origins of human culture. In *Mind the gap: Tracing the origins of human universals*, Kappeler PM, Silk JB (eds), 429–450. Heidelberg: Springer.

Whiten A (2011) The scope of culture in chimpanzees, humans and ancestral apes. *Philosophical Transactions of the Royal Society B* 366:997–1007.

Whiten A, Spiteri A, Horner V, Bonnie K, Lambeth S, Schapiro S, De Waal F (2007) Transmission of multiple traditions within and between chimpanzee groups. *Current Biology* 17:1038–1043.

Wiessner P (2009) Parent-offspring conflict in marriage: Implications for social evolution and material culture among the Ju/'Hoansi bushmen. In *Pattern and process in cultural evolution*, Shennan S (ed), 251–263. Berkeley, CA: University of California Press.

14

Population and Food Sustainability

Alejandro de las Heras

CONTENTS

Revisiting the IPAT Framework

In the early 1970s a population-only explanation of anthropogenic environmental impacts was adduced, the "population bomb" (Ehrlich 1968). A discussion ensued as this model was ill-adapted to the impacts observed in affluent countries. A different explanation emerged, called IPAT, whereby impacts equal a function of population, affluence, and technology (formulated by 1972 by Commoner on the one hand and Ehrlich and Holdren on the other hand; York et al. 2002): large populations, high consumption, and low technology are posited to entail the largest encroachment on nature. While a suggestive simplification, this fails to capture the current relative weight of population impacts and future possibilities to curb population growth.

Accordingly, this chapter aims at reassessing the role of population as the root cause of the current environmental and social unsustainability. The roles of economy and technology are dealt with in the following chapters.

Human Population Biometrics and Food Requirement Impact

Population Transitions

Two thousand years ago, the human population was 133 million to 300 million and at the beginning of the Anthropocene in 1760 it was 750 million (Kremer 1993). As of 2012, the human population was 7.05 billion persons and growing at a 1.1% annual rate (Greene et al. 2012), meaning that 14.1 billion humans may be living on the planet by 2075. This situation emerged from an epidemiological transition (Omran 2005) brought about by vaccines and simple hygiene measures, foremost access to safe water, the chief driver of human increase in life expectancy since 1840 (BMJ 2007). Additionally, during a time lag called the demographic transition, the large difference between high birth rates (pertaining to a high mortality regime) and descending mortality rates (during the epidemiologic transition) was responsible for most of the population increase during the last century. However, the global population growth started to decelerate by the turn of the century as a result of contraceptive use using both hormonal and mechanical methods. Fertility control is now practiced in most societies, including those with predominant Muslim or Catholic creeds commonly said to oppose such control. This is due to the diffusion of knowledge on the efficacy and relative innocuousness of contraceptives whereby they became acceptable to small groups at first and increasingly larger ones thereafter.

The current world population slowdown means that future generations will be smaller and that gradually, population inertia due to still large numbers of women of child-bearing age will decrease, leading eventually to a convergence with the population regime of developed countries (high life expectancy, low fertility). If such convergence occurs, the world population might attain 10 billion by 2100 (United Nations 2011). In developed countries, fertility is already below the generation replacement level, meaning that each woman will be replaced by less than one woman of child-bearing age. The upshot is that populations in the 21st century in developed countries will likely grow based solely on immigration, whereas in developing countries population growth rates will continue to decrease.

Human Height and Human Biomass

Human adult height seems to depend on the genetic potential height and net nutrition (total food consumption minus energy, and nutrients deviated to physical efforts and recovery from disease) (Alter 2000). As a consequence of the Industrial Revolution and urbanization, the height of army recruits or the general populations of Europe and the United States decreased but almost linearly increased thereafter. The Dutch recruits gained 8 cm between 1843 and 1918 (Steckel 2001). The Swedish recruits gained more than 12 cm between 1820 and 1965 (Sandberg and Steckel 1997). Both countries are among the tallest in the world nowadays. By contrast, the American men

BOX 14.1   THE ENVIRONMENTAL IMPACT
OF HUMAN POPULATION

In biological terms, humans impact the environment to fulfill their food requirements. Population size, that is, the number of humans, is therefore only one of the factors. Prolonged life expectancy (72 years for women and 67 for men; Greene et al. 2012) is an additional factor. But as human nutrition improves, the height of individuals increases. Increased height and body mass index (weight in kilograms divided by height in meters squared) augment food requirements.

For any given age and sex, the food requirement also depends on the basal metabolism (i.e., energy requirement at rest) and the level of physical exercise. Food requirements are maximum for breast-feeding mothers and adolescents; these are two groups typically large in young populations. Conversely, in aging populations, longer life augments food requirements while older individuals, women outnumbering men, and lesser fertility diminish food requirements.

The current obesity pandemic (see Chapter 15) is a cultural compounded effect of excess food supply, attributable to technological changes in the course of the Green Revolution and diet changes during the nutritional transition (excessive food, sugar, and fat intakes from early in life; Popkin 2004). In the meantime, physical labor has given way to office work and exaggerated automobile use. This cultural effect is ascribable to consumption, not to human population in a biological sense, although both cause excess demand for foods that require agricultural land and water, fragmenting and reducing ecosystems, outcompeting and displacing wildlife.

were the tallest in the 19th century but their height increase stalled for those born after 1950 (Ogden et al. 2004). There is evidence suggesting that height increases in all human populations when conditions are favorable, such as is the case globally since the onset of the Green Revolution. The environmental downside is that a taller population has a higher nutritional requirement and may also gain considerably more weight. Global biosocial unsustainability is most visible in the 165 million stunted children under 5 years of age (very low height for age), 101 million underweight (very low weight for age), and 52 million wasted (very low weight for height) (UNICEF, World Health Organization [WHO], and World Bank 2012).

No other animal the size of *H. sapiens* comes even close to the human population size. This is why humans probably outweigh the biomass of any other species. A recent estimate of human biomass is 287 Mt (million metric tonnes) out of which 15 Mt correspond to overweight and 3.5 Mt to obesity (Walpole et al. 2012). North America, with only 6% of the world's population (average 80.7 kg weight), contributes 34% of the human obesity biomass. Meanwhile, Asia has 61% of the population and only 13% of the excess biomass. The food requirement of the global human biomass, and the way the large human necromass (total mass of the deceased per annum) is disposed of, have direct environmental impacts.

Need and Greed

To understand the environmental impacts of biological food needs, they must be separated from greed. In a useful convention in biomedicine, energy requirements are mostly met with carbohydrates and tissue replacement with proteins. Lipids are an essential component of cell walls but require lesser amounts and can be processed metabolically from carbohydrates and protein-containing food. Agriculture and fisheries supply virtually all the demand and cause most of the environmental impacts related to human biological needs. Today, hunting supplies a minuscule fraction of human needs and causes disproportionate impacts.

Land in today's human food chain is simplified to include mostly cropland and pastureland (Figure 14.1). During a year a certain amount of land is kept under these uses, some of it being nutritionally required (need) and some being transformed into waste or excess consumption (greed). The area devoted to nutritional need is the product of the extant human population (the initial population plus population growth), the individual nutritional need, a yield factor (production of nutritionally useful organic matter and energy per unit area of land accounting for plant disease, drought, and insect herbivory), and a trophic efficiency factor (to account for loss of matter and energy due to land change, respiration, excretion, decay, combustion).

Paraphrasing the Mnong Gar people in the 1940s Vietnam, "we have eaten the jungle" (Condominas 1957) and produce more food than needed by the entire mankind (Hazell and Wood 2008). We fail to provide enough vegetal food for some human populations while at the same time trading excessive amounts of meat (Figure 14.1). Greed, more than need, motivates this situation and is a factor of the current nutrition transition. In middle-income countries like Brazil, there seems to be since the second half of the 20th century a chronic state where in breadbasket areas, like the states of Rio Grande do Sul and Mato Grosso, there is on the one hand undernourished children (Victora and Vaughan 1986) or undernourished women (Demographic and Health Surveys (DHS) 2006), and on the other hand exports and excess domestic supply of unnecessary foodstuff from large exports-oriented landholdings. These are not isolated cases of ill-distributed agricultural wealth. Although undernourishment in all the developing world has decreased as a proportion of the total population, and now affects 870 million people worldwide, it has grown in absolute terms in Sub-Saharan Africa over the last 20 years (Food and Agriculture Organization of the United Nations [FAO], World Food Programme [WFP], and International Fund for Agricultural

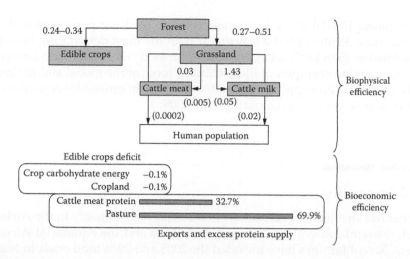

FIGURE 14.1

Efficiency in the human food chain. Forest is converted into cropland and grassland with 0.24–0.34 and 0.27–0.51 mass efficiencies, respectively (meaning that, for instance, the mass of forest is replaced by a mass of crops smaller by a 0.24 factor). Feed–milk conversion is the most efficient food source with a 1.43 efficiency (as cows add water and virtually no waste is generated) and a 0.05 feed–milk protein efficiency. The feed-to-meat edible protein efficiency is 0.005. Finally, feed is turned into human bodily proteins with 0.02 and 0.0002 efficiencies, via milk and meat, respectively. More inefficiencies are related to the food system (Brazil in 2000 in this case) wherein crop carbohydrate energy and cropland just meet the population requirements (0.1% deficiency). Meanwhile, meat and pastures present 32.7% and 69.6% surpluses, which contribute to exports and the domestic and global nutrition transition toward too much meat and fat. (Efficiencies based on Smil V, 2002, *Population and Development Review* 28:599–639; Mooney H et al. 2001, *Terrestrial global productivity: Past, present and future*, San Diego, CA: Academic Press; Zhao M et al., 2005, *Remote Sensing of Environment* 95:164–176; Van de Werf G et al., 2003, *Global Change Biology* 9:547–562.) (Average energy requirement in Brazil in 2000 based on the age and sex distribution [U.S. Census Bureau, n.d., International Data Base, http://www.census.gov/ipc/www/idbprint. html], height and body mass index [Coelho CA et al., 2004, Health, human capital and economic growth in Brazil. European Regional Science Association Conference Papers. http://www-sre. wu-wien.ac.at/ersa/ersaconfs/ersa04/PDF/490.pdf; Shetty P, James W, 1994, Body mass index: A measure of chronic energy deficiency in adults. FAO Food and Nutrition Paper. Rome: FAO], and basal metabolism requirements [FAO, 2001, Human energy requirements. Report of a Joint FAO/WHO/UNU Expert Consultation. FAO Food and Nutrition Technical Report Series 1. Rome: FAO].) (Protein requirements were based on Scrimshaw N, 1996, Human protein requirements: A brief update. The United Nations University Press Food and Nutrition Bulletin, 17. http://archive.unu.edu/unupress/food/8F173e/8F173E02.htm#Human protein requirements: A brief update; weight and population per age; Centers for Disease Control and Prevention, 2005, CDC growth charts: United States. Percentile data files with LMS values. http://www. cdc.gov/growthcharts/percentile_data_files.htm; Shetty P, James W, 1994, Body mass index: A measure of chronic energy deficiency in adults, FAO Food and Nutrition Paper, Rome: FAO; and Coelho et al., 2004, Health, human capital and economic growth in Brazil, European Regional Science Association Conference Papers, http://www-sre.wu-wien.ac.at/ersa/ersaconfs/ersa04/PDF/490.pdf.) (Nutritional content of crops and meat, agricultural output in Brazil, agricultural yields, and extant land cover were based on FAO 2005, FAOSTAT: Food balance sheets, Brazil, year 2000, http://www.faostat.fao.org; Global Forest Resources Assessment 2000: Main report. Forestry Report 140. Rome: FAO.)

Development [IFAD] 2012). Undernourishment is also a persistent malady causing more deaths in India every year than the most devastating famines known to date (Sen 1997). Moreover, the acute food crises examined immediately are extreme examples of ill-distributed food, at the global and regional levels, coupled with supply deficit and soil depletion caused by exports agriculture in poor countries (Goulding et al. 2008).

Food Insecurity

Hunger has affected several nations in the last decade, mostly in the African Sahel, the semiarid ecotone between the Sahara and the equatorial African forests. Recent famines have included the 2005 and 2006 food crisis in Niger and the Horn of Africa, and the 2010 and 2012 Sahelian emergencies. These events demonstrate several critical failures: the global governance system has unsuccessfully channeled agricultural surplus, which keep being produced in affluent countries. Climate factors have also been adduced (Figure 14.2);

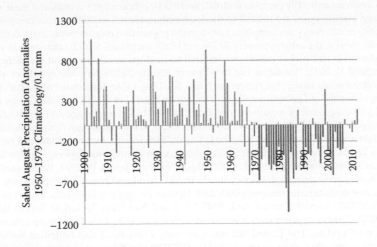

FIGURE 14.2
Sahelian August precipitation anomalies, 1900–2012, average over 20-10N, 20W-10E. (Data from National Oceanic and Atmospheric Administration [NOAA], National Climatic Data Center [NCDC], and Joint Institute for the Study of the Atmosphere and Ocean [JISAO], 2013, Sahel precipitation anomalies in 0.1 mm, 1900–2012, 1950–79 climatology, Global Historical Climatology Network, http://jisao.washington.edu/data/sahel.) The month of August was chosen here to illustrate the soudure critical phase in which the crop of last year has to last as food and the rain season has to establish itself lest seeds have to be used as emergency food. It is precisely during this phase that the deepest rainfall deficit has been observed in recent decades (in dark gray). Recent models explain this deficit as a result of increased aerosol loadings and greenhouse gases; they forecast a drier Sahel in the 21st century and beyond (Held et al. 2005).

these may explain the trend and timing of famines along with associated natural factors, such as the chronic desert locust threat (FAO 2013a). But the social organization—for example, a negligent/callous government versus a democratic one—commands a set of factors that turn a drought or flood into a famine (Drèze and Sen 1989). Ethnically biased governments stifling revolts through food deprivation and elites supposed to distribute food aid actually reexporting it to more profitable markets are some of the conducts observed in crises including the 1967 to 1970 Biafra and current crises in Somalia and Sudan. International crop prices and supply may contribute to famine (FAO 2013b) since in an emergency flaring prices make hunger more likely.

Transitioning to Low Population and an Adequate Food Supply: A Framework

Acute food issues currently are equity issues. However, a lasting solution to food quantity problems in a finite planet imposes limits to the total human population. In addition, food quality is already mired in an industrialized food model with consequences such as the pandemics of obesity (Walpole et al. 2012), the BSE (bovine spongiform encephalitis), MRSA (multiply resistant *Staphylococcus aureus*), avian flu (Morse 2004; Omenn 2010), and mycotoxin issues augmented due to trade and climate change (Paterson and Lima 2011; Wu and Guclu 2012), as well as declining micronutrient content in diets, hormone disruptors in food, and the fact that gluten peptides in cereals are inflammagens and much of human nutrition is cereal-based (Finch 2012).

Future food issues will be quantitative, qualitative, and distributive, and affect both industrialized and developing countries. Dealt with separately, these facets may not reach adequate solutions. The Green Revolution seemingly had solved quantitative issues but its side effects include an ongoing pollination and seed dispersal crisis in domestic and wild plant species, and surplus production, which decreased prices and challenged agricultural livelihoods.

Envisioning possible futures thus warrants a two-pronged sustainable framework of demographic change and equitable food security.

Population Degrowth, Cyclicity, and Age Structure

Conventionally, pandemics, hunger, and war have been considered the only means to lower the human population. Family planning policies can at best reduce the population growth rate. More boldly, a recently proposed strategy for a sustainable future included "negative population growth for a number of generations, followed by zero growth" (Nekola et al. 2013).

This might be possible within the theory of stable populations (Dublin and Lotka 1925) defined as closed populations, such as humans on Earth, with an age structure and survival constant over time, such as expected if medical progress slows down. It follows that birth, death, and growth rates are constant over time. The latter is a function of the net reproductive rate (nrr) and the mean age of fertility; in particular, when nrr = 1, the growth rate is zero. The age structure is completely determined by the mortality schedule and the growth rate.

In terms of human behavior, achieving a theoretical stable population is rather restrictive: the offspring of all women must be kept constant over time at a little above two children, or balanced so as to achieve this result. Also, a constant age structure occurs when the mean age of fertility is constant. Medical progress must be frozen and death at old age must comply with a set schedule. These conditions have to be obtained voluntarily and/or compulsorily, which poses large bioethical questions. The following describes a variant, how it can come about, and what its consequences might be.

Negative growth can be achieved in a sequence of low fertility slimming down the bottom of the population pyramid (between 1950 and 2050) (Figure 14.3) followed by attrition at old ages at the pyramid's top (between 2050 and 2100). This is already what affluent countries with negative natural growth are experiencing.

As a result of old age attrition, the 2100 population is fairly young (the population is concentrated at the pyramid's bottom) and even with constant fertility may lead to a higher birth rate and consequently a population increase. Fertile women in 2100 ought to have a larger variance in their descent but the result should be a replenishment of the age groups at the top of the pyramid, as in 2200. Over time, the population and mean age of population describe symmetric cycles driven by the fertility cycles (bottom of Figure 14.3).

The conditions for a population degrowth and further birth and death control toward a stable or cycling population are as follows. First, a reproductive health approach, coupled with the exercise of reproductive rights, is needed to help diffuse birth control technologies in large sectors of the developing world, and adolescent and marginalized populations in affluent nations. Contraception is an immediate factor that has driven the global fertility transition. But the underlying causes of contraceptive use are on the one hand women empowerment through education and market labor access, and on the other hand low infant mortality via adequate sanitation technology so that high fertility rates become unnecessary. Second, policies like one-child restrictions or freedom of euthanasia may raise bioethical or fairness concerns. More stringent policies than these seem more controversial and also impractical owing to the lack of global governance mechanisms and because of national and religious competition. Some countries or groups would then pursue a separate policy diverging from the global sustainable

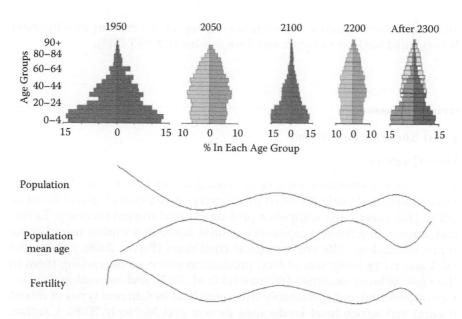

FIGURE 14.3
Adapting human population to the planet's carrying capacity. Left to itself the human population would probably behave like any natural population, with cyclical growth and degrowth, similar to boom and bust cycles described from the archaeological and ethnographic records (Steele and Shennan 2009). An aging-population period would have low fertility and a population decrease. This would be followed by a period of slightly heightened fertility as opportunities augment and competition lapses, leading to a small recovery of population size. Population inertia makes the return to an aged population relatively slow (many young women can give birth to few children and still produce a large offspring).

population objectives. Third, hunger and pandemics and most particularly wars would probably be even more damaging than today because some age groups could be more affected (jags in the 1950 pyramid) and jeopardize future population recovery. And fourth, as degrowth and aging may entail lower numbers of workers and less production, more technological inputs may be required to fulfill basic food and health needs.

The benefits of population degrowth and stabilization or cycling might be, first, to prevent a catastrophic collapse of human population following massive deterioration of the planetary life-support systems. Second, a smaller population would decrease human consumption thereby lifting pressure on natural habitats. Third, a further consumption and production decrease might stem from population aging (National Research Council 2012). Fourth, alternating periods of population rejuvenation might alleviate the consequences of aging in terms of funding welfare states: higher taxes or less entitlements; higher health expenditures as a consequence of higher chronic disease prevalence. And fifth, population degrowth might achieve a decoupling of socioeconomic development and population size, similar to Norway and other small countries consistently on top of the human development lists (United Nations Development Programme [UNDP] 2013).

Food Security Solutions

Food Quantity

It is yet unclear whether we might soon reach a global peak of land encroachment or just local peaks followed by recovery (Ausubel et al. 2013; Ellis et al. 2013). This uncertainty compels a prolonged effort toward recovery. To this end, augmenting natural capital is essential, through adequate use of genotypes adapted to different ecological conditions (Pretty 2008); halting the land and ocean footprints of food production and even decreasing them in the more affluent countries (Weinzettel et al. 2013); and increasing the surface of protected and sustainably cultivated land in different types of mixed natural and agricultural landscapes (Scherr and McNeely 2008). Caution, however, is warranted with eco-efficiency yield gains (Wilkins 2008), which might decrease prices and subsequently augment the amount of production to maintain agricultural income.

Food Quality

Micronutrient deficiencies in all countries call for a diversity of foods and stress the importance of information on food diversity and nutrition for all. Synergies are required between the food, health, sanitation, and education sectors to foster better nutrition, gender equity, and environmental sustainability (FAO 2013c).

To improve food quality, a transition must occur toward micronutrient-rich, contaminant-free food. Human diets in the past introduced some anti-inflammatory factors: salicylates (some fruit skins, willow bark), polyphenols (e.g., curcumin), and omega-3 polyunsaturated fatty acids, but the literature on the topic is incipient. Anti-inflammatory food is important as many adults suffer from chronic intestinal inflammation from reactions to cereal gluten peptides, which resist digestion; most diets are cereal-based (Finch 2012).

Cereals such as maize, spices, peanuts, tree nuts, and milk also are colonized by mycotoxin-producing fungi. Aflatoxin B1 is the most powerful known natural carcinogen; it can also have many less acute effects. More than 5 billion persons could be chronically exposed to food aflatoxins, especially as more than 1 billion Mt of maize were traded in the last decade (Wu and Guclu 2012) and milk proteins are much more bioavailable and digestible, and should replace meat, which is much more wasteful to produce. Natural control uses native atoxigenic strains of *Aspergillus flavus* fungi, which out-compete and exclude aflatoxin-producing *A. flavus* (Anon. 2010).

A mushroom-based nutraceutical–nutriceutical expansion must take place. Mushrooms are largely underused in the West despite their being protein rich and their growing global use as nutraceuticals, nutriceuticals, and pharma products. Nutraceuticals are foods with medical or health benefits. Mushroom nutriceuticals are (partially) refined food extracts consumed as a dietary supplement with health benefits, from known species without long-term toxicity. In the early 21st century global annual sales of mushroom nutriceuticals were US$6 billion with the United States accounting for 0.1%, and Asia and Europe for 99%. The U.S. expansion of this market has been 20% to 40% per year. Nutraceuticals in the United States were a US$14 billion annual market (Chang and Miles 2004).

Food Distribution

An interdependent global food system should be pursued that minimizes energy costs such as food production and transportation. Nonemergency food aid, often conditional to political clauses and privileged access to markets, should be eliminated and replaced by compensation for lost development (see Box 14.2), mostly in kind through technology transfers in the education and health sectors. The Doha Round of negotiations of the World Trade Organization has pointed to the detrimental role of agricultural tariffs and subsidies in temperate countries in generating unfavorable terms of trade for agricultural products. Tariffs and subsidies in industrialized countries are a waste of public monies to sustain an agriculture that is less competitive than sun-subsidized tropical countries, deliver expensive food to temperate markets, constitute unfair competition with fledgling economies, produce large surpluses that are underutilized, and degrade soils and biota.

Food insecurity must be addressed with wealth redistribution, not only the pervasive discourse of economic growth. There is a large potential in investing in smallholdings to foster sustainable intensification, encourage environmental preservation, address recent price volatility, and obtain higher prices for agricultural products (FAO, WFP, IFAD 2012). Similar actions may palliate both food deficit and excess: enhancing the availability of nutritious food; reducing agricultural losses and waste along the food chain; and fostering healthier diets to make more sustainable use of resources (FAO 2013c).

Sustainable Agriculture

The technological memory in agriculture should be deep and wide. Ten thousand years of agricultural innovations around the globe still have

BOX 14.2   A COLONIAL PATTERN UNDERPINS FAMINES AND MALNUTRITION

Adam Smith was one of the first to relate the Bengal Famine of 1770 to one of the first capitalist corporations, the East India Company, "the mercantile company which oppresses and domineers in the East Indies" (Smith 1776; Smith and Garnier 1852, e.g., Book 1, Chapter 8; Book 4, Chapter 5). Subsequently, the British viceroy oversaw a peak of wheat exports during the Great Famine of 1876 to 1878 (Waldorf 1879).

Nowadays, neocolonialism relies on financial multilateral institutions to forcibly open Third World agricultural markets to competition from developed economies protected by subsidies and tariffs. Neocolonialism in this case distorts the economic competition and favors the less efficient agriculture of northern countries, less endowed with sunlight. The opening of the Indian food market to external competition raised its agricultural imports from US$1 billion to over US$4 billion between 1995 and 2000. In 2000, the World Trade Organization argued for further opening. The new imports came in a large measure from the United States and Europe, who were poised to heavily subsidize millions of farmers but more particularly the largest ones. This helped surplus production bring prices down and diminish farming income. Coconut, coffee, and pepper prices dropped 45% to 80%. Seventy percent of the edible oil now is imported. Farming protests to this situation were repressed with firearms. In this context, it is difficult to understand stances such as held by Nobel Prize winner Amartya Sen recommending increased trade opening and exports as the solution to hunger, and castigating food self-sufficiency (Shiva 2002).

to be incorporated in sustainable agroecosystems defined as hybridizing the global indigenous knowledge and current scientific understanding.

Similar to the scientific viewpoint, indigenous agricultural systems rely on knowledge and experiments on process changes and interrelated processes, which substantially differ from the one-year, nature-excluding conception of industrialized agriculture (Gómez-Pompa 1987; Alcorn 1989). Plots dedicated to fair trade are possibly the only large-scale application so far of agroecology. There is also growing knowledge on the best agroforestry management practices leading to enhanced richness and abundance of different natural taxa, which contribute to biological pest control (Wanger et al. 2009). Other biological advances include the isolation of semiochemicals to repel pests from the crop and attract them into traps. This, along with companion crops usable as feed and cattle pest biological control, helps obtain higher yields, retain water, and reduce erosion (Hassanali et al. 2008).

A plethora of combinations can expand the limits of agroecosystems with principles from emergent sustainable agricultural practices. Bioreserves, for instance, use the all-important edges to cater to pollinators and plague predators. This principle is now a legal mandate in Brazilian Amazonia legal reserves (riparian corridors, hilltops, and slope conservation areas), which promotes *in situ* conservation of native germplasm. Similarly, bio-intensive agriculture intends to change consumers into producers (every-one can recover and sow seeds from eaten greens) and augments the soil organic matter. As to permaculture, it takes the notions of local food and autonomy to the design of backyard food production. Organic agriculture in turn stresses the importance of quality food and recommends minimal or nil use of agrochemicals. Other principles of sustainable agriculture are microirrigation, and the reuse of yellow and greywater. More generally, an overarching principle is that no biogeochemical or biological cycle should be interrupted.

Learning from and improving on the Green Revolution is essential and dictates a systematic precautionary approach. Such is the case for geoengineering emergency proposals to mitigate climate change through stratospheric aerosol injections; these should be considered in light of the role of (tropospheric) aerosols in the ongoing mechanisms of drought emergence in the Sahel. Alternatives have to be developed through subsidiarity mecha-nisms that include inputs from all the relevant scientific communities (Hulme 2012). The same precautionary principles should be applied to genetically modified organisms as relates to sustainable food systems (Jacobsen et al. 2013; Lynd 2013).

A transition to petroleum-deprived agricultural systems worldwide should initiate now (The Soil Association 2007). Energy savings may require more extensive production systems such as organic farming (Goulding et al. 2008). In relation to tillage and its high energy cost, no-tillage agriculture has been investigated since the inception of the agroecosystem concept (Crossley et al. 1984). Animal plows may be required but animal well-being and the sustainable management of genetic resources and diversity have come to the fore under the concept of precision husbandry (Flint and Woolliams 2008).

A mushroom-based non-Green Revolution must take place that taps the vast biosynthetic capacities and immense diversity of fungi. The non-Green Revolution is the conversion of the lignocellulosic biomass waste materials by mushrooms enhancing food supplies, health, and environmental quality. The use of agricultural waste is vastly underdeveloped, thereby squandering vast lignocellulosic resources and releasing pollutants. Secondary or tertiary materials can be produced out of the most common agricultural waste, including cereal straw, coffee grounds, and sisal substrate used in mush-room production. Animal and earthworm feed and soil improvers are other uses. No waste is generated consistent with the zero emission and total productivity principles (Chang and Miles 2004).

References

Alcorn JB (1989) Process as resource: The traditional agricultural ideology of Bora and Huastec resource management and its implications for research. *Advances in Economic Botany* 7: 63–77.

Alter G (2000) Stature, survival and the standard of living: A model of the effects of diet and disease on declining mortality and increasing stature. Paper presented at the Population Association of America, March 29–31, 2001.

Anon. (2010) Higher quality maize with Aflasafe. *Appropriate Technology* 37:39–41.

Ausubel JH, Wernick IK, Waggoner PE (2013) Peak farmland and the prospect for land sparing. *Population and Development Review* 38:221–242.

BMJ (2007) BMJ readers choose the "sanitary revolution" as greatest medical advance since 1840. *British Medical Journal* 334:111.2.

Centers for Disease Control and Prevention (2005) CDC growth charts: United States. Percentile data files with LMS values. http://www.cdc.gov/growthcharts/percentile_data_files.htm (accessed May 30, 2005).

Chang ST, Miles PG (2004) *Mushrooms: Cultivation, nutritional value, medicinal effect, and environmental impact.* Boca Raton, FL: CRC Press.

Coelho CA, Monteiro C, Conde W, Sarti MS (2004) Health, human capital and economic growth in Brazil. European Regional Science Association Conference Papers. http://www-sre.wu-wien.ac.at/ersa/ersaconfs/ersa04/PDF/490.pdf (accessed June 20, 2013).

Condominas G (1957) *Nous avons mangé la forêt de la pierre-génie Gôo.* Paris: Mercure de France.

Crossley DJ et al. (1984) The positive interactions in agroecosystems. In *Agricultural ecosystems*, Lowrance R, Stinner B, House G (eds) 73–82. New York: John Wiley & Sons.

Demographic and Health Surveys (DHS) (2006) DHS Stat Mapper. Brazil. Women with BMI <18.5 (underweight) as per the 1996 Brazilian Demographic and Health Survey. http://statmapper.mapsherpa.com/index.phtml (accessed June 23, 2012).

Drèze J, Sen A (1989) *Hunger and public action.* Oxford: Oxford University Press.

Dublin L, Lotka AJ (1925) On the true rate of natural increase. *Journal of the American Statistical Association* 20:305–339.

Ehrlich PR (1968) *The population bomb.* New York: Ballantine.

Ellis EC, Kaplan JO, Fuller DQ, Vavrus S, Klein Goldewijk K, Verburg PH (2013) Used planet: A global history. *PNAS* 110:7978–7985.

Finch CE (2012) Evolution of the human lifespan, past, present, and future: Phases in the evolution of human life expectancy in relation to the inflammatory load. *Proceedings of the American Philosophical Society* 156:9–44.

Flint APF, Woolliams JA (2008) Precision animal breeding. *Philosophical Transactions of the Royal Society of London. Series B, Biological Sciences* 363:573–590.

Food and Agriculture Organization of the United Nations (FAO) (2001) Human energy requirements. Report of a Joint FAO/WHO/UNU Expert Consultation. FAO Food and Nutrition Technical Report Series 1. Rome: FAO.

Food and Agriculture Organization of the United Nations (FAO) (2005) FAOSTAT: Food balance sheets, Brazil, year 2000. http://www.faostat.fao.org (accessed October 28, 2005).

Food and Agriculture Organization of the United Nations (FAO) (2013a) Desert Locust Bulletin. 3 June 416:1–9.

Food and Agriculture Organization of the United Nations (FAO) (2013b) Global food price monitor. The Global Information and Early Warning System on Food and Agriculture May:1–17.

Food and Agricultural Organization of the United Nations (FAO) (2013c) The state of food and agriculture 2013. Food systems for better nutrition. Rome: FAO.

Food and Agriculture Organization of the United Nations (FAO), World Food Programme (WFP), and International Fund for Agricultural Development (IFAD) (2012) *The state of food insecurity in the world 2012: Economic growth is necessary but not sufficient to accelerate reduction of hunger and malnutrition.* Rome: FAO.

Gómez-Pompa A (1987) On Maya sylviculture. *Mexican Studies* 3:1–17.

Goulding K, Jarvis S, Whitmore A (2008) Optimizing nutrient management for farm systems. *Philosophical Transactions of the Royal Society of London. Series B, Biological Sciences* 363:667–680.

Greene M, Joshi S, Robles O (2012) *The state of world population 2012. By choice, not by chance. Family planning, human rights and development.* New York: UNFPA.

Hassanali A, Herren H, Khan ZR, Pickett JA, Woodcock CM (2008) Integrated pest management: The push-pull approach for controlling insect pests and weeds of cereals, and its potential for other agricultural systems including animal husbandry. *Philosophical Transactions of the Royal Society of London. Series B, Biological Sciences* 363:611–621.

Hazell P, Wood S (2008) Drivers of change in global agriculture. *Philosophical Transactions of the Royal Society of London. Series B, Biological Sciences* 363:495–515.

Held IM, Delworth TL, Lu J, Findell KL, Knutson TR (2005) Simulation of Sahel drought in the 20th and 21st centuries. *PNAS* 102:17891–17896.

Hulme M (2012) Climate change: Climate engineering through stratospheric aerosol injection. *Progress in Physical Geography* 36:694–705.

Jacobsen SE, Sørensen M, Pedersen SM, Weiner J (2013) Feeding the world: Genetically modified crops versus agricultural biodiversity. *Agronomy for Sustainable Development.* doi: 10.1007/s13593-013-0138-9.

Kremer M (1993) Population growth and technological change: One million BC to 1990. *The Quarterly Journal of Economics* 108:681–716.

Lynd J (2013) Gone with the wind: Why even utility patents cannot fence in self-replicating technologies. *American University Law Review* 62:663–700.

Mooney H, Roy J, Saugier B (2001) *Terrestrial global productivity: Past, present and future.* San Diego, CA: Academic Press.

Morse SS (2004) Factors and determinants of disease emergence. *Revue scientifique et technique (International Office of Epizootics)* 23:443–451. http://www.ncbi.nlm.nih.gov/pubmed/15702712 (accessed October 15, 2011).

National Oceanic and Atmospheric Administration (NOAA), National Climatic Data Center (NCDC), and Joint Institute for the Study of the Atmosphere and Ocean (JISAO) (2013) Sahel precipitation anomalies in 0.1 mm, 1900–2012. 1950–79 climatology. Global Historical Climatology Network. http://jisao.washington.edu/data/sahel (accessed June 6, 2013).

National Research Council (2012) Aging and the macroeconomy: Long-term implications of an older population. Committee on the long-run macroeconomic effects of the aging U.S. population. Board on Mathematical Sciences and their Applications, Division on Engineering and Physical Sciences. Washington DC: The National Academies Press.

Nekola JC, Allen CD, Brown JH, Burger JR, Davidson AD, Fristoe TS, Hamilton MJ, et al. (2013) The Malthusian–Darwinian dynamic and the trajectory of civilization. *Trends in Ecology and Evolution* 28, 127–130.

Ogden CL, Fryar C, Carroll MD, Flegall KM (2004) Mean body weight, height and body mass index, United States, 1960–2002. *Advance Data from Vital and Health Statistics*, 347:1–20.

Omenn GS (2010) Evolution and public health. *PNAS* 107:1702–1709.

Omran AR (2005) The epidemiologic transition: A theory of the epidemiology of population change. *The Milbank Quarterly* 83:731–57.

Paterson RRM, Lima N (2011) Further mycotoxin effects from climate change. *Food Research International* 44:2555–2566.

Popkin BM (2004) The nutrition transition: An overview of world patterns of change. *Nutrition Reviews* 62:140–143.

Pretty J (2008) Agricultural sustainability: Concepts, principles and evidence. *Philosophical Transactions of the Royal Society of London. Series B, Biological Sciences* 363:447–465.

Sandberg LG, Steckel R (1997) Was industrialization hazardous to your health? Not in Sweden! In *Health and welfare during industrialization*, Steckel R, Floud R (eds), 127–159. Chicago: NBER-University of Chicago Press.

Scherr SJ, McNeely JA (2008) Biodiversity conservation and agricultural sustainability: Towards a new paradigm of "ecoagriculture" landscapes. *Philosophical Transactions of the Royal Society of London. Series B, Biological Sciences* 363:477–494.

Scrimshaw N (1996) Human protein requirements: A brief update. The United Nations University Press Food and Nutrition Bulletin, 17. http://archive.unu.edu/unupress/food/8F173e/8F173E02.htm#Human protein requirements: A brief update (accessed June 5, 2005).

Sen A (1997) Hunger in the contemporary world. Discussion paper 8. London: The Suntory Centre, Suntory and Toyota International Centres for Economics and Related Disciplines, London School of Economics and Political Science.

Shetty P, James W (1994) Body mass index: A measure of chronic energy deficiency in adults. FAO Food and Nutrition Paper. Rome: FAO.

Shiva V (2002) The real reasons for hunger. *The Guardian*, June 23:1–3. http://www.guardian.co.uk/world/2002/jun/23/1 (accessed July 16, 2012).

Smil V (2002) Eating meat: Evolution, patterns and consequences. *Population and Development Review* 28:599–639.

Smith A, Garnier G (1852) *An inquiry into the nature and causes of the wealth of nations. With a life of the author. Also, A view of the doctrine of Smith, compared with that of the French economists; with a method of facilitating the study of his works; from the French of M. Garn* (Vol. L, p. 633). London: T. Nelson and Sons.

Steckel RH (2001) Health and nutrition in the Preindustrial era: Insights from a millennium of average heights in northern Europe. NBER Working Paper No. 8542. http://www.nber.org/papers/w8542 (accessed July 13, 2013).

Steele J, Shennan S (2009) Introduction: Demography and cultural macroevolution. *Human Biology* 81:105–118.

The Soil Association (2007) One planet agriculture. Preparing for a post-peak oil food and farming future. The case for action. Hopkins R, Holden P (eds). Bristol: The Soil Association.

UNICEF, World Health Organization (WHO), and The World Bank (2012) Levels and trends in child malnutrition: Joint child malnutrition estimates. New York, Geneva, Washington DC: WHO, UNICEF, The World Bank.

United Nations (2011) World population to reach 10 billion by 2100 if fertility in all countries converges to replacement level. UN Press Release. http://esa.un.org/unpd/wpp/Other-Information/Press_ Release_WPP2010.pdf (accessed July 17, 2012).

United Nations Development Programme (UNDP) (2013) Human development report. New York: United Nations Development Programme.

U.S. Census Bureau (n.d.) International Data Base. http://www.census.gov/ipc/www/idbprint.html (accessed February 14, 2005).

Van de Werf G, Randerson J, Collatz G, Giglio L (2003) Carbon emissions from fires in tropical and subtropical ecosystems. *Global Change Biology* 9:547–562.

Victora CG, Vaughan JP (1986) Land tenure and child health in Rio Grande do Sul: The relationship between agricultural production, malnutrition and mortality. *Brazilian Journal of Population Studies* 1:123–143.

Waldorf C (1879) The famines of the world: Past and present. Being two papers read before the Statistical Society of London in 1878 and 1879 respectively, and reprinted from its journal. London: Edward Stanford.

Walpole SC, Prieto-Merino D, Edwards P, Cleland J, Stevens G, Roberts I (2012) The weight of nations: An estimation of adult human biomass. *BMC Public Health* 12:1–6.

Wanger TC, Saro A, Iskandar DT, Brook BW, Sodhi NS, Clough Y, Tscharntke T (2009) Conservation value of cacao agroforestry for amphibians and reptiles in South-East Asia: Combining correlative models with follow-up field experiments. *Journal of Applied Ecology* 46:823–832.

Weinzettel J, Hertwich EG, Peters GP (2013) Affluence drives unsustainable consumption of land and sea. *Global Environmental Change.* doi:10.1016/j.gloenvcha.20

Wilkins RJ (2008) Eco-efficient approaches to land management: A case for increased integration of crop and animal production systems. *Philosophical Transactions of the Royal Society of London. Series B, Biological Sciences* 363:517–25.

Wu F, Guclu H (2012) Aflatoxin regulations in a network of global maize trade. *PloS one* 7:e45151.

York R, Rosa EA, Dietz T (2002) Bridging environmental science with environmental policy: Plasticity of population, affluence, and technology. *Social Science Quarterly* 83:18–34.

Zhao M, Heinsch F, Nemani R, Running S (2005) Improvements of the MODIS gross and net primary production global data set. *Remote Sensing of Environment* 95:164–176.

15

Consumption and Property

Alejandro de las Heras

CONTENTS

The human population does not directly impinge on the environment (see Chapter 14). Here again, the notion that population growth alone leads to unsustainability can be dismissed. As an example, each U.S. inhabitant emits six times more CO_2 than their Chinese counterpart. Although China's population growth rate has descended dramatically as a result of drastic fertility policies, increased production and consumption per capita have more than compensated for this descent. Even the undeniable progress in science and technology (S&T) in China has not reined in environmental damage. These elements point to the importance of the economic context in the search for sustainability.

The environmental and social impacts of economic growth are generically termed externalities meaning that they are unpaid for by the responsible party. In China, it is not only domestic consumption that is leading economic growth but also the industrial delocalization policy (or "deindustrialization") initiated by affluent countries in the 1980s that moved heavy industries and their externalities to low-wage countries. Nowadays, additional motives for industrial migration are less stringent labor and environmental regulations in poorer countries. This process also brings into question the notion, termed "environmental Kuznets curve," that development inevitably leads to clean production technologies and diminished externalities. The footprint—the amount of externalities irrespective of where production and waste are generated—of high-consumption economies that have delocalized can be traced in emerging and fledgling economies.

But consumption alone, or even excess consumption, does not explain the level of externalities. Wealth concentration has to be considered, as one person can earn the equivalent to hundreds of persons' worth of annual wages. More important, capital can be accumulated so that one person or one corporation holds more assets than entire countries (Anderson et al. 2005). Because wealth is mostly generated via (natural) resource exploitation, the quest for individual wealth is a purpose responsible for vast environmental degradation. At the core of wealth accumulation is the notion of property, which excludes others from access to certain resources. These are issues that can seldom be resolved by S&T; indeed in this economic context, sustainable S&T can prove ineffectual.

Following up on the analysis of the IPAT framework started in Chapter 14, it is possible to tease out the influence of consumption and property, the main movers of affluence concentration in the world's current socio-economic system. This analysis then extends to the assumptions (freedom, competition, rationality, and access to information) underlying the idea that we live in the best possible economic system. Whether profound changes have to occur in capitalism is a question relevant to the efficacy of sustainable S&T.

Affluence and Property

Psychologists have discovered that happiness does not accompany wealth (Csikszentmihalyi 1999) and well-being begets desirable outcomes including economic ones (Diener and Seligman 2004). People are also happier in natural environments (MacKerron and Mourato 2013). Economists have shown that waste, such as gasoline in a traffic jam, increases the production of gasoline and so is recorded by official statistics as adding to national economic growth (Stiglitz et al. 2009). With such observations they have concluded that the gross domestic product is a measure of consumption but not of national well-being. This decoupling of wealth and happiness has prompted the emergence of the happiness economics field, actually rediscovering utility as a maximization of happiness for all, not an individual quest. Happiness and the fulfillment of our potential are better sustainability measures than just income or capital. Relevant to our biological welfare, affluent nations do not seem to have time for sex (Stiglitz et al. 2009). Leisure could both augment welfare and diminish our environmental impacts (Druckman et al. 2012).

Does private property make us happy? The quintessential institution of capitalism was argued to be indispensable for the preservation of natural resources. Common goods, on the contrary, were said to lead to a race to depletion as nobody wanted to consume less of a good equally accessible to all (Hardin 1968). It took more than two decades to empirically demonstrate that this idea was not supported by observations. Elinor Ostrom showed that it is not the type of property but the interactions between people, especially monitoring and penalties for bad conduct, that lead to resource exhaustion or sustainability. So perhaps, exclusion of others, underlying private property, and leading to win–lose situations might not be as important as access to resources and diversity of property institutions.

Market competition and monopolistic tendencies drive capital concentration. The urban funnel syndrome (Luck et al. 2001) also concentrates capital, intellect, manpower, and final products in cities; only nature is excluded therefrom. Urban funnels and market competition act as pumps concentrating resources and valves externalizing waste: they take resources away from and deliver waste to rural areas. The cities are sustainability's greatest challenge (Rees and Wackernagel 1996).

Current Risk Societies

The accumulation of manmade hazards (climate change, stratospheric ozone depletion, epizootics, genetically modified organisms, nuclear plant meltdown, and so forth) seem to suggest our entry into risk societies (Beck 1992). This can be linked to the profit motive behind a large number of innovations and more precisely to the drive to accumulate private property and social capital (economic, prestige, and power capital) in capitalist economies.

Metropolises in turn can be viewed as emergent properties, self-organizing from accumulation. Resources become concentrated in the particularly large "urban funnels" of global cities that drive global trends. These cities are susceptible to hurricanes, tsunamis, earthquakes, and the social and economic disarray in their wake. Global cities such as New York, Shanghai, Tokyo, or London have stock exchanges that, although dealing with a minority share of their nation's wealth, create ripples and squalls in the domestic and international economies. The global cities are vulnerable and so are the economies that depend on them. The end result of stock exchange crises is further economic concentration and enhanced risk.

The economic risks are seemingly augmenting due to the recent confluence in global cities of a large increase in large-scale storms such as in New York; climate-change-driven sea level rise; blackouts related to solar coronal mass ejections in a large portion of northeastern America; and pandemic viral outbursts (avian flu and H1N1 flu) intensified by airplane traffic. Additional hazards are due to indefinite trends that may affect confidence in the markets. Large urban areas induce stress that correlates with mental ailments (Abbott 2012), social and political strife, and economic recessions. Acute examples of social unrest in the recent past were terrorist attacks from within a country: in Europe in the 1970s (e.g., Germany, Italy, France, Spain, and the United Kingdom), the Sarin gas attack in the Tokyo subway in 1996, or the Unabomber and school killings in the United States. International terrorism is a more recent risk.

Market Channeling of Biological Impulses

The success of capitalism stems from channeling biological impulses to turn them into profit. Biological drives have been linked to our economic behavior since before capitalism, but they assume a different combination in the latter. So for instance, neophilia, an animal preference toward a new object even if it is exactly the same, underlies some of our consumerist behavior and maintains the turnover of commodities and hence consumption. Speed relates to kinesthesia, which may drive some of our taste for ever faster vehicles. Costly signaling in the human species has been documented in symbolic objects such as jewels for at least 80,000 years, and can be linked to competitive reproductive or economic behavior. Competition for territory has also possibly been mollified by the rules of property. Consumerism in turn uses our senses and the plasticity of our tastes, and psychologically taxes us to resist temptation to consume. Arguably, arms sales benefit from the longstanding warfare social drive (see Chapter 13) unquenched by market economies. An extreme competitive drive is apparent in the following examples of corporate social responsibility failure.

Tampering with Our Biology: Consumption and Corporate Responsibility

A focus on essential "goods" produced by our bodies rather than industrialized food affecting our bodies, and the commoditization of food supplements shows that profit, the main motive for action under the mainstream conception of capitalism, negates attempts at counterbalancing deleterious effects of profits.

Human milk is a "magic potion" delivering nutrients, immunity, and safe hydration (Hinde 2013). Exclusive breast-feeding also is the par excellence contraceptive, especially in Africa (prolactin released during breast-feeding delays the return of ovulation after the delivery of a newborn and so postpones the next pregnancy). The problem is that profit making and marketing have challenged breast-feeding and infant survival (see Box 15.1) due to products

> ### BOX 15.1   MILK FORMULA
>
> As U.S. Senator Edward Kennedy cogently asked in 1978: "Can a product which requires clean water, good sanitation, adequate family income, and a literate parent to follow printed indications be properly and safely used in areas where water is contaminated, sewage runs in the streets, poverty is severe and illiteracy is high?" (Murphy 2012). Thirty-five years later, milk formula is still of particular concern where diarrhea-related infant mortality is concerned.
>
> The marketing of powder milk formulas often targets nursery personnel; this includes the provision of sterile water, as part of "manipulation by assistance" in both affluent and poor countries (Coates 2005). Meanwhile, the most basic sanitary precautions such as to boil water, bottles, and nipples (Johnston 2010) in the preparation of milk formulas are seldom advertised, especially in illiterate populations. Plentiful evidence shows that exposing women to milk formula or information thereof significantly inhibits breast-feeding even in urban contexts in affluent countries (Kaplan and Graff 2008).

of lesser quality than human milk, and despite corporate social responsibility, the response of companies to their own unsustainability. Milk formula is a global problem that has prompted a long-standing international boycott.

Similarly, the metabolic syndrome of conjoined obesity, diabetes, and high blood pressure, already a large issue in the United States and countries with similar lifestyles, is extending fast where fatty and sweet products and excess meat are marketed. Obesity, now considered a pandemic, is bound to affect 51% of all adult U.S. residents by 2030, with an additional 36% overweight. By 2048, all American adults and 30% of the adolescents aged 12 to 19 and children aged 6 to 11 might become overweight or obese if the 1976 to 2004 trends continue (Wang et al. 2008).

Lifestyle traits leading to metabolic syndrome include fast food, excessive car and TV use, and lack of exercise. Lifestyle is advertised and spreads among people in the same network; having a friend who became obese augments a person's chance of becoming obese by 57%. The increased probability is 40% considering siblings and 37% considering spouses. Neighbors are spared an increment in probability (Christakis and Fowler 2007). Some products, soft drinks in particular, and their manufacturers bear an exceedingly large burden of responsibility in unwittingly promoting diabetes and obesity (Ventura et al. 2009).

Merchandized medications also have dramatic consequences. Excess vitamin supplementation among babies born in 1987 to 1992 in the United States has been linked to increased incidence of autism (Zhou et al. 2013). Moreover, the credibility of pharmaceutical corporations has come under criticism in relation to their assessments of product innocuousness. This is why the European Union is taking steps to apply freedom of information to the laboratory and clinical trials of drug companies (Cressey 2013).

Social Sustainability

Three dimensions underpin social sustainability: equity, peace, and cultural diversity. Peace and cultural diversity warrant more space than available here. Equity posits that all humans are entitled to develop their inherent potential or capacities (as proposed by Maslow, and then Sen 2005). This is very different from equal access to consumer goods. Equity cuts across gender, generation, nation, and class barriers but an even more wide-spanning notion of equity can apply to all living organisms. Equity is about the distribution of energy, matter, and information resources to fulfill one's vital needs and potential. In this sense, it is concerned with welfare, a concept first developed by Pigou (Sen 2009), and so differs from a maximization of aggregate output, the backbone of mainstream economics. Fairness has both quantitative and qualitative expressions. The former can be traced back to Bentham in 1789 (1907), who argued in favor of maximum welfare for all. The latter draws from Rawls (1999) and implies a resource-distribution decision-making process by all and the possibility to ignore one's economic position to decide upon resource distribution. Equity and efficiency are not competing goals; equity and equal opportunities in early childhood education promote economic efficiency (Heckman 2011). With regard to peace, it has recently been shown to be less damaging to the environment and societies than warfare (Glew and Hudson 2007).

Economic Sustainability

Compared to social sustainability, the formal debate on economic sustainability seems clear-cut: weak sustainability is the version supported by mainstream economics and upholds present value over future returns on investments and over future generations; it also claims that technological progress is endogenous, that is, it does not require investments and therefore is to an extent, unavoidable; and finally, it posits all resources to be substitutable, meaning that all the trees in the world can be substituted by, say, algae or scrubbers to capture atmospheric carbon. Strong sustainability, on the other hand, maintains the value of future resources, especially those delivering vital environmental services that cannot be substituted.

When present value is much larger than future value, interest rates rise (equivalently, discount rates are applied that diminish the value of expected future profits). This reflects low confidence in the future and ensuing high rates of environmental and labor exploitation. And so where sustainability is concerned, signals in the economic system like interest rates should reflect not only the risks associated with a possible loss of invested capital, but also how any given investment is able to undermine or replenish natural and social capitals. Under strong sustainability, a less damaging investment should command a lower interest rate since it preserves the possibility of future investments and diverts immediate consumption, which lowers nature's reserves and at the same time undermines its regeneration capacity. Some countries have chronically had more confidence in the future, such as Japan, which has been translated in high research and development investments and rates of patents. Substantial investments in technologically advanced economies belie the inevitability of technological progress. Moreover, the efficacy of technology to solve the economic and political drivers underlying environmental and social issues is far from obvious. In light of this discussion, the application of the weak sustainability framework seems to explain current unsustainability.

Can the Tenets of Capitalism Lead to Sustainability?

There is arguably not a natural economic regime for humans because of the diversity of their political institutions and inclinations. Capitalism is a great system in theory: it posits all-around efficiency, unfettered freedom, and promises unending material satisfaction. At present, however, capitalism has a very inefficient metabolism: consuming and wasting too much environmental and human resources, and achieving no balance with its environment. The concentration of resources by elites and push for mass consumption beyond needs further reduce its efficiency.

First, freedom refers to freedom of enterprise or simply the possibility of choice between two brands of the same product. However, excess of choices in market economy correlate with less happiness (Schwartz 2004). Unfettered freedom to access cheap or even free labor ended up reviving slavery and child labor in the early Industrial Revolution and perhaps more episodically to date. Scientists and engineers at the time did little to resist this voracious drive for human and natural resources. And currently, the endless quest for material satisfaction and unabated access to natural resources are poised to deplete world reserves. Colonization until the mid-20th century, conditional financial aid from multilateral organizations thereafter, and (economic) war waged to control key natural resources (e.g., copper and oil) have force-fed capitalist solutions. This has run counter to the extolment of freedom (see Box 15.2).

BOX 15.2   ARRESTED TECHNOLOGICAL DEVELOPMENT

Adam Smith already mentioned that Bengal exported manufactured products by 1776 (Smith and Garnier 1852). Colonial rule halted the technological progress in India, a move best exemplified by the prohibition of high-quality exports to the empire and their replacement by low-quality calico exports. Gandhi later advocated for autonomous technological development. Similarly, the aperture to capitalism was imposed upon Japan by the U.S. Perry Expedition, on Korea by Japanese invasion, and on China by multinational intervention.

Nowadays, China, India, Korea, and Japan epitomize the importance of autonomous technological development. China, India, and Brazil represent 30% of the world's economic output (United Nations Development Programme [UNDP] 2013), while Korea and Japan lead the way in consumer appliances, robotics, and the automobile industry. The environmental toll of these products, however, calls for the next wave of value generation based on existing sustainable technologies. This will likely depend on pressure on corporations, which, of their own accord, do not engage in technology transfer or sustainable innovation (Westley et al. 2011).

Second, the ideal of free competition is lost on monopoly tendencies (market domination) exhibited by virtually all large corporations. This hinders sustainability-oriented innovations, such as renewable energies, which have had to confront vested interests in petroleum.

Third, rationality is bounded (Kahneman 2003), diverse in small sustainable societies and experimentally proven to differ from the Homo economicus hypothesis underlying a discourse of capitalism as the *summum bonum* or the natural state of mankind.

And fourth, access to information, a precondition of free and rational choice, is often costly. The clearest case has probably been the tobacco issue in which information on toxicity has had to permeate through a barrage of advertisement. As they stand, these tenets of capitalism have little relation to the imperatives of social and strong economic sustainability. They also lack grounding in biological drives and a universal character across cultures.

Solutions

Corporations have vast financial power and political influence. They have grown more powerful than many a nation, escaping state and citizen supervision. This is why they are addressed first in the following sections: corporate tools are presented and then contrasted with alternative value-creation processes, core sustainability goals, and the slow ongoing dematerialization process in advanced capitalist economies. Turning to other actors, alternative value creation suggests new roles for a highly skilled labor force and consumers. As to the state, new sustainability metrics and economic tools are briefly addressed.

Eco-Innovation

Eco-innovation is narrowly defined as any improvement that reduces resource consumption and pollution over the life cycle of a good (Eco-Innovation Observatory 2010). This definition can further be expanded to include all actions of relevant actors (households, firms, politicians, unions, associations, and so forth) who develop and apply novel ideas, behaviors, processes, and products toward reducing the environmental burden and goals of enhanced environmental functioning (Rennings 2000).

Eco-innovation has been proposed to help fight the current resource lock-in, which stems from a combination of consumer behavior, market forces, regulations, and obsolete technologies, and which is maintained in spite of cheaper, more equitable and less environmentally harmful solutions. What allows for a resource lock-in to continue is a techno-institutional complex (TIC) or set of "systemic interactions among technologies and institutions." For a TIC to persist, discrete technologies fail to be accurately seen as obsolete and they are presented as "complex systems of technologies embedded in a powerful conditioning social context of public and private institutions." The TIC's vested interests "lock out alternative technologies" (Unruh 2000).

The principles for eco-innovation at the enterprise level to succeed on a large enough scale are:

- Limits to economic scale below the carrying capacity of the ecosystems and harvests of renewable resources below regeneration rate
- Negative waste rates (decontamination) in natural ecosystems and maximum throughput in artificial biodegradation microcosms
- Value creation that minimizes resource use rather than maximizing throughput

An eco-innovative product or service corresponds to a level of commitment with sustainability. This level depends on the assessment of current and anticipated drivers (Figure 15.1), impacts on natural resources, necessary consumer behavior changes, necessary infrastructure changes, potential lock-ins, and the relationships with partners and stakeholders.

Depending on the strength of the drivers, different levels of eco-innovation can be sought (Table 15.1). Higher benefits require deeper changes and investments as well as risk analyses involving markets. These levels require

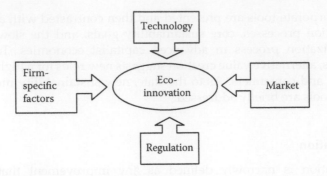

FIGURE 15.1
The drivers of eco-innovation. The market drivers include reputational risks (the cost of looking like a greedy, polluting corporation weighed against the costs of investing in resolutely more environmental products), and the growing but still elusive market of greener consumers. Internal drivers are executive team values, and available design and economic appraisal tools. (After Horbach J et al., 2012, *Ecological Economics* 78:112–122.)

TABLE 15.1

Eco-Innovation Levels and Corresponding Socioeconomic Changes

Level	Purpose	Investment	Consumer Lifestyle Change	Infrastructure Change Needed
1	Incremental improvement	–	–	–
2	Complete concept redesign	+	+	–
3	Alternative functionality	+++	+++	++++
4	Functionality completely fitting into sustainable society	+++++	+++++	++++

Sources: Stevels A, 1997, *Journal of Sustainable Product Design* 3:47–55; Bocken NMP et al., 2012, *Technovation* 32:19–31.

Notes: –, minimal; +++++, maximal; R&D, research and development.

changes in consumer behavior, infrastructure, and institutions. The benefits of level 1 innovations are exploratory because changes at this level are easily reversible and costs can be recovered in the short term. But level 1 can also be misleading and create an impression of commitment and sufficiently deep changes. Up to level 2, no substantial changes are needed to the business basis. Considering the current environmental crisis, only level 3 and 4 innovations can elicit effective changes. But they represent a steep learning curve for every organization, especially large, all-too-often unwieldy corporations. Therefore, eco-innovation levels can best be approached as a necessary learning sequence gradually involving more inputs from different departments (design and marketing at first, then research and development as well as operations) and finally external input from technical universities. Increasing institutional involvement corresponds to gradually more comprehensive

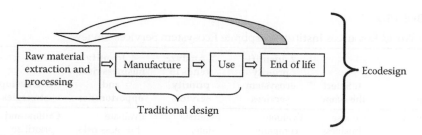

FIGURE 15.2
The value-chain links of eco-design. Reductions of materials, energy, and waste are obtained throughout the cradle-to-cradle value chain. (After Knight P and Jenkins JO, 2009, *Journal of Cleaner Production* 17:549–558.)

tools of environmental validation of the change process (from initial life-cycle analysis of one product to the complete life-cycle analysis of the brand and complete life-cycle cash flow).

Eco-Design

Any level of eco-innovation requires eco-design that considers recycling at the end of life (Figure 15.2) and reductions in environmental impacts in all the stages of the cradle-to-cradle cycle. Among the increasingly long list of eco-design tools, material, energy and toxicity balances, environmental impact assessment matrices, life-cycle cost analysis, and alternative function-fulfillment tools cannot be ignored (Knight and Jenkins 2009). A complementary tool is the water footprint of a product.

Corporate Valuation of Ecosystem Services

The foregoing approaches are product-centered. A more comprehensive approach situates a whole corporation in its global environmental setting. The Corporate Ecosystem Services Review (CESR) developed by the World Resources Institute helps businesses manage ecosystem services, identify risks, and analyze priorities, facilitating strategy creation (Table 15.2). Effective CESR appraisals involve managers and analysts, local stakeholders including labor force, suppliers, researchers, NGOs, and customers from around the world.

The following are water-related examples of eco-innovation, eco-design, and corporate valuation of ecosystem services conducted by small, medium, and very large corporations. The Water & Sewage Metropolitan Enterprise in Ecuador along with The Nature Conservancy developed a scheme that collects payments from downstream water users for watershed management and watershed protection by farmers upstream. A water conservation reward scheme was launched in 2011 in Sao Paulo, Brazil, by brewer AmBev, water utilities Sabesp and CODAU, and retailers Submarino, Americanas.com, and

TABLE 15.2

The World Resources Institute Corporate Ecosystem Services Review

Step	1. Select the scope	2. Identify priority ecosystem services	3. Analyze trends in priority services	4. Identify business risks and opportunities	5. Develop strategies
Activity	Choose business units involved	Evaluate company dependence and impact on ecosystem services Identify ecosystem services most important to performance	Evaluate state, drivers, and trends in priority ecosystems	Evaluate business risks and opportunities due to trends	Outline and prioritize strategies to manage risks and opportunities
End product	Boundary of analysis	List of 5 to 7 priority ecosystem services	Short paper and dataset on trends for priority ecosystem services	List and description of risks and opportunities	Set of strategies
Completion time (weeks)	1–2	1–2	2–5	1–2	1–2

Source: Hanson C, et al. 2012, The Corporate Ecosystem Services Review: Guidelines for identifying business risks and opportunities arising from ecosystem change, version 2.0. Washington DC: World Resources Institute.

Blockbuster. Together they created a bank account that monitors water use. Retailers offer discounts and rewards to customers for water savings. Water shortages and contamination are serious in Sao Paulo, and brewer and water utilities benefit from keeping water sources safe and available (WWF 2011). Nike has ventured in water-related issues by gaining expertise in recycling plastic bottles into apparel. Puma has conducted an analysis of value chain impacts on ecosystem services, not just water (see Box 15.3).

Alternative Value Creation

The eco-innovation, eco-design, and corporate valuations of ecosystems all appear to lack a social sustainability dimension. This seems a direct contradiction with the social responsibility notion. An equally fundamental issue is that no substantial decreases in consumption are proposed; this will eventually annul any redesign and efficiency gains.

BOX 15.3   WATER USE AT PUMA

In 2011, Puma announced an environmental profit and loss statement (Puma 2011). It is one of a small number of companies doing so. What is unique is the application of an ecosystem services approach. Water was valued on the basis of reductions in services (provisioning, supporting and regulating, same as MEA, see Figure 5.1). They developed a value for water based upon a relationship between value and scarcity: the withdrawal of water as a percentage of actual renewable freshwater resources. The average value by location is €0.81/m³ with a range of €0.03 to €18.45/m³ depending on local scarcity. Puma looked at water consumption across their entire value chain as shown in Figure 15.3.

Water appeared to be the chief environmental concern at Puma (it accounts for 33% of total impacts) but what is significant is that Puma was trying to measure its impacts (directly or through suppliers) on the entire ecosystem (in terms of greenhouse gases, land use, air pollution, and waste). These data tell management where major costs and water-related risks exist: water in Asia-Pacific, footwear, and raw materials are more costly to Puma and the environment. If water becomes scarcer in certain areas, suppliers could be unable to deliver raw materials to Puma. The company and its suppliers are faced with a choice of proactively managing the risks or passively waiting to see what happens.

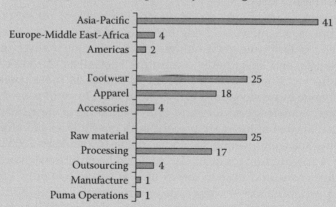

FIGURE 15.3
Puma's impact on water ecosystem services, expressed as production costs (million euros).

Reducing consumption without entering an economic depression is a conundrum in market economies first expounded by Keynes in 1936. One workaround is to create value chains that do not overexploit nature and recycle waste while allowing for nonmaterial consumption. These value chains should conform to overarching sustainability goals attainable with

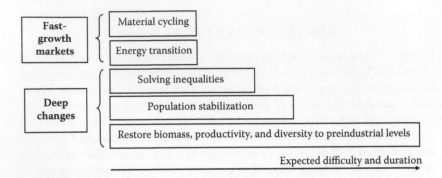

FIGURE 15.4
Overarching sustainability goals as related to current markets and deep changes to capitalism. These goals are overlapping and urgent. Material cycling and energy transition to renewable sources are fast solutions to environmental unsustainability since they are fast-growing markets; they call for the participation of entrepreneurs. Deeper changes will take longer since maintaining inequalities and large populations actually help create large, low-wage, labor markets. The deep changes are amenable to very large, low-profit, adequate technology markets, which call for the participation of the local population with government support. Both advanced and basic science and technology solutions are required to progress toward these overarching goals.

currently fast-growing markets or very large, untapped, adequate technology markets (Figure 15.4).

The foregoing value chains will provide employment for sustainability lawyers, activists, scientists, and engineers to train new professionals: sustainability accountants and criminologists, actuaries, journalists, marketing ethics specialists, and sustainable computing specialists. A long-term multiplicative effect will spill over the whole economy as these professionals gradually change our conceptions and aspirations. This will pave the way to a fully fledged knowledge society and professional Earth stewardship.

In addition, market failures are fraught with possibilities of value generation in helping solve environmental and social issues (Luksha 2008). Markets have failed at solving hunger, war, chronic health conditions entailing massive monetary costs, and human losses for the world economy. But health, happiness, peace, and conservation of environmental services can cancel out the monetary costs of physical illness, psychological maladies, war, and nature depletion.

Dematerialization

Since indefinite material growth is impossible due to the limited planetary carrying capacity, we are only left with a set of options consistent with strong sustainability, chiefly the emergent practices of postmaterialism, postconsumerism, and degrowth (Schneider et al. 2010) (Figure 15.5). Anticonsumerism and cultural jamming are forms to contend with the

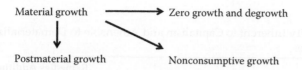

FIGURE 15.5
Three sustainable future economic pathways. Carrier bags can exemplify their complementarity: zero growth means no increase, degrowth means less bags, postmaterial growth means no more bags, and nonconsumptive growth implies recycling or biodegradability and reuse by microorganisms to recirculate nutrients into plants and the trophic chain. Sustainability is best achieved using all pathways.

conformism to consumerism. Although dematerialization warrants participation from all economic actors, it should at the very least involve a movement of citizens, concerned scientists, and engineers collectively garnering foresight intelligence (Ehrlich and Ehrlich 2013).

Capitalism is already very much dematerializing: finance is not materialistic at all but electronic. Luxemburg or the Cayman Islands are wealthy, although materially speaking unproductive. Money itself is symbolic and the paper or plastic it is made of has very little value. These symbols, as all things cultural are arbitrary, the best proof being that for most human history, societies have survived without them. Quality has in fact become more valuable than size or quantity. So further dematerializing the economies would only follow the trend of deindustrialization and increasing value of technology or knowledge; this is why the largest petroleum companies have focused on prospection and exploration and are moving away from exploitation, refinery, and transport. By doing so, they have largely decontaminated their operations if not their past environmental record.

Consistent with the aforementioned, future value creation may best be served by renewable resources exploited at a rate below natural regeneration, investments rather than consumption, and savings rather than credit, economic degrowth to or a little above biological needs, and infinite nonmaterial personal growth. Dematerialization along with other actions can help solve unsustainability issues deeply connected to capitalism (Table 15.3). Other inherent unsustainabilities are discussed later.

Changing Macroeconomics

Alterations in the economic metrics (national accounting) and steering instruments (interest rates, taxes, and incentives) are needed to move the world economy toward sustainability. Official statistics are in crisis in several important countries where 30% of the citizens trust them and the policies (Stiglitz et al. 2009). One reason is that politicians are seen as self-interested and committed to short-term, not long-term policies (Acemoglu et al. 2007). There is, however, a movement initiated by hybrid

TABLE 15.3

Unsustainability Inherent to Capitalism and Amenable to Dematerialization and Other Remedies

Cause	Effect	Remedies Additional to Dematerialization
Obsolescence and fashion	↑ Consumption ↑ Waste	Value creation (not demand creation) Satisfy biological needs and protect health
Prices not reflecting long-term scarcity or value added by labor	↑ Consumption ↑ Waste ↑ Inequalities	Complement current development metrics with inclusive wealth, health, happiness, peace, natural regeneration indicators; assign weight to indicators and value to goods and services in subsidiary processes
Cheap labor, low education, population growth	↑ Consumption ↑ Inequalities	Equity, education, birth control
Privatization, concentration of political and economic resources and power	↑ Inequalities	Reclaim cooperation, information, diversity and subsidiarity

accounts following United Nations recommendations, based on money and environmental costs. Physical input–output tables go one step further and account for matter and energy flows; as implemented by Austria and Scandinavian countries (Mulalic 2007; Statistics Austria-SERI 2011). To match social demand and natural biocapacity, efforts are carried out following the environmental footprint framework (Best et al. 2008; Galli et al. 2012). The latest, incipient shift recognizes the broader sense of development and aims at integrating social, environmental, and economic indicators (United Nations University International Human Dimensions Programme on Global Environmental Change [UNU-IHDP] and United Nations Environment Programme [UNEP] 2012).

National accounting determines the size of the budget, taxation, incentives, and interest rates. In sustainable national accounting, input and feedback from society is required (obeying the principle of subsidiarity or decision at the lowest possible hierarchical level), as in the Canadian Alternative Federal Budget (Elson and Cagatay 2000). Incentives ought to be apportioned considering that, similar to excess consumption and savings, excess investments may lead to environmental degradation, as businesses vie for the fastest possible returns on their investments. As to taxes, the Thoreau principle states that they have to be ethically spent and only citizens can and must, through civil disobedience if necessary, assess the ethics of tax expenditure (Thoreau 1849).

Incentives and taxes must be proportional to pollution and restoration, according to the "polluter pays" or "abuser pays" principle, which reflects the depletion of natural and social capital. Conversely, restoration, natural regeneration, and recycling are to be incentivized. Government initiatives, like the

2012 bioeconomy initiatives in the European Union and the United States, ought to be operated according to these principles to move toward a sustainable economy.

Satellite accounts now reflect the situation of a country with regard to, for instance, gender equity. Similarly, restoration and natural regeneration, health, happiness, quality of life, peace, planetary stewardship, knowledge (rather than information), and equity are to be given a larger weight than consumption and production in satellite and then national accounts.

Is Capitalism a Fertile Ground for Sustainable Science and Technology?

From a material viewpoint, capitalism excels at economic growth based on consumption of natural resources. Destructive creation further enhances growth. When as a result, natural resources tend to zero, profits go to infinity. From a values viewpoint, all tangible and intangible values (e.g., life, beauty, and variety) are ultimately related to money (hedonic valuation) and so nothing is irreplaceable as long as money is available. Present value is always larger than future value, especially as technology and fashion render goods and ideas obsolete. These and those listed in Table 15.3 are not the only deep issues. They are rather the most tractable and so amenable to solutions through dialog. Others are more value-laden and could only be tackled as sustainability makes wider inroads in our economies.

This means that although capitalism displays inherent unsustainabilities, deep changes can redirect the economies. Material and financial growth have to be replaced as the overarching goals. Values have to change in three substantial ways according to Georgescu-Roegen: first, reversing hedonic valuation—"If our values are right, prices will be OK." Second, a society should be valued according to how much choices it allows at home and abroad. And finally, a society must decide how much production beyond biological needs it can sustain (Gowdy 1998).

Scientists and engineers left to their own devices cannot achieve sustainability for all. The positivist illusion that technical progress could solve all problems (Comte [1830–1842] 2001) is in fact largely responsible for the current environmental and socioeconomic issues. This illusion helps maintain the economic, political, and social status quo that thwarts the transition to sustainability. S&T play a key role in massive production, the supply side of consumerism, and also in patents that buttress the institutions of (intellectual) property. Programmed obsolescence distracts technological progress from solving substantial human and environmental issues. This contradicts the illusion of independent science. It is, however, possible that the foregoing solutions might prepare the ground for a more definite contribution of science and technology to sustainability, as exposed in Chapter 17.

References

Abbott A (2012) Urban decay. *Nature* 490:162–164.

Acemoglu D, Golosov M, Tsyvinski A (2007) Political economy of mechanisms. Research Notes #2007/02. State University–Higher School of Economics, Moscow.

Anderson S, Cavanagh J, Lee T (2005) *Field guide to the global economy*. New York: New Press.

Beck U (1992) *Risk society: Towards a new modernity*. London: Sage.

Bentham J (1907) An introduction to the principles of morals and legislation. Library of Economics and Liberty. http://www.econlib.org/library/Bentham/bnthPML1.html (accessed July 19, 2013).

Best A, Giljum S, Simmons C, Blobel D, Lewis K, Hammer M, Cavalieri S, et al. (2008) Potential of the ecological footprint for monitoring environmental impacts from natural resource use: Analysis of the potential of the ecological footprint and related assessment tools for use in the EU's Thematic Strategy on the Sustainable Use of Nature. Brussels: Report to the European Commission, DG Environment.

Bocken NMP, Allwood JM, Willey AR, King JMH (2012) Development of a tool for rapidly assessing the implementation difficulty and emissions benefits of innovations. *Technovation* 32:19–31.

Christakis NA, Fowler JH (2007) The spread of obesity in a large social network over 32 years. *New England Journal of Medicine* 357:370–379.

Coates MM, Riordan J (2005) Tides in breastfeeding practice. In *Breastfeeding and human lactation*, 4th ed., Riordan J (ed), 3–29. Sudbury, MA: Jones & Bartlett.

Comte A (2001) *Cours de philosophie positive*, 4 volumes. BookSurge Publishing.

Cressey D (2013) Drug-company data vaults to be opened: European agency will publish firms' clinical-trial results. *Nature* 495:419–420.

Csikszentmihalyi M (1999). If we are so rich, why aren't we happy? *American Psychologist* 54:821–827.

Diener E, Seligman MEP (2004) Beyond money: Toward an economy of well-being. *Psychological Science in the Public Interest* 5:1–31.

Druckman A, Buck I, Hayward B, Jackson T (2012) Time, gender and carbon: A study of the carbon implications of British adults' use of time. *Ecological Economics* 84:153–163.

Eco-Innovation Observatory (2010) Methodological report. http://www.chamberofecocommerce.com/images/EIO_Methodological_Report_2010.pdf (accessed May18, 2012).

Ehrlich PR, Ehrlich AH (2013) Can a collapse of global civilization be avoided? *Proceedings of the Royal Society* 280:20122845.

Elson D, Cagatay N (2000) The social content of macroeconomic policies. *World Development* 28:1347–1364.

Galli A, Wiedmann T, Ercin E, Knoblauch D, Ewing B, Giljum S (2012) Integrating ecological, carbon and water footprint into a "Footprint Family" of indicators: Definition and role in tracking human pressure on the planet. *Ecological Indicators* 16:100–112.

Glew L, Hudson MD (2007) Gorillas in the midst: The impact of armed conflict on the conservation of protected areas in sub-Saharan Africa. *Oryx* 41:140–150.

Gowdy J, Mesner S (1998) The evolution of Georgescu-Roegen's bioeconomics. *Review of Social Economy* LVI:136–156.

Hanson C, Ranganathan J, Iceland C, Finisdore J (2012) The Corporate Ecosystem Services Review: Guidelines for identifying business risks and opportunities arising from ecosystem change, version 2.0. Washington DC: World Resources Institute.

Hardin G (1968) The tragedy of the commons. *Science* 162:1243–1248.

Heckman JJ (2011) The economics of inequality. The value of early childhood education. *American Educator* Spring:31–47.

Hinde K (2013) Lactational programming of infant behavioral phenotype. In *Building babies: Primate development in proximate and ultimate perspective*, Clancy KBH, Hinde K, Rutherford JN (eds), 187–207. New York: Springer.

Horbach J, Rammer C, Rennings K (2012) Determinants of eco-innovations by type of environmental impact: The role of regulatory push/pull, technology push and market pull. *Ecological Economics* 78:112–122.

Johnston L (2010) Infant formulas unravelled. *South African Pharmacist's Assistant* 10:23–26.

Kahneman D (2003) Mapping bounded rationality. *American Psychologist* 58:697–720.

Kaplan DL, Graff KM (2008) Marketing breastfeeding: Reversing corporate influence on infant feeding practices. *Journal of Urban Health* 85:486–504.

Knight P, Jenkins JO (2009) Adopting and applying eco-design techniques: A practitioners perspective. *Journal of Cleaner Production* 17:549–558.

Luck MA, Jenerette GD, Wu J, Grimm NB (2001) The urban funnel model and the spatially heterogeneous ecological footprint. *Ecosystems* 4:782–796.

Luksha P (2008) Niche construction: The process of opportunity creation in the environment. *Strategic Entrepreneurship Journal* 2:269–283.

MacKerron G, Mourato S (2013) Happiness is greater in natural environments. *Global Environmental Change.* http://dx.doi.org/10.1016/j.gloenvcha.2013.03.010 (accessed April 22, 2008).

Mulalic I (2007) Material flows and physical input-output tables: PIOT for Denmark 2002 based on MFA. Copenhagen: Statistics Denmark.

Murphy RC (2012) The facts of INFACT: How the infant formula controversy went from a public health crisis to an international consumer activist issue. MA dissertation, University of Minnesota.

Puma (2011) PUMA completes first environmental profit and loss account which values impacts at €145 million. http://about.puma.com/puma-completes-first-environmental-profit-and-loss-account-which-values-impacts-at-e-145-million (accessed May 4, 2012).

Rawls J (1999) *A theory of justice*, Rev. ed. Cambridge: The Belknap Press of Harvard University Press.

Rees W, Wackernagel M (1996) Urban ecological footprints: Why cities cannot be sustainable—And why they are a key to sustainability. *Environmental Impact Assessment Review* 16:223–248.

Rennings K (2000) Redefining innovation: Eco-innovation research and the contribution from ecological economics. *Ecological Economics* 32:319–332.

Schneider F, Kallis G, Martinez-Alier J (2010) Crisis or opportunity? Economic degrowth for social equity and ecological sustainability. *Journal of Cleaner Production* 18:511–518.

Schwartz B (2004) The tyranny of choice. *Scientific American* April:71–75.

Sen A (2005) Human Rights and Capabilities. *Journal of Human Development* 6:151–166.

Sen A (2009) Capitalism beyond the crisis. *The New York Review of Books.* http://www.nybooks.com/articles/archives/2009/mar/26/capitalism-beyond-the-crisis/?pagination=false&printpage=true (accessed February 12, 2012).

Smith A, Garnier G (1852) *An inquiry into the nature and causes of the wealth of nations. With a life of the author. Also, A view of the doctrine of Smith, compared with that of the French economists; with a method of facilitating the study of his works; from the French of M. Garn* (Vol. L, p. 633). London: T. Nelson and Sons.

Statistics Austria-SERI (2011) Physical input-output tables for Austria 2005. Final report on technical implementation. Vienna: Statistics Austria-SERI.

Stevels A (1997) Moving companies towards sustainability through eco-design: Conditions for success. *Journal of Sustainable Product Design* 3:47–55.

Stiglitz JE, Sen A, Fitoussi JP (2009) Report by the commission on the measurement of economic performance and social progress. www.stiglitz-sen-fitoussi.fr (accessed June 20, 2010).

Thoreau HD (1849) Resistance to Civil Government. http://www.gutenberg.org/cache/epub/71/pg71.txt (accessed March 7, 2012).

United Nations Development Programme (UNDP) (2013) Human development report. New York: United Nations Development Programme.

United Nations University International Human Dimensions Programme on Global Environmental Change (UNU-IHDP), United Nations Environment Programme (UNEP) (2012) Inclusive Wealth Report 2012: Measuring progress toward sustainability. Cambridge: Cambridge University Press.

Unruh GC (2000) Understanding carbon lock-in. *Energy Policy* 28:817–830.

Ventura EE, Davis JN, Goran MI (2009) Sugar content of popular sweetened beverages based on objective laboratory analysis: Focus on fructose content. *Obesity* 19:868–874.

Wang Y, Beydoun MA, Liang L, Caballero B, Kumanyika SK (2008) Will all Americans become overweight or obese? Estimating the progression and cost of the U.S. obesity epidemic. *Obesity* 16:2323–2330.

Westley F, Olsson P, Folke C, Homer-Dixon T, Vredenburg H, Loorbach D, Thompson J, et al. (2011) Tipping toward sustainability: Emerging pathways of transformation. *Ambio* 40:762–780.

WWF (2011) Green game-changers. WWF Report UK. http://www.wwf.org.uk/what_we_do/working_with_business/green_game_changers/50_innovations_to_inspire_business_transformation_.cfm (accessed May 12, 2012).

Zhou SS, Zhou YM, Li D, and Ma Q (2013) Early infant exposure to excess multivitamin: A risk factor for autism? Autism Research and Treatment. Volume 2013, Article ID 963697. http://dx.doi.org/10.1155/2013/963697 (accessed May 24, 2012).

16

Built Environment

Alejandro de las Heras

CONTENTS

Issues with Building Materials

Cities have accumulated 75% of all the materials ever extracted by humans. Cities started to grow in the 1700s as manpower in-migrated in search of livelihood following the Enclosure Act in Scotland and similar land privatization in England. The number of large buildings augmented thereafter and high-rises appeared following the invention of steel and the rediscovery of concrete, forgotten after the Antiquity. Despite enhanced durability thanks to steel and concrete, it is not utility but change that governs the lifetime of buildings. The high turnover of materials is due to the destruction of buildings (Fernandez 2006), in line with creative destruction in the search for profit that seems essential to capitalism (Schumpeter 1943). The obsolescence discourse has evolved in the past century from "financial decay" (explaining the demolition of very large buildings in very good conditions in Chicago and New York), to the urban renewal discourse (and the complete replacement of entire districts), to progressive or planned obsolescence and creative waste by 1930 (Fernandez 2006).

By the end of the 20th century, out of 2.5 billion metric tonnes of non-fuel material circulating in the United States each year, 70% were destined to construction. Due to the fast turnover of building materials, the construction industry is responsible for 40% of U.S. CO_2 emissions (Williams 2007).

The dominant life cycle is construction–demolition–waste, with little refurbishment or renovation and considerable postbuilding activities (decommissioning, dispersal of residuals, landfill management, incineration, and land reclamation). In the United States 15% to 30% of the municipal waste is due to this cycle, especially since renewable materials went down from 42% to 5% in all U.S. industries in the 20th century.

The real estate market also dictates a different life cycle in which abandonment of vast buildings in good conditions can be observed, as in Detroit. Midtown Manhattan in contrast buoys with "hyperlocalized expectation of boundless prosperity" (Fernandez 2006).

Metals and Concrete

Globally, smelting is responsible for 13% of the SO_2 emissions (the main cause of acid rain). Metal production at the turn of the century was 900 Mt leaving 6 billion tons of waste behind. Waste is expected to increase as the grade of ores decreases until complete exhaustion of proven reserves for even some of the most common metals in the Earth's crust: Al, Fe, and Si have between 95- and 125-year reserves; copper has 70-year reserves (Fernandez 2006).

Concrete, along with glass, stone, and clay are ceramics. Concrete has a high embedded energy and represents 6% to 8% of the global CO_2 emissions, and cement 3% (Joint Research Centre of the European Commission (JRC)/PBL Netherlands Environmental Assessment Agency 2011). Innovation is required in concrete structures, previously thought to be durable, to prevent creep and corrosion. As to glass, its use is soaring but only partly due to double glazing.

Polymers and Composites

The fastest growing materials are polymers (plastics) and adhesives. Plastics already surpass the combined production volume of steel, aluminum, and copper, and they account for 4% of the petroleum consumption.

The exhaustion of petroleum reserves raises concern about the resumption of coal use as a precursor of polymers such as used in Europe until the 1950s (calcium carbide reaction to form acetylene). Plastic waste is immense: it amounts to 9% of the U.S. solid waste and large volumes persist in the oceans and on continental Africa. Excess use of off-gassing polymers, adhesives, and volatile organic hydrocarbons in interior finishing leads to the sick building syndrome.

PVC, the most widely used polymer in architecture, is also the most conducive to health hazards. PVC manufacturing produces dioxins, furans, and PCBs (Chapter 12). Burning PVC produces hydrochloric acid and many caustic and toxic substances. Upon recycling, PVC may be confused with PET, which melts at a higher temperature, and releases hydrochloric acid. Landfilling PVC may also release toxic compounds. The Georgia Gulf PVC factory spillage into the aquifer led to the relocation of an entire town in Florida.

As to composites, they are combinations of different materials that capture a new set of properties (concrete is technically a composite of steel, cement, and other minerals). Today, glass and carbon fibers are widely used to reinforce epoxy coatings on concrete structures. Many smaller structures also substitute metal with fiberglass or with metal-reinforced epoxy or polyester matrices, which show outstanding energy absorption. However, embedding several materials makes composites difficult to recycle (Fernandez 2006).

Recycling

Recycling depends on the metal and the ease of salvage but is on average below 50% in the United States. The pursuit of lighter materials while reducing the mass of a construction has augmented the volume of waste: one unit of aluminum requires the consumption of 300 units of other materials; deforestation is also associated with bauxite mining for aluminum production. Northern Europe recycles 90% of the aluminum and the United States 42%, despite the fact that the recycled metal requires only 5% of the energy required for its primary production. Copper is most efficiently recycled in China and the Third World, entailing toxicological (Chapter 12) and child labor issues. Similar to metals, recycling depends on the class of polymers: thermoplastics are more readily recycled as they liquefy upon heating, unlike thermosets, which should target durability and reuse (Fernandez 2006).

Notwithstanding recycling, a material bottleneck may soon appear: petroleum-based polymers will have to be replaced at around the same time as metal reserves become depleted. This may increase the use of silicones and ceramics. The latter can be easy to recycle (glass, clay) or may end up producing smaller fraction sizes in the process. This down-cycling to a lower grade explains why stone is increasingly used at ever smaller sizes or as thinner layers.

Transport: Compounded Sustainability Issues

The dependence of cities on cars is responsible for a good share of the accelerated depletion of oil reserves. The continuation of this problem will make the energy transition to renewable energies very difficult as this transition will require oil energy to be carried out. Cars have replaced transit or public transport, the demise of which can be traced to conspiracies involving oil and car companies (Register 2006). Oil wars to control more of the extant reserves also unwittingly contribute to depletion. They are waged to fuel the continuation of the car-dependent economy and suburban culture (*The End of Suburbia* 2004). As a result urban growth now takes place around suburbs and roads, in the form of gated and bedroom communities, edge cities, back offices, and malls (Johnson 2001; Register 2006), which are only accessible by car and require extensive parking lots.

The environmental consequences of urban areas are vast impervious areas, which hinder aquifer recharge, and urban heat islands. The urban approach to water is often in terms of dams and desiccation. Urban sewage and runoff pollute contiguous rivers and hinterlands. Water becomes channelized and does not distribute in the entire watershed as before. A lower infiltration area thus reduces aquifer recharge and causes subsidence, a lower water table, and hence lower pressure limiting the use of gravity for water transport. Consequently the costs of pumping augment. Agriculture is continuously displaced to ever more remote and larger areas as land becomes less suitable than the original places of urbanization. Air pollution is also mostly driven by car use and its death toll has now surpassed the number of car crash fatalities (Williams 2007). This is a global trend with small differences, a feature of the global homogenization exemplified by Beijing, which started closing streets to bicycles in 1998 (Register 2006). Rush hours make for more lanes; parking lots make it easier to use a car; this vicious circle is closed by more efficient cars deluding us into not solving root issues.

As a result, good city livability—considering housing, health, transport, water, air, and green spaces—is not sustainable in the largest cities. The living standards of the most livable cities, if they were to be replicated throughout the world, would require three or more planets (Australia and New Zealand), or in the best cases more than one and up to two planets (German cities). The largest American cities are livable to a moderate-to-good extent but are seldom sustainable (three planets or more). Currently, models of cities and technology affordable by one planet Earth can only deliver moderate-to-low livability (Casablanca, Manila) (Newton 2012). With regard to transport, a review of 300 studies on public attitude in the United Kingdom seems representative of increasing global awareness but at the same time reluctance to undergo transformations in our lives. A majority of people see traffic congestion as a national problem, although the impact on families is somewhat less clear-cut. The large majority support public transport improvements, speed reduction, and less traffic in residential areas, although road building and pricing are controversial topics (Goodwin and Lyons 2010).

Supply Regions

Supply regions are far from the cities that they supply, and human populations cannot thrive there without subsidies. The Great Plains are such a region, with the shortest growing season in the United States and recurrent natural disasters (the 1880s intense blizzards, 1890s droughts and financial panic; the 1930s Dust Bowl, the early 21st century water reserves depleting in the vast Ogalalla aquifer) (Register 2006). Siberia is another archetypal case that might grow as a result of climate change and the inroads leading human expansion into central Asia, visible from nighttime light satellites for a decade. In Africa, the land grab phenomenon (Chapter 5) points to another costly, large-scale colonization definitely more oriented to exploitation than livability.

Sustainable Built Environment: Principles and Metrics

Until now, most built environments have been self-organizing, rarely stemming from a planned layout, and have emerged from collections of individual artifacts (buildings and cars) increasing the use of energy to deliver suitable indoor conditions. Sustainable built environments call instead for temperatures and hygrometry in bioregions as the starting point, followed by planning of the exchanges with the rural surroundings into an industrial ecology with green-blue infrastructure and minimal waste generation and transport. The design of buildings comes last, aiming at maximum matter and energy efficiency and recyclability, use of local material, and local knowledge. In keeping with sustainability, the built environment has to maximize well-being and functionality through a series of measurable steps.

1. The size of the settlement layout must correspond to natural limits, in particular: hinterland and watershed size, water reserves and recharge, soil availability, sun supply, and net primary productivity. Most often densification and surface degrowth are necessary, but richness in terms of abundance of diverse species and human cultures should be augmented.

2. The share of local inputs should be maximized from the design onward: native/indigenous and local knowledge, vernacular design, and local materials. Local ecosystem functions are constantly replenished (restored, let regenerate, and enhanced). Participative, decentralized planning, and management cater to the diversity of approaches to sustainable life in common.

3. Connectivity and proximity, rather than transport, guide the spatial organization. This means that green (vegetation) and blue (water) spaces are mixed with built structures. Green-blue corridors and networks pervade the layout, as well as green-blue infilling or addition of fine texture interspersed with buildings. This translates into maximal green and blue area, maximal native plant diversity to adapt to small niches, diminished spatial autocorrelation (the measure of spatial sameness in neighboring places), maximum exchanges by contiguity, and maximal diversity of uses and ecological functions throughout, at every scale.

   Maximum net primary productivity is achieved through restoration, defined as human-driven ecosystem recovery after anthropogenic disruption. Restoration implies networking what buildings, pavement, and freeways sever: natural and cultural flows. Diversification via disturbance management increases resilience of the built-natural environment. To this end, channeled rivers are unearthed, embankments are left to decay and are bioremediated. And consequently, the following functions of the surrounding natural environment are enhanced: stormwater absorption, water depuration, and water recharge.

   Maximum system functionality is achieved through cycles of natural disturbance and recovery. This requires human disturbance to be minimized. If restoration cannot reclaim buildings, artificial wetlands, green walls, and roofs provide layout and design opportunities.

4. Carrying capacity in terms of sustainably hosting human population is dependent on maximum system functionality, that is, maximum renewable energy and matter gain, minimal losses and disturbances, minimized human consumption, and recycling. Energy and matter losses in maintaining the built system are minimized through maximal use of natural biodegradation processes, and green and blue infrastructure rather than waste generation, management, and confinement.

5. A diversified economy enhances its resilience. Specialization can only be pursued as long as resilience, autonomy, and food security are not compromised. Value creation in the new diversified urban economy stems from innovations in and trickling effects of green-blue infrastructure, landscape architecture, large-scale industrial ecology (including but not limited to recycling), construction of long-lived buildings, adaptive reuse of buildings, renewable energy, the science and technology of sustainable materials, bioclimatic and off-the-grid technologies, human powered vehicles, permaculture, and restoration ecology. New activities in ecocities also call for new interior design and new but less appliances.

6. Designs for future generations coalesce in 100-year plans. Sustainable design charts the development of connections with nature, fosters stewardship of nature, plans diverse property rights—commons, private property, and other innovative forms—uses renewables only, and collects and stores local renewable energy.

Accumulated experience has culminated in a set of layout, design, and material use principles that can be creatively and productively articulated in a diversity of local solutions (Box 16.1).

BOX 16.1 ARTICULATION OF SUSTAINABILITY PRINCIPLES IN COUPLED BUILT-AND-NATURAL SYSTEMS

Bioregions and Biourbanism

Studies on the bioregion inform participatory processes to designate areas, nodes, and vertices; their primary functions; and exchanges that occur among them. Areas include stormwater, rain harvest, green, unplugged areas, as well as brownfields to be cleaned up. Nodes, or hubs, are transit access points, facility locations, and long-lived public structures. Vertices are natural conveyor belts facilitating matter and energy exchanges: green and blue infrastructure, channels for rural–urban and industrial ecology exchanges, and networked connections rather than single-strand (Williams 2006).

Elements of Industrial Ecology Applied to Sustainable Buildings

Dematerialization is the process whereby less material and waste are handled; costs diminish but economic returns are maintained. This makes for resource-light societies that reduce their culturally induced need for space. A decoupling of money and destructive use of resources is warranted, as well as considering that not all resources are substitutable or always renewable (Fernandez 2006).

Useful waste, similar to biological waste in nature, is a useful resource. To benefit from it the biogeochemical cycles must remain closed; networks of matter and energy exchange must be preserved or created.

Renewables—within the limits of their natural regeneration should replace single-use materials. Otherwise, material reuse and recycling as well as adaptive reuse of extant buildings is to be sought.

Economy, sobriety of design, inventiveness, and consideration of opportunity costs (what could be done with saved resources) are key to sustainable material design. Very long-lived buildings prolong the useful life of nonrenewables and reduce extraction; these can coexist with naturally degradable structures. According to this view, buildings live and breathe, have a metabolism, repair and die, but their materials are always reused or recycled.

Top 10 Green Projects Awards

Lessons from the American Institution of Architects and U.S. Department of Energy Committee Top 10 Green Projects Awards include: simplify (nonmoving parts, deautomatize, reduced liability and costs, low maintenance), educate users, create flexible spaces (with double or multiple uses, evolvable over time), and assess (postoccupancy tests, integrate lessons into collective knowledge) (Williams 2006).

The Barcelona Declaration

The Barcelona Declaration proposes a change in professional culture so that architects and associated professionals become natural and cultural stewards. Cooperation and knowledge sharing are the means to bridge professional gaps. The construction professionals have to educate and motivate clients, owners, and users. The ultimate goal is to cater to the needs of future generations of all species (Williams 2006).

Layout Solutions

To move toward organic (integrated into a recycling loop) rather than mechanic (waste-generating and power-consuming) built environments, the initial focus has to be not on the form of a built structure, but on the surrounding ecosystem functions and their limitations.

Bioregions and Biourbanism

Biourbanism adapts to the bioregion, merging nature and culture within cities and changing our activities. Restoration and regeneration are needed to recoup natural resources degraded or destroyed during the life cycle of the built environment. Beyond the usual site characterization by mechanical and fertility characteristics of its soils, biourbanism plans for water–land–vegetation systems, which limit the type of construction and guide land uses.

As recent tsunamis and hurricanes suggest, the use of topography helps cope with the return of very large events. This also implies an intimate knowledge of climatic averages, seasonality, and extreme events. Watersheds and water tables limit the ecological distribution of water and primary productivity, and thus are both ecological management and biourbanism units. Wetlands with mosquitoes, rather than desiccated, should be out of bounds due to the possibility of (re)emergent diseases; they should be part of the natural elements in the urban hinterland and separated from urban places by rural uses. As to littorals, they naturally function as storm buffers; water extraction there is to be avoided to limit saltwater intrusions into the freshwater table.

Wastewater and solid waste infrastructure and utilities have to be replaced by landscape architecture applied to artificial wetlands (Chapter 3) and recycling, respectively, thereby reducing much-needed and costly infrastructure repairs. For the most part, utilities are costly means to generate externalities, our getting rid of unwanted matter. These externalities can turn into agricultural and renewable energy resources (Chapter 10). The Catskills range in the state of New York captures water, a solution chosen over huge wastewater treatment plants, similar to the Cedar River watershed in Seattle whose protection has gone on for a century so that trees help replenish the city's water supply.

The blue infrastructure includes wetlands, duck ponds, prairie waterways, and floodplains that capture and stock stormwater and help oxidize (degrade) organic matter. It replaces traditional storm-sewer pipes, which propitiate the growth of pathogens. The green infrastructure includes greenways, public orchards, and corridors; more vegetation means less runoff, more infiltration, more recharge, and lesser urban heat islands. Other environmental functions of urban vegetation are modulation of incident solar radiation and temperature as well as reduction of low intensity radiation (noise) and atmospheric particulates. The blue-green infrastructure is an instance of material degrowth that enhances the ecological, cultural, recreational, and economic value of urban places.

Rural–Urban Industrial Ecologies

Virtually all urban economies are based on natural resources processed by agriculture, fishing, or mining that take place in the hinterland surrounding a city. This was demonstrated during the revival of the European cities. After the demise of the Roman Empire, the Dark Ages witnessed a downturn in Europe's urban life, which ended in the Middle Ages thanks in part to the fertilization of the rural areas by urban waste (De Landa 1997). In other words, the hinterland can turn a waste into a resource to generate development.

Today, intended flows of matter and energy between hinterland and city would be called industrial ecology and attempts are made at understanding and promoting them (Guest et al. 2009; Ramaswami et al. 2012). The emergence of new rural–urban industrial ecologies is best exemplified by the reclamation of the Hammarby Sjöstad brownfield in Stockholm based on a closed loop system at the district scale, where all water and solid flows are recycled or treated. Hammarby's decade-old experience has been followed by Dockside Green in Victoria, Canada; Yesler Terrace in Seattle; and the Royal Seaport in Stockholm (Svane et al. 2011; Moddemeyer 2012). Keys to the efficiency of such systems are innovative waste and resource transport networks, and reduced transport distances. More fundamentally, progress in ecological engineering will depend on the utilization of natural energy and self-organizing processes (Mitsch and Jørgensen 2003).

The rural–urban industrial ecologies (Figure 16.1) also hold the key to agricultural recovery in the hinterland and ecological restoration farther afield in supply regions. Such isolated rural areas may in the future relent to agricultural pressure and costly long-distance freight to distant urban markets. They can then relive their own natural vocation and a "cultural–natural metaphor," such as the Buffalo Commons in the Great Plains owned by the state, private landholders, and nongovernmental organizations (NGOs) like The Nature Conservancy. Eventually, industrial ecologies and natural regions may fulfill the expectation of urban landscapes within seas of mixed natural-agricultural landscapes and natural regions (Register 2006).

Transport Solutions

Urban planning has privileged cars causing intractable traffic congestion. Compact, dense, vertical, multiuse cities, devolved to pedestrians are the

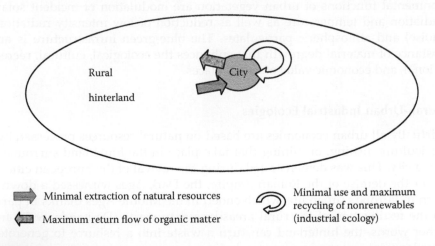

FIGURE 16.1
Schematic of rural–urban industrial ecology (the three-way teapot): A dense but diversified urban area in a large rural hinterland. Densification allows for devolution of land to agricultural and natural uses. Organic resources and waste flows sustain the city over the long run, while nonrenewable minerals are recycled within the industrial estates and recirculate within the city. These social-metabolism flows drive layout and design of built environments. A similar idea is being developed in the urban harvest approach: cities and their hinterland have overlooked productive capacities, which could make them self-reliant (Agudelo-Vera et al. 2012).

proposed antidote to car-dependent urban sprawl. Compact development reduces metropolitan miles traveled, energy consumption, and CO_2 emissions through diversity and accessibility even more than through density (Ewing et al. 2011).

Over time, sprawl should be broken down into smaller, denser settlements separated by crops and wilderness. Slow streets with buses and parking lots dismantled to accommodate for gardens and urban orchards would reduce the need for freeways to escape from the stifling city. Thus, roads at the urban edge would be absorbed by the greening of the city, becoming pedestrian–commercial parkways (Register 2006). Further road traffic and pollution reductions can be obtained by planning for or retrofitting bikeways. Amsterdam's bicycles have claimed the streets on an equal footing as pedestrians and cars. Human-powered vehicles are even more efficient than bicycles owing to their enhanced aerodynamics.

Rush hours have forced an ever-increasing number of road lanes. Twenty-four-hour use of streets would take advantage of nighttime lights, which at present represent a sunken cost. Tokyo's freight traffic occurs preferably at night. Less intense day traffic would go a long way toward reducing photochemical smog. More transport reduction is to be achieved through the design of ecocities, rural–urban industrial ecologies, and public transport systems. Compared to the United States, the greater demand for public transport in Europe, in people from all walks of life, relates to better and more frequent service, low fares and easy ticketing, integration of short- and mid-distance services, restrictions to car use, and policies promoting compact and mixed land uses. In Switzerland, each person makes annually 240 public transport trips as compared to 24 in the United States and 51 in the Netherlands due to the extraordinary importance of cycling there (Buehler and Pucher 2012).

Design Solutions

In sustainable design, height saves land area, and a smaller shape factor—the external area to inner volume ratio—saves material. Creating the illusion of space through the use of light saves both material and energy. Sustainable design should be the prime mover in the transition from energy-consumptive to energy-producing buildings. Sustainable buildings operate with or without electricity, reducing their life-cycle cost. Their design is cradle-to-cradle, meaning that their decay leads to the reuse of their material. They fit and function with the ecosystem (in terms of size, rate of decay, and restoration needed to compensate for environmental degradation over their life cycle). These buildings incorporate place-based knowledge of culture and nature. This contrasts sharply with both green design, which is fossil-fuel dependent, and with certifications fraught with timid energy efficiency improvements and use of nonrenewables.

Ecocities and New Synthesis Architecture

Ecocities are intended as extremely low-energy-consumption and energy-generating cities (Register 2006). Ecocities depart from the stand-alone building and the green buildings, which are only one or two stories high. Buildings suit the human body and its activity, and restore a balance with nature. Mixed land uses diversify the economy and make it more resilient while providing a wider array of possibilities for human endeavors.

The first step is restoration and regeneration of biotopes, natural cycles and circulation of wildlife in and around cities. This goes beyond setting aside remote natural reserves. Cities with crisp limits shrink and merge with nature, in "fractal landscapes," thanks to infilling, hedges, and urban boscage. Green corridors allow for animal and vegetal migration, while discontinuous boulevards and a diverse network of different street types morph the metropolis into several pedestrian towns linked by bicycles, human-powered vehicles, and public transit (rail mostly) and some animal power. Bikeways stretch in the countryside.

The second step integrates the green and blue intrusion, brownfield reclamation, harnesses natural energy sources in buildings and streets, and achieves low-maintenance bioclimatic structures. In-built flexibility allows for retrofitting (Williams 2007).

The third step is to make cities compact and diverse, to eliminate the need for cars. Biogas fuels smaller and more efficient buses. The roads are made narrower and traversable. Pedestrian areas display more 3D features with less and lighter material: arcades, awnings, porches, verandahs, gardens, walkways, gallerias, cafés, sports centers, water flows, green roofs, green walls, greenhouses, urban orchards, and roof arboretums. Blind walls and desolate roofs disappear. Semi-interior spaces and largely open spaces take over. The city becomes multilayered with shops, offices, and housing in successive vertical order (Register 2006).

The outcome is a set of built islands within diverse cultural and working landscapes. The city is designed to conserve, accumulate, and distribute energy and matter; turn waste into resources; and restore nature. The city may restore more soil than it consumes. The negative footprint (i.e., restoration) offsets positive footprint (i.e., deterioration), and the overall footprint is negative or at least neutral.

Downtown boomburbs, tall buildings, and arcologies (architectural ecologies) might already be ecocities in the making. Boomburbs are fast-growing suburbs, but in response to build-out (scarceness of available land) and rising land prices some are showcasing ecocity solutions: infilling, proximity to rail, transit-oriented development, compactness, mixed use, pedestrian-oriented neighborhoods, historic preservation and revitalization, greyfield development (adaptive reuse of malls and redevelopment of parking lots into retail districts), and lifestyle centers (standalone retail mixed with multifamily housing) (Lang et al. 2008). Arcologies, one of the boldest ecocity ideas, emerged because the Asian economic powerhouse cities (Shanghai, Hong Kong, Tokyo) are land starved. During the 1985 to 1995 economic bubble in Japan, architects started envisioning increasingly taller and larger buildings, culminating with arcologies able to mix all urban uses in the densest fashion and accommodate up to 1 million residents (Al-Kodmany 2011). As the economic bubble deflated, these designs were filed away. Their sustainability might have been faulty to begin with. But they remain as examples that ecocity ideas permeated into prime mainstream architecture and development projects.

Bioclimatic Houses and Buildings: Reduced Embedded and Operation Energy

In addition to the use of geometry to save energy embedded in construction materials, bioclimatic design is concerned with the following principles to save energy during the operation of a building (Gonzalez 2004).

Well-being is a function of organ perception of natural and artificial parameters (climate, light, sound, air quality).

Hygrothermal well-being depends on the body energy exchange, driven mainly by the temperature difference between body and environment. Gender and activity level also intervene, as do the exchange area and material resistance to heat loss (for instance, thermal resistance clothing). Fast (slow) energy loss gives a cold (warm) sensation. But the effective temperature depends on the relative moisture. Air movements also influence the thermal sensation. Comfort is lost if the feet–head temperature difference is more than 3°C.

The air quality depends on oxygen content, absence of combustion products, absence of odors, elimination of overheated air, and condensation avoidance.

Acoustic designs modulate low-intensity energy in the form of sound, vibrations, and impacts. These are propagated by exterior structures and air conditions (wind direction and speed). Vegetation has a low attenuation effect because effective barriers must be heavy.

Maximum passive and active use of solar input through daylighting, solar heating, air circulation, solar dials, photosynthesis, food drying, photovoltaic panels, and wind turbines allow for energy self-sufficiency in every building.

Passive solar energy use relies on the greenhouse effect: sunlight has a 300 to 3500 nm wavelength and glass is permeable to rays below 2500 nm (only 3% are less energetic and do not traverse the glass). Sunlight loses energy upon absorption by objects inside a building, and the resulting low-energy radiation (infrared) only poorly traverses glass panes back out.

The many light conduction (through glasshouses, conservatories, galleries) and modulation solutions (through window size, orientation, glass pane treatment, textile and rigid shutters, photosensitive diaphragm shutters) combine to make users feel safe inside and connected with the outside world, an important source of well-being.

The foregoing principles save on heating and air-conditioning, and hence mitigate climate change and stratospheric ozone depletion.

Existing design strategies include the following.

Vernacular houses and archeological buildings store knowledge on local conditions and time-tested solutions.

Sunlight provides very cost-effective light, warmth, and air circulation. However, cloud cover and the shape of the building can distribute heat irregularly in the building, making the following adaptations necessary.

Conservatories regulate the energy entering the house. An additional inner mass stores solar energy and slowly releases it after sunset (Trombe principle).

Double and triple glazed windows retain an air cushion especially if the outer glass is treated to lower its emissivity.

Hollow buildings are those that have deep rooms far away from sunlight. These are only useful in hot climates. In higher latitudes hollow buildings require reflecting floor surfaces to propagate sunlight in the inside.

Berms (earthen embankments) reduce convective energy losses to the air, while allowing sunlight on the sunny side. Cellars and buried constructions, as in the Kasbah belt from Morocco to Afghanistan, protect from both sun and cold.

Oval-shaped buildings and venturis help improve air circulation. Recycling kitchen and (dried) bathroom air also saves energy. Dark or light shades modulate the sunlight emissivity of a building (how much energy its outer walls store).

Convection can circulate warm and cool air around the building. Air distribution is important to dissipate moisture that otherwise condensates inside the building. In summertime, nighttime evacuation of warm air saves cooling energy the following day, especially using materials that remain cool (thermal inertia given by high volume and density and low specific heat). Underground air can be drawn into and circulated inside a building through buried ducts that cool air captured outside the house.

Ventilated roofs allow the passage of air between two layers of roof. Shadows (tensile structures, blinds, trellises, pergolas, or trees) help cool a building.

The heat sensation can also be moderated with cool inner surfaces, a reduction of relative moisture, and an increase in speed of air flows. Ventilation can derive from wind entering a window or circulating between two windows on different walls, which then have different wind pressures. In addition, differences in height enhance the convective circulation. Around high-rises wind is augmented since the friction of air with lower buildings is reduced. Therefore, high-rises can be wind and solar energy providers. Solar chimneys work on convective movement enhanced by allowing the sun to overheat the outer part thanks to its dark color and sorptive material. The wind can also improve heat extraction by creating a leeward depression. Factories make extensive use of rotating air extractors based on these principles. The reverse principle is to force the entry of wind down a chimney, which is then cooled through contact with stored water; it is traditionally used in Middle Eastern houses.

Water circulation may cool or warm a building. Warming requires geo-thermally or solar-heated water. Water at lower temperature than the air absorbs heat. Evaporative cooling has also traditionally been used in fountains or water trickling down facades. Water cooled in recipients during the night and stored inside during the day has also been traditionally used.

Green roofs (whose history is 4600 years long) insulate from extreme temperatures. They compensate for vegetation loss due to the presence of buildings. They are useful in climate modulation (attenuation of urban heat islands, photosynthetic CO_2 capture, reduction of land surface temperature), sound proofing, protecting sun-sensitive waterproofing materials, and harvesting rainwater flows. The density and size of the vegetation on the roof augments the structural weight and requires light soil, mixtures of soil with lighter natural or artificial materials, which also improve soil aeration and water retention. Irrigation must be minimal to reduce the green roof's mass; this dictates the choice of vegetation species. Evapotranspiration is also able to locally mitigate hot air conditions. Food is increasingly being grown on green roofs.

The Sustainable Rural Home as a Closed System

If urban politics and economics do not evolve, concerned citizens around the world can still develop their own sustainable houses, the principles of which have been studied in closed systems or sun-subsidized systems (Figure 16.2). Desert life, space travel, and autonomous habitats like the Biosphere 2 project

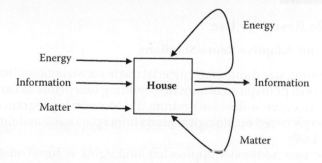

FIGURE 16.2
Flows in self-sustainable households. Solar subsidies, postconsumerism, and the absence of outflowing waste make them asymptotically autonomous. They are not an information sink but a source of information for other households. The house is a bio-social-technological system or biosoma, which ought to minimize the machine component and optimize the biological and social components (Bugliarello 2011).

(Morowitz et al. 2005; Nelson et al. 2008; Nelson et al. 2009) should guide rural autonomy. If cities do not change, there is the possibility that the peri-urban hinterlands change and gradually transform the city by contagion.

Sustainable rural housing is a hybrid of autonomous habitats and vernacular house design, using bioclimatic principles. Each household is a scaled-down industrial ecology wherein material and energy flows, in particular water, carbon, and nitrogen are closed, combined, cycles, with no externalities (waste exports) nor subsidies, except for the sun.

Sustaining the biotic side of this household ecology requires maximizing photosynthetic and fungal production of edible, nutra- and nutriceuticals, medicinal, and other useful plants (see Chapter 14). Green hedges, green roofs, green walls, no mowing, native species, and no invasives are requisites, because they host pollinators, create microclimates, and their diversity contributes to biological plague control. Beekeeping, vermicompost, and other microhusbandry species complements microagriculture. Larger animals and humans provide animal power. Organic waste is recycled into fertilizers via dry toilets (Chapter 3) and biomethane (Chapter 10).

The abiotic side is based on rain harvest, dripping irrigation, and infiltration to replenish the aquifer. Incoming organic matter is limited, reused, or recycled on site. Solar stoves make for safe, cost-effective reduction of everyday chores, or garden parties under more cloudy climes. Fridges are replaced by food drying and preserves because coolants are the fastest-growing climate change agent (Chapter 11).

Human-powered vehicles replace bicycles, which replace cars for commuting. Large households, intentional communities, or cooperatives may spare enough waste to generate biomethane and power a shared truck.

Sustainable Resource Use

Long-Lived and Adaptive Reuse Solutions

Reduced building size diminishes material use in constructions: the "existenz minimum" concept inspired minimalist building blueprints as early as 1929. And by 1976, Georgescu-Roetgen's minimal bioeconomic program expressed how an economy based on biological needs underpins sustainability (Gowdy and Mesner 1998).

Nowadays, two additional approaches undergird reduced material consumption and turnover. First, prolonged building life, using the system ductility of stone (stone buildings adapt to land movements). Buildings in Los Angeles and London have recently been built with planned lives of 500 and 700 years, respectively. And second, it is necessary to make the buildings reusable and allow them to "migrate" from one purpose to another (Fernandez 2006).

Material Solutions

The long-term behavior of polymers and composites is not well known, making them difficult to predict in long-lived systems. Corrosion and decay in building skins (provoked by chemical and fungal activity, bird droppings, and acid rain) diminish their useful life and entail maintenance costs in skins and basements. These issues have led to material innovations. New ultra-light alloys allow for air-supported or inflated roofs, which pursue the dematerialization trend of the economy. Cellular or metal foams are also being developed for high-performance sound or light applications. Composites as well as fire- and fracture-resistant steel alloys (especially in seismic regions) are used for long-life structures along with HPS (high-performance steel). Emerging ceramics include improved concrete, with fine-particle filling and binding to replace Portland cement; pozzolanic materials to increase strength, toughness, and water resistance; reactive powder concrete with extremely high strength and ductility with very fine steel whiskers; or smart concrete with carbon fibers. Emerging fired clays add ceramic or metal to enhance energy absorption. Glass ceramics are improving mechanical and thermal properties, and are easily machined with metalworking tools. Foamed silicates are superior insulators. Responsive glass changes its optical properties (Fernandez 2006). The costs of these new materials are likely to decrease as demand for them grows.

The foregoing materials are targeted to long-lived buildings: they use soon-exhausted resources, are difficult to recycle, command hitherto high prices, and have low biodegradability. This contrasts with biopolymers, biodegradable plastics from renewable resources. They depend mostly on starch, harvested from corn, wheat, or potato, and transformed by microbes into lactic acid whose molecules are then chemically processed to link into polymers. Polyhydroxyalkanoates (PHAs) are microbial polyester thermoplastics with properties very similar to petroleum-based plastics, which can further be customized as is the case of PHBV (with remarkable flexibility and toughness and very close to polypropylene and polyethylene). Four emergent biopolymers are capturing interest: PLA, PHA, PCL (a biodegradable synthetic polymer), and PGA. The copolymer of 3-hydroxybutyrate and 4-hydroxybutyrate (P(3HB-co-4HB)) is also being investigated, as are chitin from insects, shells, bones, and spider silk (Fernandez 2006); the latter has much more resistance than a steel strand of the same weight.

Less technologically advanced, biomaterials comprise possibly the only veritable, albeit limited, renewables. Emergent biomaterials include hybrid wood composites and laminated woods reinforced with textile layers (for instance, natural fibers such as sisal, jute, flax, or denim waste) to reduce wood anisotropy, that is, the fact that wood does not behave homogenously in all directions. Wood is the epitome of renewables. The multifarious uses of wood still lead to deforestation; this is how Europe lost most of its forests (20% to 95% depending on the country). In wood-limited areas, rammed earth has produced outstanding historical buildings still preserved today. Rammed earth, like concrete beforehand, is being rediscovered. Lime is also a very versatile biomaterial quarried from limestone deposits formed by the accumulation of calcite in the shells of marine animals. Lime embeds much less energy than Portland cement; it can reabsorb 100% of the CO_2 used during its manufacture; and it can be recycled or naturally decay without toxicity. In addition, lime is hygroscopic: it absorbs moisture and can release it later (Denison and Halligan 2010). Many wood–fiber–mineral composites have been traditionally employed and will increasingly be used (in the form of adobe, wattle, daub, cob), as tapping the underutilized agricultural waste becomes a necessity.

References

Agudelo-Vera CM, Mels A, Keesman K, Rijnaarts H (2012) The Urban Harvest Approach as an aid for sustainable urban resource planning. *Journal of Industrial Ecology* 16:839–850.

Al-Kodmany K (2011) Tall buildings, design, and technology: Visions for the twenty-first century city. *Journal of Urban Technology* 18:115–140.

Buehler R, Pucher J (2012) Demand for public transport in Germany and the USA: An analysis of rider characteristics. *Transport Reviews: A Transnational Transdisciplinary Journal* 32:541–567.

Bugliarello G (2011) Critical new bio-socio-technological challenges in urban sustainability. *Journal of Urban Technology* 18:3–23.

De Landa M (1997) *A thousand years of non-linear history*. New York: Zone Books, Swerve Editions.

Denison J, Halligan C (2010). *Building materials and the environment*, vol. 2. Leicester, UK: Stephen George & Partners.

Ewing R, Nelson AC, Bartholomew K, Emmi P, Appleyard B (2011) Response to Special Report 298 Driving and the built environment: The effects of compact development on motorized travel, energy use, and CO_2 emissions. *Journal of Urbanism: International Research on Placemaking and Urban Sustainability* 4:1–5.

Fernandez J (2006) *Material architecture: Emergent materials for innovative buildings and ecological construction*. Oxford: Architectural Press.

Gonzalez Neila FJ (2004) *Arquitectura bioclimatica en un entorno sostenible*. Madrid: Munilla-Lería.

Goodwin P, Lyons G (2010) Public attitudes to transport: Interpreting the evidence. *Transportation Planning and Technology* 33:3–17.

Gowdy J, Mesner S (1998) The evolution of Georgescu-Roegen's bioeconomics. *Review of Social Economy* LVI:136–156.

Guest JS, Skerlos SJ, Barnard JL, Beck MB, Daigger GT, Hilger H, Jackson SJ, et al. (2009) A new planning and design paradigm to achieve sustainable resource recovery from wastewater. *Environmental Science and Technology* 43:6126–6130.

Johnson S (2001) *Emergence: The connected lives of ants, brains, cities and software.* New York: Scribner.

Joint Research Centre of the European Commission (JRC)/PBL Netherlands Environmental Assessment Agency (2011) Emission Database for Global Atmospheric Research (EDGAR), release version 4.2. http://edgar.jrc.ec.europa. eu (accessed November 15, 2011).

Lang RE, Nelson AC, Sohmer RR (2008) Boomburb downtowns: The next generation of urban centers. *Journal of Urbanism: International Research on Placemaking and Urban Sustainability* 1:77–90.

Mitsch WJ, Jørgensen SE (2003) Ecological engineering: A field whose time has come. *Ecological Engineering* 20:363–377.

Moddemeyer S (2012) Understanding the nature of change: Building resilience into urban life. *Water 21* August:14-18.

Morowitz H, Allen JP, Nelson M, Alling A (2005) Closure as a scientific concept and its application to ecosystem ecology and the science of the biosphere. *Advances in Space Research* 36:1305–1311.

Nelson M, Dempster WF, Allen JP (2008) Modular biospheres: New testbed platforms for public environmental education and research. *Advances in Space Research* 41:787–797.

Nelson M, Pechurkin NS, Allen JP, Somova LA, Gitelson JI (2009) Closed ecological systems, space life support and biospherics. In *Handbook of Environmental Engineering, vol. 10: Environmental Biotechnology,* Wang LK, Ivanov V, Tay JH (eds), 517–565. New York: Humana.

Newton PW (2012) Liveable and sustainable? Socio-technical challenges for twenty-first-century cities. *Journal of Urban Technology* 19:81–102.

Ramaswami A, Weible C, Main D, Heikkila T, Siddiki S, Duvall A, Pattison A, et al. (2012) A social-ecological-infrastructural systems framework for interdisciplinary an integrative curriculum across seven major disciplines. *Industrial Ecology* 16:801–813.

Register R (2006) *Ecocities: Rebuilding cities in balance with nature.* Gabriola Island, British Columbia, Canada: New Society Publishers.

Schumpeter JA (1943) *Capitalism, socialism and democracy.* London: Allen & Unwin.

Svane Ö, Wangel J, Engberg LA, Palm J (2011) Compromise and learning when negotiating sustainabilities: The brownfield development of Hammarby Sjöstad, Stockholm. *International Journal of Urban Sustainable Development* 3:141–155.

The end of suburbia: Oil depletion and the collapse of the American Dream (2004). Directed by G Greene. Canada: Electric Wallpaper Company. Documentary.

Williams DE (2007) *Sustainable design: Ecology, architecture, and planning.* Hoboken, NJ: John Wiley & Sons.

17

Sustainable Science and Technology

Alejandro de las Heras

CONTENTS

Issues in Science and Technology

Mankind is unwittingly experimenting with vital planetary processes, with challenges for which conventional science and technology (S&T) are not prepared. These irreplaceable natural processes have been valued at trillions of U.S. dollars every year but the collapse of any natural process is likely not reversible to the same state nor on a human time scale. The S&T communities have to take responsibility for their inadvertent contribution to these changes (Lubchenco 1998), which was only statistically verified in 1971 as environmental degradation in the United States could not be explained by population growth. Production technologies determined by profit-seeking economic elites were blamed. Along came the realization that, first, much knowledge and energy were needed to solve environmental issues; second, technological fixes without lifestyle changes would be popular but insufficient; and third, most emerging technologies are burdened with unintended environmental effects (York et al. 2002).

Several biases accompany technological fixes. The reductionist approach typical of S&T is a tendency to overspecialize, which fails to encompass sufficient information on the coupled human–nature systems (CHANS). These require a systemic approach since they have evolved from simple local interactions to a global entanglement (Liu et al. 2007).

Even the flexibility and capacity of S&T to contribute to solving global unsustainability has come under question. This is because pollution remediation and renewable energies are bound by the second law of thermodynamics to fix local issues at the cost of importing energy and matter in exchange for entropy in the surroundings. One unit of restored order creates more than one unit of disorder in the adjacent environment. This begets expanding needs for restoration. As a consequence, even modest economic growth rates based on ecoefficiency technologies would eventually double the size of the economy and require further technology refinements, until none could be gained due to their increasing high costs, associated to diminishing returns typical of technological innovation and the second law. Therefore, long-term solutions to unsustainability can only stem from decreases in population, per capita consumption, or both. These are not technological problems but social and ethical ones (Huesemann 2001).

Issues in Science

Billions of dollars are expended annually on allegedly policy-relevant research, without evidence that this substantially changes decision-making capabilities or the course of environmental degradation. About $25 billion was spent on climate research in the United States between 1989 and 2003, which improved fundamental understanding and moved forward the institutions of global climate governance but narrowed the knowledge outcomes to expert users; by doing so, the more fundamental behavioral changes that citizens and consumers need to effect were probably sidelined. Little knowledge was generated on the need of citizens for information. Obtaining such information is not part of current research portfolios inclined to believe that science is automatically usable (Sarewitz and Pielke 2007).

The main reasons for the gap between knowledge generation and usable knowledge are, first, that the current metrics to assess science (production of peer-reviewed papers and citations) leave outreach out of the desirable focus (Dilling and Lemos 2011). Second, self-interest exists in individual scientists, funding institutions, and societies, and it is the main cause of bias in environmental research (Huesemann 2002). And third, even though scientists are translating their results for policy makers and the media at an increasing rate, facts, figures, and future projections have had far less effect on policy than expected by the scientists, simply because science is one among many factors that can influence policy. Policy-relevant science is carried out in too small a network to exert sufficient influence (Palmer 2012).

Issues in Technology

Other issues relate to professional engineers. Forty-four percent of the respondents to surveys from the UK Institution of Chemical Engineers (IChemE) believed that the United Nations were just alarming people about future water shortages or that sunspots, not humans, caused climate change. This lack of awareness was proportional to the duration since graduation. In the meantime the public image of engineers has deteriorated, as they are held responsible for key problems such as climate change. The self-image of engineers has also deteriorated, as a fraction of them seem content to fit as experts in the technocratic structures of governments and private organizations; they view themselves as guns for hire or paid hands who have been resisting sustainability in the engineering curriculum.

In the opposite direction, engineering associations have committed to sustainability: Engineers Australia in 1994, 18 signatories of the London Communiqué in 1997, and then 20 more signatories of the Melbourne Communiqué in 2001. Thereafter, the U.S. National Academy of Engineering developed the 2020 Engineer as a Leader in Sustainable Change vision, with a grounding of multidisciplinary design with social scientists in higher education. The Royal Academy of Engineering has taken steps to embed sustainability in the engineering courses, not just add another topic. IChemE has followed suit, and Engineers Australia issued a sustainability charter stating that sustainability should be part of all human endeavors (Byrne and Fitzpatrick 2009).

Patents and Copyrights

Between 1990 and 2007, patent applications in the 90 most relevant patent offices in the world have gone from 500,000 to 800,000 per year. The United States, Germany, United Kingdom, France, and Japan are the key players in this arena. Only 1% were international applications (at least two inventors residing in different countries). The foregoing countries retain their supremacy in international cooperation networks (applications with at least two countries of residence), although China, India, and South Korea are important regional hubs (De Prato and Nepelski n.d.). So, there is a Matthew effect of accumulated advantage (the poor get poorer and the rich richer): knowledge production leads to unequal capital accumulation. A brain drain also penalizes poorer countries and adequate technologies more adapted to Third World issues. The problem is not new: the Sussex manifesto (1970) called for actions in all countries as high S&T investments occurred only in high-income countries, and external and internal (away from some disciplines) brain drains affected poorer countries (Kaplinsky 2010, 2011).

Science journal articles are not often accessible to the general public. Green open access, that is, versions of papers prior to the final published version, is almost exclusively used in mathematics, physics, and biomedical research despite the fact that open access publication is now compulsory for all U.S.-based research (Van Noorden 2012, 2013). Open data would similarly augment the usefulness of science and mitigate still prevalent lacks of data transparency and connectivity between individual research pieces. Despite a growing number of data repositories (e.g., Dryad, Pangeae, and GenBank), many scientists still do not store their data in sharable ways, due to time constraints and fear of data misuse or discovery scooping (Nelson 2009). There are also mounting concerns that publishing copyrights might extend to prohibiting text mining and derivative works (Wilbanks 2013).

Emergent Solutions: Knowledge Systems

Not all science is usable, and some is bound to vested interests, so that science needs to be replaced by a different knowledge production system (Sarewitz and Pielke 2007). Knowledge systems (KS) are fulfilling this requirement; they are networks of actors linking knowledge, know-how, and action. Diverse players interact across user–producer boundaries to design and harness knowledge in support of sustainability (McCullough and Matson 2011). In KS, decision makers are all science users, not only policy makers. KS comprise users, policy makers, scientists, government agencies, and nongovernmental organizations that share information.

Usable knowledge is emerging as requisite since most science is publicly funded around the world. Successful use of knowledge involves coproduction or iteration between knowledge producers and users, as well as negotiation between science push (researchers and information providers set the science agenda) and demand pull (priorities set by users and stakeholders). The following foster iterative cycles of knowledge production and use: information brokers; boundary organizations working between the worlds of research and science users; and "collaborative groups" wherein decision making is highly distributed and many groups are vested in the outcome. It is imperative in KS to admit flexible research agendas to promote an iterative coproduction of knowledge (Dilling and Lemos 2011).

Effectively harnessing S&T sustainability in KS means managing knowledge–action boundaries through enhanced salience, credibility, and legitimacy of the information, its communication, translation, and mediation. Communication involves eliminating assumptions as to what other stakeholders may hold as valuable information. Translation means eliminating jargon and differences in what a persuasive argument is. And mediation implies legitimacy through transparency, accountability to all relevant parties, as well as approving rules and criteria. Porous knowledge-action boundaries allow for efficient information flows and keep politics out of the scientific process. An additional strategy to harness S&T for sustainability involves coproduction of results valid across boundaries (Cash et al. 2003), a process called participatory sustainability science. These solutions help avoid self-interest biases.

Evolving Responsibilities

Scientists and engineers have personal and collective responsibility regarding public funding, S&T decision making, technological usability, and ethical use. This entails a continuous reexamination of goals. Sustainable S&T should help rethink and reshape the interactions within the CHANS through improved understanding, social and environmental precautions and protection, and new technologies to minimize ecological footprints.

Sustainable S&T helps foresee the consequences of social choices rather than providing solutions. Anticipation of CHANS trends requires a deeper understanding of feedbacks, collective behavior, emergent properties, and system nonlinearity. Many intuitive or unconscious leads can be sought in cultures more sustainably related to nature rather than the Western exploitative model.

S&T should feed subsidiary not technocratic decision making to solve complicated and contentious technological or scientific issues related to, for example, bioengineering, animal tests, and genetically modified organisms. Science is answerable to public funding but this has meant too often communication of results to state agencies. Scientists and engineers should communicate widely with the public so that decision making can be devolved to the widest social sphere wherein, one must assume, possible victims are made aware, and other scientists can contribute to reduce nonanticipated effects.

The S&T agenda-setting process has to cater to local dialogs, as global and regional scales overlook large local variability. The WEHAB (water, energy, health, agriculture, and biodiversity) targets set forth internationally in the last decade have to be prioritized; they are the easiest to achieve with current

S&T and indeed suffer from high-tech solutions. The S&T search for solution pathways should replace the science of processes leading to problems; this would likely trigger novel science (Lubchenco 1998; Clark and Dickson 2003; Hulme 2012; Palmer 2012).

How to (Re)train for Sustainability

In 1997 the report of the Joint Conference on Engineering Education and Training for Sustainable Development in Paris recommended sustainability to be integrated into engineering education, ongoing projects, research, and retraining of all faculty members (Byrne and Fitzpatrick 2009).

Sustainability problems require new sets of knowledge, skills, and attitudes that must be acquired in real-life KS situations. Project- and problem-based learning along with information technologies gradually replace textbook and classroom exercises and rigid settings, which thwart empowerment, and multi- or transdisciplinary work. The diversity of group and individual interests, values, and definitions of sustainability are catered to. Starting points (ideologies) and end points (expected future) are reexamined. Cooperation trumps competition in face-to-face and virtual settings. Groupware allows for enhanced organizational learning. This transforms the student experience into a junior consulting experience (van-Dam Mieras et al. 2008).

Competencies functionally link knowledge, skills, and attitudes in future problem solvers, change agents, and transition managers. A set of competencies too large to be developed in higher education calls for lifelong learning and includes systemic thinking; anticipatory thinking; interdisciplinary work; cosmopolitan perception; planning and implementation; empathy, interpersonal thinking and acting, and solidarity; (self-)motivation; and the capacity to critically reflect on one's own cultural influences. In particular, systemic thinking applies to model building and validation. The anticipatory competence involves collectively developing rich scenarios, trajectories, and transitions. The interpersonal competence is the ability to encourage participatory problem solving (Wiek et al. 2011).

Sustainability courses are being taught in higher education since little more than 15 years ago; undergraduate and graduate programs have emerged in the last few years and PhD programs initiated around 2007. There is as yet little accumulated knowledge as to whether the underpinnings of these courses have allowed alumni to effect deep changes. However, institutional support from the sustainability and outreach offices has proved essential to continue fieldwork over several years in the same location. In addition, transformational sustainability research now applied in leading sustainability programs includes collaborative goal setting and implementation, and has augmented relevance (Brundiers and Wiek 2013). The end result of sustainability training should be an involvement in the evolution of knowledge systems.

Harnessing Ongoing S&T Revolutions

Presumptions of sustainability without deep changes in human population size, consumption per capita, and technology have lost credibility (Huesemann 2001; York et al. 2002). Sustainability transitions are likely to be accompanied by new, possibly abrupt, imbalances, which can require human equilibrating actions at several time and spatial scales (Holling et al. 2002). An essential test of the willingness and abilities of the S&T communities will be to harness emerging and fast-growing technologies. This means to subsume them to the stabilization of natural trajectories; to building equity and resilience in the human subsystems through meeting WEHAB goals; and to sustained S&T development using the principles of industrial ecology to recycle scarce resources (Figure 17.1).

Scientists and engineers can no longer ignore the consequences of the technologies they develop and help bring to the markets. Hopes and concerns have been expressed on public and corporate policies in the emerging bioeconomy (launched in 2012 in the European Union and the United States). The terminator seed technology (genetically modified to force the

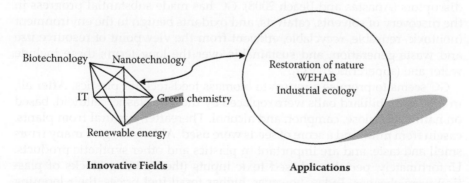

FIGURE 17.1
Interactions in fast-growing S&T fields (left) can lead to sustainability-enhancing applications (right). The 3rd Nobel Laureate Symposium on Global Sustainability points to a reachable efficiency revolution through decoupling economic growth from natural resources, as required by industrial ecology (Stockholm Memorandum 2011). A U.S. National Research Council report urges the information technology and computer science communities to bring to bear their efforts on global sustainability issues (Millet and Estrin 2012).

yearly purchase of seeds, even in the poor areas where crops were origi-
nally domesticated) has generated much disquiet. In connection to the nano-
bio-info-cognitive convergence, the idea of "interconnected virtual brain of
the Earth's communities" geared toward the "conquest of nature" (Roco and
Bainbridge 2003, p. 93) has to be turned into a sustainable counterproposal.
Much research on nanostructures is pending to ensure their innocuity, in
particular due to the ease with which arsenic, cadmium, and many other
nanoparticles cross the blood-brain barrier and cell membranes, possibly
giving rise to mutagenicity and other effects (Anastas and Beach 2008). It is
only through the simultaneous application of the precautionary principle
and long-term trials of the new materials that doubts can be overcome.

There is no doubt, however, that progress is taking place in the use of green
chemistry and nanotechnology, including energy applications, and in green
computing. These are reviewed next.

Green Chemistry and Sustainable Nanotechnology

Green chemistry (GC) is the epitome of sustainable S&T, as it looks for compounds and reactions devoid of environmental hazards over entire life
cycles. GC principles aim at preventing the release of hazardous compounds.
In GC, degradability and low toxicity are explicit design criteria. GC could
eliminate contaminant products and byproducts, for example, DDT, PVC,
the (organo)chlorine chemistry, perfluorooctanoic acid, polybrominated
diphenyl ethers, phthalate softners in plastics, bisphenol A, and hormone
disruptors (Anastas and Beach 2008). GC has made substantial progress in
the discovery of solvents, catalysts, and oxidants benign to the environment
(nontoxic, reusable, recyclable, efficient from the viewpoint of resource use
and waste generation, and sustainable over the long term); these include
water and (supercritical) CO_2.

GC seems to promise a return to biomass feedstock for plastics. After all,
in 1860 ivory billiard balls were replaced by the first plastic, celluloid, based
on natural cellulose, camphor, and alcohol. Thereafter, furfural from plants,
casein from milk, and a score of seeds were used. Aldehydes give many fruits
smell and taste, and are important in plastics and other synthetic products.
Unfortunately, petroleum-based toxic inputs (the building blocks of plastics) were cheaper. Today, however, higher fossil fuel prices, their looming
exhaustion, the overwhelming use of plastics in our society, as well as carcinogenic and otherwise toxic monomers (e.g., benzene for nylon, also one of
the most widely used solvents; and acrylamide for polyacrylamides) call for
other feedstocks. In relation to plastics, green chemistry has advanced in the
use of benign pathways involving benign catalysts, oxidants, and alcohols.
Olefin metathesis catalysts will likely lead to many polymer applications

based on renewables like unsaturated fatty acids and natural oils. Natural and artificial polymers can be produced with enzyme systems to obtain high-performance and benign antioxidant polymers. The use of the DNA base thymine may enhance the durability of biodegradable polymers, many of which can be derived from cellulose and starch. Heterogeneous catalysts reduce the use of solvents, corrosive reagents, aggressive oxidants, and acid catalysts in applications such as thermoplastics. In turn, catalysts on polymer supports (e.g., chitosan films from crustacean shells and starch-based materials) enhance their recyclability and performance. Cashew nut shell liquid waste can form a composite with jute and produce nanotubes. Defatted soy flour can form composites with flax, hemp, or bamboo, or form adhesives replacing toxic compounds in plywood. Terpenes are used for thermoplastics. Vegetable oils can be converted into polyols in the manufacture of polyurethane. CO_2 reacts efficiently with epoxides to produce cyclic carbonates (Varma 2007) able to replace polystyrene. Essential uses of plastics may therefore go on sustainably, but superfluous ones ought to be eliminated.

Other natural products are already in the development phase after demonstrating their usefulness and innocuity. Natural compounds allow the use of degradable functional groups; an enlarged range of enzymatic activity; the use rather than waste of renewable biomass; the replacement of hazardous corrosion inhibitors by plant extracts (Bammou et al. 2010); the usefulness of waste such as fermented restaurant oil with cosmetic and therapeutic activity; and the use of rice, nut, and fruit waste antioxidant phenols to replace food additives suspected of being hazardous. Also, gamma-valerolactone from fruits, a food additive, might have notable sustainable fuel properties.

In general, enzymes produced by living organisms decrease waste, reagents, and solvents. Engineered enzymes and aldol reactions (common in nature) help obtain chiral products (which include many pharmaceuticals). Enzymes in the remediation of xenobiotics can be derived from "directed evolution" (mutation and screening of enzyme-producing organisms) as a replacement for genetically modified organisms. Hazardous reagents could soon be replaced by metal catalysts and H_2 gas hydrogenation. Hazardous chlorine used in paper pulp bleaching now could be substituted by iron-based catalysts and benign catalysts. Polymers of N-vinylformamides can now substitute polyacrilamides. The neurotoxic tributyltin hydride and antifoulants containing organotin (banned because it bioaccumulates) can be replaced. CO_2 is already used as solvent in industrial decaffeination instead of toxic solvents; in dry cleaning and spray painting it has eliminated bioaccumulating pollutants (Anastas and Beach 2008).

GC principles are essential conditions being used in harnessing nanotechnology. Green nanocatalysts (typically 2 to 3 nm long) are synthesized with green solvents. Many are metals (gold, palladium, platinum, silver, copper, iron, etc.), which due to ore exhaustion are best used in the smallest possible quantities. Fischer-Tropsch catalysts have been used in the production of fuels. Most interesting, iron nanoparticles catalyze hydrogen peroxide during decontamination of organic pollutants. Atmospheric oxygen is a green oxidant at room temperature involved in transforming alcohols to aldehydes and ketones. Esterification of alcohols in one step using oxygen and gold nanoparticles avoids damaging oxidants, strong acids, and excess reactants. Despite growing interest, less than 2.5% of the literature on nanoparticle catalysts addresses their green obtention (Narayanan 2012).

Nanotechnology is also poised to operate radical changes in energy and digital information applications. An application in wind turbine blades is carbon fiber coating, as protection against sand and water (hyperhydrophobicity) for enhanced durability. The improved vibration damping results in aerodynamic stability, and power generation efficiency, position control, and durability of the blades (Liang et al. 2011).

Batteries and fuel cells can hold high energy densities but they deliver power slowly, unlike supercapacitors, which can take up or release up to 10 kW kg^{-1} in a few seconds, with a full charge often obtained in around 1 second. Supercapacitors store hundreds or thousands more charge than conventional capacitors. Supercapacitors can also be endowed with long lifetimes thanks to nanoporous or nanogranular electrodes, which improve electrolyte ion circulation and thus capacitance and power density. Hybrid vehicles with combustion engines and fuel cells have used a supercapacitor to recover braking energy. Hybrid buses with supercapacitors are 25% to 30% more energy efficient than diesel buses or hybrid buses with batteries (Lu et al. 2012).

Graphene, an atom-thick carbon structure, was isolated in 2004 and is a candidate to replace silicon in electronic devices due to very high mobility of electrons (10 times that in silicone), very low resistivity (35% less than copper), and greater strength than diamond. Computing can be done 1000 times faster than in silicon. It is transparent and has been used on photovoltaic cells, touch screens, LCDs, or next-generation light devices. Graphene has a high surface-to-mass ratio (2630 m^2 kg^{-1}), which makes it ideal for supercapacitor cells. Graphene is also ideal for sensors, as a single molecule can be detected on its surface. It has good sensitivity to pollutants such as NH_3, NO_2, dinitrotoluene, or CO (Basu et al. 2010).

Green Information Technology and Sustainable Computing

Science and engineering depend heavily on information technology (IT). In turn green IT would largely benefit from S&T solutions to reduce the IT industry's environmental footprint, and contribute to enhanced economic and environmental performance. Green IT is an initial hardware step toward sustainable computing, mostly software solutions encompassing environmental, business (Guyon et al. 2012), and labor solutions. In sustainable IT, the scale and complexity of issues call for a fast evolution of solutions, real-time assessment and optimization of resources, and modeling of consequences, all the while adapting solutions to natural and human needs. Higher education should include life-cycle analysis, agriculture, ecology, natural resource management, economics, and urban planning courses (Millett and Estrin 2012). Integrated hardware and software solutions are still beyond the horizon. The remainder of this chapter reviews IT issues and green IT solutions.

IT Material and Energy Consumption

Manufacturing amounts to 70% of the natural resources used during the life of a PC. Despite miniaturization and convergence (the inclusion of several functions in a device, which were formerly performed by several appliances), mobile devices appear and become obsolete at a fast rate, augmenting the use of natural resources. Laptop sales are increasing fast and in 2011, 472 million smartphones with computer capabilities were sold globally (almost a twofold increase from 2009).

Only 40% of the computers and 10% of mobile devices are recycled. A large fraction of the e-waste accumulates in countries like China and Pakistan. The increasing use of digital media will not diminish paper consumption, expected to increase from 370 to 504 million tons between 2005 and 2020 (CSS 2011). Fast innovation rates have made predictions of obsolescence difficult, resulting in surplus physically destroyed to avoid price drops. Despite PCs being designed for motherboard and memory upgrading, the industry has taken little or no steps to make this a more attractive or affordable option: selling entirely new PCs generates many times more profit than selling a small internal item. To this effect, standardization by engineering associations should contribute to more upgradability. Finally, the ITC industry will face considerable challenges as some manufacturing compounds are both toxic and bound to hugely increase in price as their reserves deplete within 20, 27, and 30 years for lead, tin, and copper, respectively (Kounatze 2009).

Processors and storage account for 86% of the ITC energy consumption in the use phase, a proportion growing 18% annually despite steady decreases in energy consumption per unit of processing (Floyer and Miniman 2011). A fast graphic card can be the largest energy expense in a computer. Huge numbers of computers in the world remain idle (s0 state) but powered around the clock, so that only 34% of the energy consumption occurs while working (CSS 2011). The U.S. Environmental Protection Agency (EPA) Energy Star program and server manufacturers recommend the sleep (s3) mode for power management because it allows for faster work resumption but uses more energy than the hibernate (s4) mode (CSCI n.d.).

Emerging Green IT Solutions

Very simple measures can be hugely beneficial. For example, if U.S. college students allowed power management on their computers, more than 2.3 billion kWh per year would be saved (CSS 2011). For many users, most savings would stem from not installing a video card and using instead the motherboard video output. Businesses and government agencies can implement hoteling or hot desking to use office resources only when needed by an employee.

Another large set of innovations seems to depend on improved data center and server use. Software can now direct data to those data centers where energy is less expensive. Virtual labs can be created with thin clients. Virtualization is the use of one computer running software for many other computers, so each server can operate at 80% capacity instead of 10%. Energy Star servers consume 30% less energy and if they replaced all new servers being sold, they would save $800 million and the equivalent of 1 million vehicles' worth of CO_2 (CSS 2011). Until 2010, the IEEE (Institute of Electrical and Electronic Engineers) standards required networks to be powered up regardless of the data amount being transmitted. Today, the Energy Efficient Ethernet standard provides a low power idle for periods of low data transmission, which occur most of the time (Diab 2010). Power switching at the chip level, disk stopping depending on the input/output demands, hardware offload protocols, and decreasing server power have gathered attention (Floyer and Miniman 2011).

Manufacturers have faced the sustainability challenges in different ways. Intel started voluntarily reporting and reducing perfluorocompounds in 1994 to 1996. Recently, it committed to switching 85% of its U.S. energy consumption to renewable sources to abate its 3.9 million tons of CO_2 emissions in 2007. Other ongoing solutions are server refresh (decommissioning and upgrading) and virtualization; heat generation decreases through managing geographic location (that determines the air-conditioning requirements of a data center and the technology used to produce electricity). Desktop-based videoconferences and telepresence for larger groups have saved $73 million and 65,000 tons of CO_2 in business trips. Solid state drives consume less power in laptops and are being considered for data centers for energy and heat reductions with enhanced performance (Guyon et al. 2012).

Others are providing carbon-neutral processors and computers. Customers pay for tree plantation or wetland conservation to offset CO_2 emissions over a 3-year-lifetime energy consumption (VIA processors). This is more viable if the energy consumption in the lifetime is low (15 W or less as opposed to typical 175 W desktops) and computers are returned to the manufacturer at the end of life (Zonbu computers). In addition, social sustainability targets are within reach: VIA looks for pay-per-use markets for heat- and dust-resistant PCs in an attempt to connect the next billion people, following a philosophy of information as the driver of "social mobility, economic equality [...] and global democracy." Computer centers already apply this lemma in South Africa and Vietnam. VIA has partnered with Motech to build solar-powered devices and it has established the first solar cyber community in the South Pacific (Gingichashvili 2007).

More is to be gained by scientists and engineers designing and purchasing for longevity, upgradability, modularity, miniaturization, and battery life improvement.

Open Science and Technology

Open S&T augments the accessibility of results and products, and is probably the cheapest way to enhance the socioeconomic usefulness of research and development budgets. Open S&T will reach its full potential when a majority of citizens can partake in the supply and demand of knowledge. Japan and South Korea have taken the lead in science communication and research to enhance citizen access to science results, through improved museums, the mass media, online outreach, and training of science communication and outreach specialists (Sipp 2013).

Open communication requires adaptations of property rights. New types of Creative Commons licenses have been chosen by entire scientific communities to enhance collaboration. Other Creative Commons licenses used to share and freely remix music samples while keeping a continuous credit record might migrate to scientific data sharing (Nelson 2009). The Web could allow for aggregation and "collective judgment of the communities themselves," circumventing slow journal editing processes. This has prompted many researchers to share their results as they are generated or to use new journal types to incrementally build their papers or automatically knit them (Priem 2013).

Open engineering is perhaps even more important for social sustainability, as technology benefits are more directly related to human livelihoods and well-being. Open engineering now encompasses systems for the design stage such as computer-assisted design, 3D printing (Pearce et al. 2010), personal manufacturing machines (Lipson and Kurman 2010), free forming machines, and design reusing hardware blueprints. Of particular importance are demand-driven innovation tools, such as open source appropriate technology design (Pearce and Mushtaq 2009), and peer production. Equally important is the legal protection of designs against exclusion of potential users, with mechanisms such as the MIT and GNU licenses.

Collective Intelligence in Sustainable Knowledge Societies

S&T alone cannot solve socioeconomic issues causing the current global unsustainability but sustainability politics are doomed without the involvement and actions of most people (Beck 2010). S&T can help bring about knowledge societies more apt to solve their own issues. The virtues of KS are hitherto limited to an industry or an agricultural area. The Internet, however, loosely focused at present offers possibilities to extend KS tenets to societies at large. However, integrating Internet infrastructure and KS architectures may not suffice to circumvent deeply ingrained global and national governance systems functioning as information black boxes whereby decision making swerves away from the global citizenry.

A foresight intelligence can emerge from academia and civil society (Ehrlich and Ehrlich 2013), but it may need tools to establish itself. Policy engineering redesigns the mechanisms of participatory decision making successful in local environments and KS but expands them to large societies so that the populace in a knowledge society partakes of its responsibility in turning it sustainable. Policy engineering is inspired more by social networks than technocratic proposals submitted by small groups to the powers that be. The Arab Spring of 2010 was the first large-scale freedom movement to successfully use the Internet and mobile technology, showing that in societies with relatively modest access to the Internet, direct democracy is possible and trumps centralized information at a fraction of the cost of elections.

Policy engineering, similar to any sound political system trying to effect urgent as well as deep changes, allots incentives to foster sustainable technologies; changes in persons, rules, and organizations; and legal instruments. Of critical importance are the income and capital distribution consequences of these policies. Other important effects of a policy are changes in economic competition, the global and local environment, and employment (Ashford 2004).

What is different in policy engineering is that what needs to be changed and how to make it sustainable is decided in polycentric governance systems that go beyond the state–market dichotomy (Ostrom 2010). Proposals and implementation are made in a KS environment with synergies between S&T knowledge and local knowledge, with success in mind instead of vested interests. The implementation of engineered policies may use nature-inspired algorithms like swarming, neural networks, cellular automata, and genetic algorithms, currently being used by S&T to solve a wide variety of problems. These tools can use organizations, latent groups (loose groups with similar interests), and every person as nodes and networks, or perceptrons, of a larger cognitive system, which in turn connects to perceptrons in the biosphere and the planet.

Knowledge democracies (KD) can draw on neural systems to gather a mass of parallel (same-weight) inputs such as all the citizens' changing preferences and choices. Artificial neural networks (ANN) were thoroughly described by the 1990s. They mimic the control of an organism immersed in an environment via "massively parallel interconnected networks of simple elements" similar to a biological nervous system (Kohonen 1988). The inputs of millions of persons in a political process becomes possible in ANN because operations are not carried out sequentially as in a usual computer but instead "many competing hypothesis are explored simultaneously" (Lippmann 1987).

The aforementioned can only be done with the incorporation of machine computing power. Pioneered in 1960 by Licklider, the idea of human-centered man–computer symbiosis has now morphed into the quest for ubiquitous computing able to explore and adapt to the environment with limited human participation, and to use the computing power of billions of humans and computers (Tennenhouse 2000). However, an estimate of the fast-growing global computer power is 6.4×10^{18} instructions per second, similar to 10^{17} the maximum number of nerve impulses in one human brain per second (Hilbert and Lopez 2011), which may suggest the following. First, human brains outperform computers. Second, ubiquitous computing may help monitor and make simple decisions. And third, it may prove advantageous to combine the global human brain and machine computing capacities into sustainable computing, instead of delegating sustainability decision making to the comparatively minuscule set of scientists and policy makers.

This general architecture can evolve from existing Web 2.0 features but would take off with the Web 3.0, still in the horizon. The Web 2.0 is already a participative digital culture, powered by tools like JavaScript, which have allowed for Web page modifications by nonprogrammers, blogging, RSS or Atom aggregation, and scrapers, permitting the collation and synthesis of specialized information by anyone. Web 2.0 is viewed by some as the "harnessing of collective intelligence turning the web in a kind of global brain" having already produced very large collaborative works like Wikipedia, which have demonstrated mechanisms to overcome diverging views for over a decade (Chatfield 2011). Web crowdsourcing is a way to explore issues that subverts the technocratic or expert approach to problem-solving since it caters to stakeholder diversity. Large crowdsourcing examples have included in 1999 the use of idle private PCs to look for astronomical patterns in data more efficiently than supercomputers or the 2009 scrutiny by volunteers of the British MPs expenses.

Bottom-up decision making in knowledge societies may have to rely on corresponding decentralized wherewithal and market information.

Crowdfunding allows small initiatives to be directly funded by a disperse set of interested cybercitizens. Local intelligence is the collective name of hyperlocal Web sites collating neighborhood-relevant data (on shops, schools, events, etc). Web 2.0 tools have been instrumental in real-life experimentation of more sustainable lifestyles like collaborative consumption or share economy. To this citizen impulse, the states have responded by e-government allowing for citizen feedback and input via online petitions and participatory budgeting.

The next evolutions are likely to include the Web 3.0, intended to go beyond aggregation by collating all (semantically) related information automatically to permit human–machine collaboration, with a key application being access and analysis of open government databases (Chatfield 2011). Human–machine communication ultimately might require machine biomimicry of essential human features: autonomy, imitation, intrinsic moral value, moral accountability, privacy, and reciprocity (Kahn et al. n.d.) as well as biophilia, reciprocal altruism, and cooperation.

Science, Conscience, and Practical Technology

Science and technology have evolved observation tools that have raised our awareness of the universe and our understanding of our subatomic constituents. However, somewhere in the middle, at the planetary scale, we struggle to understand our place as a species. We seem to be trapped by human rules that would have us believe we have to give precedence to the political system over the climate system: "We have to work with the institutions that already exist and in ways that respect parliamentary democracy" (Lord Giddens 2009). But what happens if the parliaments of the world fail to alter the course of unsustainability we are now following? To find the answer we might run the experiment of relinquishing responsibilities and rights as humans and entrusting our survival to a group that may have more than the average share of responsibility in the current predicament.

Alternatively, in the course of global transitions to sustainability, the exact mix of multicentered governance able to draw the best out of humans beyond markets and states (Ostrom 2010) and repoliticization (Žižek 2004) would vary, as collective intelligence acquires planetary awareness and conscience. As in all systems, collective intelligence depends essentially on human beings as subsystems. The guiding principle of collective intelligence, if it is to differ from current, unsustainable political systems is that these subsystems cannot be in cybernetic parlance "slave" but rather controlling subsystems. For the latter to lead the global collective intelligence to a sustainability tipping point, a critical mass of human beings have to undergo a transition themselves. This can only occur with technologies that they can control: adequate ones rather than high tech, leading nowhere fast. If thereafter, high tech became sustainable, we might gain further consciousness. As a species we seem to have to choose between on the one hand freezing the evolution of political and technological systems and on the other hand redefining our place among living beings. Science will have to choose whether to fuel the accelerated depletion of nature or to propitiate the emergence of a global consciousness.

References

Anastas PT, Beach ES (2008) Green chemistry: The emergence of a transformative framework. *Green Chemistry Letters and Reviews* 1:9–24.

Ashford NA (2004) Major challenges to engineering education for sustainable development: What has to change to make it creative, effective, and acceptable to the established disciplines. *International Journal of Sustainability in Higher Education* 5:239–250.

Bammou L, Chebli B, Salghi R, Bazzi L, Hammouti B, Mihit M, Idrissi H (2010) Thermodynamic properties of *Thymus satureioides* essential oils as corrosion inhibitor of tinplate in 0.5 M HCl: Chemical characterization and electrochemical study. *Green Chemistry Letters and Reviews* 3:173–178.

Basu J, Basu JK, Bhattacharyya TK (2010) The evolution of graphene-based electronic devices. *International Journal of Smart and Nano Materials* 1:201–223.

Beck U (2010) Climate for change, or how to create a green modernity? *Theory, Culture & Society* 27:254–266.

Brundiers K, Wiek A (2013) Do we teach what we preach? An international comparison of problem- and project-based learning courses in sustainability. *Sustainability* 5:1725–1746.

Byrne EP, Fitzpatrick JJ (2009) Chemical engineering in an unsustainable world: Obligations and opportunities. *Education for Chemical Engineers* 4:51–67.

Cash DW, Clark WC, Alcock F, Dickson NM, Eckley N, Guston DH, Jäger J, et al. (2003) Knowledge systems for sustainable development. *PNAS* 100:8086–8091.

Center for Sustainable Systems, University of Michigan (2011) Green IT Factsheet. Pub. No. CSS09-07. http://css.snre.umich.edu/css_doc/CSS09-07.pdf (accessed February 15, 2012).

Chatfield T (2011) *50 digital ideas.* London: Quercus.

Clark WC, Dickson NM (2003) Sustainability science: The emerging research program. *PNAS* 100:8059–8061.

De Prato G, Nepelski D (n.d.) Global technological collaboration network. Network analysis of international co-inventions. http://mpra.ub.uni-muenchen.de/38818/ (accessed March 22, 2012).

Diab WW (2010) Broadcom Energy Efficient Networking. White Paper. http://www.broadcom.com/collateral/wp/EEE-WP102-R.pdf (accessed April 2, 2013).

Dilling L, Lemos MC (2011) Creating usable science: Opportunities and constraints for climate knowledge use and their implications for science policy. *Global Environmental Change* 21:680–689.

Ehrlich PR, Ehrlich AH (2013) Can a collapse of global civilization be avoided? *Proceedings of the Royal Society* 280:20122845.

Floyer D, Miniman S (2011) Networks Go GrEEN. Power. http://wikibon.org/wiki/v/Networks_Go_GrEEN (accessed February 12, 2011).

Giddens A (2009) *The politics of climate change*. Cambridge: Polity Press.

Gingichashvili S (2007) Green computing. The future of things. www.thefutureofthings.com/articles/1003/green-computing.html (accessed March 12, 2013).

Guyon B, Sheridan C, Donnellan B (2012) Developing a sustainable IT capability: Lessons from Intel's journey. *MIS Quarterly* 11:61–74.

Hilbert M, Lopez P (2011) The world's technological capacity. *Science* 332:60–65.

Holling CS, Gunderson LH, Ludwig D (2002) In search of a theory of adaptive change. In *Panarchy: Understanding transformations in human and natural systems*, Gunderson LH, Holling CS (eds), 3–22. Washington DC: Island Press.

Huesemann MH (2001) Can pollution problems be effectively solved by environmental science and technology? An analysis of critical limitations. *Ecological Economics* 37:271–287.

Huesemann MH (2002) The inherent biases in environmental research and their effects on public policy. *Futures* 34:621–633.

Hulme M (2012) Climate change: Climate engineering through stratospheric aerosol injection. *Progress in Physical Geography* 36:694–705.

Kahn PH, Ishiguro H, Friedman B, Kanda T (n.d.) What is a human? Toward psychological benchmarks in the field of human-robot interaction. http://citeseerx.ist.psu.edu/viewdoc/download?doi = 10.1.1.97.7638&rep = rep1&type = pdf (accessed May 11, 2012).

Kaplinsky R (2010) In what ways might China be a source of appropriate innovation for low income countries? Implications for the environment. http://risingpowers.open.ac.uk/Giles/WS%20Summaries%20-%20part%202/Kaplinsky%20%20Rising%20Powers%20Nov%202010.pdf (accessed August 4, 2011).

Kaplinsky R (2011) Schumacher meets Schumpeter: Appropriate technology below the radar. *Research Policy* 40:193–203.

Kohonen T (1988) An introduction to neural computing. *Neural Networks* 1:3–16.

Kounatze CR (2009) Towards green ICT strategies: Assessing policies and programmes on ICT and the environment. www.oecd.org/sti/ict/green-ict (accessed April 22, 2013).

Liang F, Gou J, Kapat J, Gu H, Song G (2011) Multifunctional nanocomposite coating for wind turbine blades. *International Journal of Smart and Nano Materials* 2:120–133.

Lippmann RP (1987) An introduction to computing with neural nets. *IEEE ASP Magazine* 3:4–22.

Lipson H, Kurman M (2010) Factory@Home: The emerging economy of personal fabrication. Science and Technology. Occasional Papers in Science and Technology Policy. Washington DC: The Science and Technology Institute.

Liu J, Dietz T, Carpenter SR, Folke C, Alberti M, Redman CL, Schneider SH, et al. (2007) Coupled human and natural systems. *Ambio* 36:639–649.

Lu P, Xue D, Yang H, Liu Y (2012) Supercapacitor and nanoscale research towards electrochemical energy storage. *International Journal of Smart and Nano Materials* 4:1–25.

Lubchenco J (1998) Entering the century of the environment: A new social contract for science. *Science* 279:491–497.

McCullough EB, Matson PA (2011) Evolution of the knowledge system for agricultural development in the Yaqui Valley, Sonora, Mexico. *PNAS Early Edition* 1–6.

Millett LI, Estrin DL (2012) *Computing research for sustainability*. Washington DC: The National Academies Press.

Narayanan R (2012) Synthesis of green nanocatalysts and industrially important green reactions. *Green Chemistry Letters and Reviews* 5:707–725.

Nelson B (2009) Empty archives. *Nature* 461:160–163.

Ostrom BE (2010) Beyond markets and states: Polycentric governance of complex economic systems. *American Economic Review* 100:1–33.

Palmer MA (2012) Socioenvironmental sustainability and actionable science. *BioScience* 62:5–6.

Pearce JM, Blair CM, Laciak KJ, Andrews R, Nosrat A, Zelenika-Zovko I (2010) 3-D printing of open source appropriate technologies for self-directed sustainable development. *Journal of Sustainable Development* 3:17–29.

Pearce JM, Mushtaq U (2009) Overcoming technical constraints for obtaining sustainable development with open source appropriate technology. TIC-STH 814-820.

Priem J (2013) Beyond the paper. *Nature* 495:437–440.

Roco MC, Bainbridge WS (eds) (2003) Converging technologies for improving human performance: Nanotechnology, biotechnology, information technology and cognitive science. NSF/DOC-sponsored report. Dordrecht: Kluwer Academic Publishers.

Sarewitz D, Pielke Jr RA (2007) The neglected heart of science policy: Reconciling supply of and demand for science. *Environmental Science and Policy* 10:5–16.

Sipp D (2013) How to hasten open access: Translocate local journals. *Nature* 495:443.

Tennenhouse D (2000) Proactive computing. *Communications of the ACM* 43:43–50.

The Stockholm Memorandum (2011) The Stockholm Memorandum. *Ambio* 40:781–785.

Van Noorden R (2012) Britain aims for broad open access. *Nature* 486:302–303.

Van Noorden R (2013) U.S. science to be open to all. *Nature* 494:414–415.

van-Dam Mieras R, Lansu A, Rieckmann M, Michelsen G (2008) Development of an interdisciplinary, intercultural master's program on sustainability: Learning from the richness of diversity. *Innovation in Higher Education* 32:251–264.

Varma RS (2007) "Greener" chemical syntheses using mechanochemical mixing or microwave and ultrasound irradiation. *Green Chemistry Letters and Reviews* 1:37–45.

Wiek A, Withycombe L, Redman CL (2011) Key competencies in sustainability: A reference framework for academic program development. *Sustainability Science* 6:203–218.

Wilbanks J (2013) A fool's errand. *Nature* 495:440–441.

York R, Rosa EA, Dietz T (2002) Bridging environmental science with environmental policy: Plasticity of population, affluence, and technology. *Social Science Quarterly* 83:18–34.

Žižek S (2004) *Plaidoyer en faveur de l'intolérance*. Paris: Flammarion.

Index